DR. BURGESS'S MINI-
ATLAS
OF
MARINE AQUARIUM
FISHES

Chaetodon mesoleucos. Photo by G. Marcuse.

DR. WARREN E. BURGESS

•

DR. HERBERT R. AXELROD

•

RAY HUNZIKER

Table of Contents

Introduction

Dr. Burgess's Mini-Atlas of Marine Fishes is designed as a pictorial aid for the identification of marine fishes as well as a guide to the setting up and maintaining of a marine aquarium. The thousands of photographs are all that will be needed for most aquarists to identify species that they are keeping in their tanks or that they have seen in their local marine fish store. They will also be able to find for the most part fishes that they are likely to collect along their own shores, including those that occur only in the cooler waters of temperate coasts. Further, many fishes are included that are too big for hobbyists but will be seen in the ever-popular public aquariums where large reef tanks or oceanariums house larger fishes, including sharks. Whenever possible, different aspects of a single species are shown. That is, where juveniles are different from adults or males differ from females, both (or even all three) are shown. Ichthyologists have even been fooled by some species that appear so different as a juvenile or as one or the other sex. This has prompted additional names for fishes that had already been named previously. The name applied to the fishes in these photographs is the name that as far as known is the currently accepted name for the fish and supersedes any name applied to it in our previous publications. The most recent scientific papers and books were used to determine the correct names. However, certain obstacles are always present that prevent identifying some of the species with absolute certainty. First of all, since the fish is represented in a photograph only and not "in the flesh" where some of the diagnostic counts or measurements could be made, educated guesses must be made using what available information there is. This information may be color pattern, geographic locality, or even identifications by a photographer who is knowledgeable in fish systematics and has actually seen the fish itself.

Whenever possible color proofs were sent to experts in the field for identification. Dr. Gerald R. Allen kindly identified members of the family Pomacentridae, Dr. Victor G. Springer did the same for the blennies, Dr. Guido Dingerkus identified the elasmobranchs, Dr. William N. Eschmeyer did the scorpionfishes, Dr. William F. Smith-Vaniz did the carangids and some blennies, and Dr. Douglas Hoese tackled the gobies. Unfortunately, some plates were not available to these experts at the time so these identifications were accomplished to the best of my (W.E.B.) ability and should not reflect on the expertise of these scientists.

The common names applied to these fishes were taken from current scientific literature whenever possible. Unfortunately, in some instances (as also happened in the scientific names) there was disagreement (even by the same author in two papers), so that one or the other name was selected. Space does not permit the inclusion of alternate common names or scientific synonyms.

In keeping with the intent of this book to provide the best means to identify the marine fishes, a section is devoted to presenting an outline drawing of a representative of many of the families. As a further aid to aquarists, a short description of the best means of keeping fishes of each family accompanies each drawing. Carrying this one step further, along with the scientific name under each photograph can be seen numerical and pictorial symbols giving a brief account of the species's basic requirements for living and a family number so that each fish can be easily placed in its systematic position. An aquarist recognizing his fish in the photographs can utilize the information in the caption or can refer to the information in the general family care write-ups in the front part of this book. The fishes are presented in systematic order according to Nelson (1984, *Fishes of the World*, John Wiley & Sons, NY), with the cartilaginous fishes first and the plectognaths last.

The symbols are approximations. Depicting the type of feeding by a single symbol is virtually impossible. An omnivorous fish that is not particular about its diet should be accorded every feeding symbol. But space limitations force us to indicate only the most likely food(s) of a given fish. It should be remembered that a well-balanced diet is always best.

Following the pictorial systematic section is a section devoted to the setting up and maintaining of a marine aquarium. The most up-to-date methods are covered and explained so that the keeping of marine fishes is brought within the capability of almost anyone with the desire to do so.

Like most things, there are sure to be changes, especially in the nomenclature of the fishes. Even so, this book should prove to be the standard for identifying and keeping marine aquarium fishes for some time to come.

Aquaristic Section

This section is provided as an introductory guide to the maintenance of marine fishes in the aquarium. Habits and requirements are given for each family, or in some cases related groups of families. Included are data such as temperament and compatibility, water conditions, relative hardiness, size, aquascape type, and feeding habits. The family listings (names and numbers) are based on the systematic listing by Nelson (1984), which is found beginning on page 23.

For families that have representatives common in the aquarium hobby, a generalized drawing is included to help with identification. Each drawing immediately precedes the corresponding written entry. After narrowing an unidentified fish down to the family level using these drawings, turning to the pictorial section of the book should give you a good chance of identifying it down to the species level. The selection of families illustrated by drawings may seem somewhat arbitrary, but effort was made to include only those that often appear in pet shops or are commonly collected by hobbyists in coastal areas.

Not all of the families listed here are suitable for home aquaria, and the writeups indicate the reasons why—many are simply too large, others too delicate, others too dangerous! In general, fishes that commonly reach an adult size in excess of a foot or so are suitable only for large public aquaria, and information on many of these species is included for the benefit of such aquaria.

It should be noted that these entries are only generalizations; individual species of some families vary widely with regard to their aquarium care. The butterflyfishes of the family Chaetodontidae are a good example—some species are very hardy, eat almost anything, and are good fishes for the beginner. Others eat only live coral and other hard-to-get items, and these are almost impossible for the average hobbyist to maintain. It is always best to research each individual species, preferably before a specimen is purchased. A good reference for this purpose is *Exotic Marine Fishes* (T.F.H. H-938), as well as much of the scientific literature.

The Families

1) MYXINIDAE
Hagfishes. Ravenous predators and scavengers, consuming dead and dying fishes. Usually found in groups. Produce large volumes of mucus that may be troublesome in small systems. Not compatible with any other fishes. Dark tank, cool water.

2) PETROMYZONTIDAE
Lampreys. Parasitic predators on living fishes. Cool, well-oxygenated water. Not compatible with fishes not intended as prey.

3) CHIMAERIDAE
Chimaeras. Cool water, dim light. Foods include benthic invertebrates and fishes. Good aquarium foods include chopped fish, crab, and shrimp. Most are too large for home aquaria, though young specimens are sometimes kept. Compatible with fishes too large to be easy prey

6) CHLAMYDOSELACHIDAE
Frilled sharks. Dark, cold tank. Foods include benthic invertebrates and fishes. Too large for home aquaria; rarely kept even by public aquaria.

7) HEXANCHIDAE
Sixgill sharks. Dark, cold tank. Sensitive to changes in water conditions. Eat fishes, crabs, shrimp. Will outgrow home aquaria but are unlikely to be seen outside of public aquaria anyway.

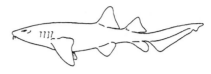

10) ORECTOLOBIDAE
Nurse sharks and relatives. Feed on benthic invertebrates. Good aquarium foods include live goldfish, chopped clam, crab, shrimp. Warm, well-lit tank. Peaceful with fishes too large to swallow, but should not be kept with any invertebrates not intended as food. Active and fast-growing, reaching a large size; large tank necessary. Relatively hardy.

11) ODONTASPIDAE
Sand tiger sharks. Aggressive fish eaters. Large tank needed, only suitable for public aquaria.

12) LAMNIDAE
Mackerel sharks. Pelagic. Very large, often aggressive. Wide variety in diet within this family; various species feed on plankton, fishes, marine mammals. Not suitable for home aquaria and not yet kept with success in public aquaria.

13) SCYLIORHINIDAE
Catsharks. Among the best sharks for home aquaria. Somewhat sensitive to adverse changes in water quality, but otherwise hardy. Many are quite attractive, and many remain moderately small. Goldfish, shrimp or prawn, chopped clam, and beef heart are good foods. Large tanks with caves preferred. Peaceful with most fishes, but may attempt to eat many invertebrates.

14) CARCHARHINIDAE
Requiem sharks. Active, aggressive predators with powerful jaws and sharp teeth. Very sensitive and generally make poor captives. Very large tanks with no sharp corners are required; these sharks swim tirelessly. Will eat live or chopped fish as well as squid, shrimp, crab.

15) SPHYRNIDAE
Hammerhead sharks. Pelagic, extremely poor survival in captivity. Extended head lobes easily

8) HETERODONTIDAE
Horn sharks. Relatively inactive, often resting on the bottom. Provide shaded grottos for shelter. Eat fishes, crustaceans. In captivity will accept live goldfish, chopped clam, shrimp, crab. Peaceful with fishes too large to swallow whole. Spines can cause injury; handle with care.

9) RHINCODONTIDAE
Whale sharks. Peaceful plankton feeders. Too large for any aquarium.

damaged. Very sensitive to any change in water conditions. Fish eaters, will also take squid.

16) SQUALIDAE
Dogfish sharks. Fairly hardy in captivity; not terribly prone to disease and not too sensitive to changes in water quality. Venomous dorsal spines in some species; handle with care. Good foods include goldfish or chopped fish, chopped clam, squid, shrimp, crab. Large tanks with no sharp edges and few obstructions. Peaceful with fishes of like size.

19) PRISTIDAE
Sawfishes. Very large, and small specimens are fast-growing, thus generally unsuitable for home aquaria but are often kept In public aquaria. Sensitive, saw is easily injured. Fish eaters, will accept chopped fish and squid. Soft substrate needed, as these rays will sometimes bury themselves.

20) TORPEDINIDAE
Torpedo rays. Sluggish bottom dwellers and sand burrowers. Not overly sensitive to water quality. Eat small fishes and benthic invertebrates—foods: chopped fish, crab, shrimp, prawn, squid. Electric organs present, which can deliver a startling but sublethal shock.

21) RHINOBATIDAE
Guitarfishes. Inactive rays; need soft substrate for burying. Fairly hardy and not disease-prone. Like dim light. Eat mostly benthic invertebrates; almost any crustacean meat and chopped fish will be accepted. Some rather large tanks needed. Peaceful with most other fishes.

22) RAJIDAE
Skates. Bottom dwellers, need soft sand for burying themselves. Eat shellfish and crustaceans; good foods include crab, shrimp, prawn, chopped clam, squid. Peaceful with other fishes. Large tanks necessary—grow large. Fairly hardy and not too sensitive to changes in water quality.

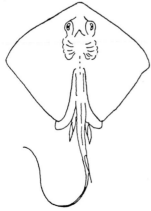

23) DASYATIDAE
Stingrays. Bottom dwellers, need soft substrate. Most grow very large and require big tanks. Barbed tail spine is dangerous—handle with extreme care. Will eat most meaty foods and will devour any tankmate small enough to swallow.

26) MYLIOBATIDAE
Eagle rays. Pelagic, often schooling. Unsuitable for all but the largest public aquaria. Foods include most shellfish and crustaceans. Warm water, very sensitive to any change in water quality. Venomous tail spines present.

27) MOBULIDAE
Mantas. Huge, plankton-eating pelagic rays. Very peaceful. Very sensitive, not suitable for aquaria.

31) LATIMERIIDAE
Coelacanth. Never kept in aquaria. Dim light preferred. Eats fishes, custaceans.

43) ELOPIDAE
Ladyfishes. Fast-moving schooling fishes. Large, spacious aquaria needed. Warm water, eat crustaceans and small fishes. Peaceful. Somewhat sensitive in captivity.

44) MEGALOPIDAE
Tarpons. Very active; brightly lit large tanks with no obstructions needed. Eats any fishes small enough to swallow. Hardy and long-lived. Peaceful with fishes of like size. Very large; large tanks needed.

45) ALBULIDAE
Bonefish. Large and spacious aquaria needed. Soft substrate a must; bonefish root about for benthic invertebrates. Foods: shrimp, crab, chopped clam, small live feeder fish. Peaceful, but very nervous and sensitive to the slightest environmental change.

49) ANGUILLIDAE
Eels. Very hardy, not disease prone. Will eat anything remotely edible. Relatively peaceful with fishes too large to swallow. Relish many invertebrates, however. Grow very quickly and will outgrow many home aquaria. Not sensitive to water quality, and can withstand any salinity from freshwater to full marine.

51) MORINGUIDAE
Small, wormlike burrowing eels. Feed mostly on worms and small crustaceans. Delicate; rarely seen in captivity.

153) AULOPIDIDAE

Benthic predators similar to lizardfishes. Will eat all fishes small enough to swallow; may attack larger ones. Also eat crustaceans. Good foods for captives include chopped fish and shrimp. Large tanks necessary for many species.

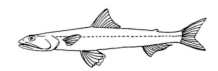

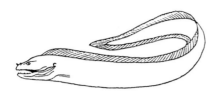

54) MURAENIDAE

Moray eels. Large, with powerful jaws and teeth. Can bite; handle with care. Warm water. Hardy, not disease prone. Large tanks needed for most species. Will eat any fishes or crustaceans that can be captured, but generally peaceful with larger fishes. Foods: goldfish, chopped fish, scallops, shrimp, crab, squid. Provide caves and rockwork for shelter.

58) OPHICHTHIDAE

Snake eels. Large but shyer than morays. Many burrow; provide soft sand or fine gravel. Peaceful but may eat small fishes. Primarily crustacean and mollusc eaters. Very hardy.

82) CONGRIDAE

Conger eels. Large, active, aggressive. Provide rockwork for shelter. Mainly fish eaters. Good foods for captives include feeder fish, chopped fish, scallops, clam, shrimp, squid. Large tanks necessary.

68) CLUPEIDAE

Herrings. Active schooling fishes. Keep in large groups In large, spacious tank. Delicate scales and fins easily damaged. Very disease prone. Eat mostly planktonic crustaceans—live brine shrimp is a good substitute for captives. Peaceful.

70) ENGRAULIDAE

Anchovies. Pelagic schooling fishes, need to be kept in large groups in large tanks. Delicate and easily stressed; difficult to keep. Eat planktonic crustaceans; feed live brine shrimp. Peaceful.

72) CHANIDAE

Milkfish. Large and active, do not take well to aquaria. Feed on crustaceans, small fishes, algae.

137) OSMERIDAE

Smelts. Large schools, large tanks. Delicate and disease prone. Will eat small fishes or chopped fish or crustacean meat.

139) SALANGIDAE

Lance-like smelt allies. Delicate in captivity. Large schools, large tanks. Eat planktonic crustaceans.

144) GONOSTOMATIDAE, 147) CHAULIODONTIDAE, 148) STOMIIDAE, 150) MELANOSTOMIIDAE

Viperfishes, dragonfishes, and allies. Deepsea fishes. Toothy predators. Because of need for darkness, cold, and pressure, never kept in captivity.

157) SYNODONTIDAE

Lizardfishes. Benthic predators, often lie partly buried in soft sand. Very aggressive; will eat most fish and crustaceans. Live fish and shrimp preferred as food—will rarely accept nonliving foods. Very delicate and subject to transport shock—acclimate as gently as possible. Water quality must be excellent.

163) ALEPISAURIDAE

Lancetfishes. Pelagic predators. Rarely kept even in public aquaria and unlikely to adapt to captivity.

166) MYCTOPHIDAE

Lanternfishes. Bioluminescent midwater fishes; superficially sardine-like. Very delicate; sensitive to water quality and rather disease prone. Plankton feeders—live brine shrimp is a good substitute.

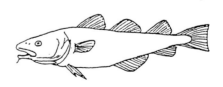

171) MORIDAE, 174) GADIDAE, 176) MACROURIDAE

Codfishes and allies. Mostly coldwater, often deepwater, often large. Spacious tanks required. Groups preferred. Foods: fish, crustaceans. Most somewhat delicate but some are kept in public aquaria.

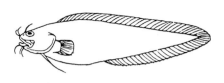

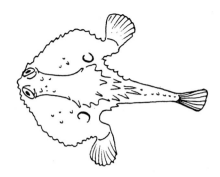

177) OPHIDIIDAE
Cusk eels and brotulids. Usually sluggish bottom dwellers. Food mostly benthic invertebrates; feed shrimp, crab, chopped clam, squid. Dim tank with caves and other hollows preferred. Usually peaceful with fishes too large to swallow.

178) CARAPIDAE
Pearlfishes. Inhabit body cavities of sea cucumbers and sometimes molluscs. Emerge to feed on small invertebrates; good foods would be brine shrimp, bloodworms, glassworms. Peaceful but delicate and shy.

179) BYTHITIDAE
Livebearing brotulas. Benthic; peaceful; fairly hardy. Will eat chopped clam, shrimp, bloodworms, brine shrimp, prepared foods. Dim tank with hollows.

186) OGCOCEPHALIDAE
Batfishes. Slow moving, benthic. Very peaceful and shy. Food mostly small crustaceans and polychaete worms but also angle for small fishes. Many refuse to feed in captivity, but sometimes live brine shrimp or glass shrimp will be accepted. Cannot compete with faster-moving fishes. Dim tank with soft substrate preferred.

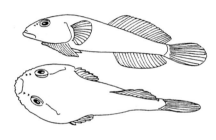

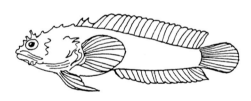

181) BATRACHOIDIDAE
Toadfishes. Big-headed benthic predators; will eat anything that can be swallowed. Large teeth and strong jaws—handle with care. Especially fond of crustaceans and mollusks. Like hollows, but place decor with care, as toadfishes like to dig and may undermine decorations.

182) LOPHIIDAE
Goosefishes. Large, toothy anglerfishes—handle with care. Will eat anything—no compatible tankmates. Very large. Need soft sub strate. Not too sensitive to water quality variations.

183) ANTENNARIIDAE
Frogfishes. Lethargic benthic anglerfishes that lure and ambush passing fishes. Can swallow prey nearly as large as themselves. Cannibalistic. Hardy, but prone to hunger strikes if overfed. Provide coral edges for perching.

198) GOBIESOCIDAE
Clingfishes. Benthic. Very hardy. Very tolerant of water quality changes; very resistant to disease. Generally peaceful, but very small fishes may be eaten. Often become tame. Foods: prawn, brine shrimp, bloodworms, chopped clam, chopped squid.

200) EXOCOETIDAE
Flyingfishes. Not easily confined, but small specimens sometimes adapt to aquaria. Jumpers; keep tank tightly covered. Peaceful, but will eat small fishes. Wide, shallow tank without obstructions. Feed bloodworms, brine shrimp, glassworms, flake foods.

201) HEMIRAMPHIDAE,
202) BELONIDAE
Halfbeaks and needlefishes. Swift surface predators. Good jumpers. Very delicate and prone to capture and transport shock. Elongate bills easily damaged. Very nervous and prone to panic in aquaria. Round tanks with no sharp corners preferable.

213) ATHERINIDAE,
214) ISONIDAE
Silversides and relatives. Surface-to-midwater planktivores. Large schools preferred. Large, spacious tanks needed. Very sensitive to any change in water quality. Deciduous scales easily damaged. Very prone to fungus. Easily killed by transport shock. Foods: brine shrimp, bloodworms, glassworms.

218) LAMPRIDIDAE
Opahs. Warmwater, very large. Feed mostly on small fishes. Poorly known.

223) REGALECIDAE
Oarfishes. Elongate, very delicate. Feed on small fishes and planktonic invertebrates. Never kept.

overly sensitive. Feed on small crustaceans, fishes. Peaceful. Feed guppies, brine shrimp, bloodworms, prepared foods.

234) BERYCIDAE
Alfonsinos. Squirrelfish-like. Peaceful but will eat small fishes and crustaceans. Foods: guppies, brine shrimp, prepared foods. Hardy.

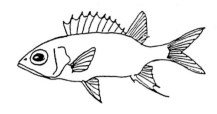

235) HOLOCENTRIDAE
Squirrelfishes. Generally peaceful, but will eat small fishes and crustaceans. Nocturnal; provide shaded hollows. Very hardy and tolerant of water quality variations. Most species school, Some become quite large. Foods: small feeder fishes, chopped clam, prawn, brine shrimp, bloodworms, all prepared foods.

238) POLYMIXIIDAE
Beardfishes. Generally similar to squirrelfishes in temperament and care.

244) MACRUROCYTTIDAE,
245) ZEIDAE, 248) CAPROIDAE
Dories, boarfishes, and allies. Rarely seen in aquaria, but are fairly hardy. Many grow quite large. Feed on fishes and crustaceans but are usually peaceful with tankmates too large to swallow.

250) AULORHYNCHIDAE,
251) GASTEROSTEIDAE
Tubesnouts and sticklebacks. Often coolwater fishes. Very easy to keep but are territorial and fight viciously with conspecifics if crowded. Peaceful with other species. Hardy. Very tolerant of salinity variations; many species can stand fresh water. Planted tank. Foods: Brine shrimp, bloodworms, glassworms, prepared foods.

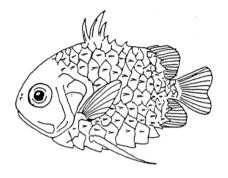

229) MONOCENTRIDIDAE
Pinecone fishes. Warmwater, very slow, peaceful, and shy. Rather sensitive to water quality. Possess bioluminescent organs; prefer dim tank. Feed mostly on small crustaceans, but often refuse to feed in captivity. Try live brine shrimp, live glass shrimp, bloodworms, glassworms.

230) TRACHICHTHYIDAE
Slimeheads. Similar to squirrelfishes; nocturnal predators. Generally peaceful but may eat small fish and crustaceans. Hardy. May be aggressive with conspecifics.

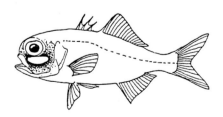

231) ANOMALOPIDAE
Flashlight fishes. Possess bioluminescent organs; dim tank. Prefer to be kept in groups. Not

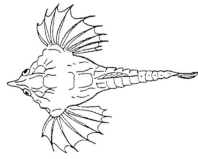

253) PEGASIDAE
Sea moths. Slow-moving bottom fishes. Peaceful, not competitive; do not keep with swifter

fishes. Not disease prone, but often difficult to feed. Prefer small living invertebrates; try brine shrimp, bloodworms.

254) AULOSTOMIDAE,
255) FISTULARIDAE
Trumpetfishes and cornetfishes. Generally warmwater. Fairly hardy. Ambush predators. Prefer planted tank for hiding. Eat small fishes, crustaceans; feed guppies, goldfish, prawn,

256) MACRORHAMPHOSIDAE,
257) CENTRISCIDAE
Snipefishes and shrimpfishes. Armored but slow and shy. Do not compete well with other species. Not prone to disease, but sometimes difficult to feed. Eat mostly small invertebrates; try live brine shrimp, bloodworms, glassworms.

258) SOLENOSTOMIDAE
Ghost pipefishes. Habits and care same as family Syngnathidae.

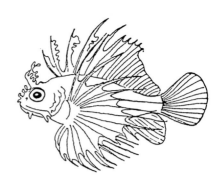

262) SCORPAENIDAE
Scorpionfishes. Venomous; handle with extreme care. Usually very hardy; not sensitive to water variations, not disease prone. Can swallow fishes nearly their own size, but peaceful with fishes too large to eat. Prefer coral-aquascaped tanks with ledges and caves. Many grow large. Foods: goldfish, prawn.

263) SYNANCEIIDAE
Stonefishes. Similar in care to scorpionfishes, but venom is deadly to humans. Best not kept at all.

264) CARACANTHIDAE,
265) APLOACTINIDAE, 266) PATAECIDAE
Velvetfishes. Benthic. Generally peaceful and small. Good for reef tanks, as plants and invertebrates supply good camouflage. Feed small feeder fishes, prawn. Generally hardy but uncommon in captivity.

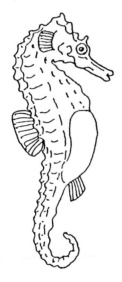

259) SYNGNATHIDAE
Seahorses and pipefishes. Very slow and shy—do not keep with competitive species. Have bred in aquaria. Branched corals, sea fans, plants needed for shelter or anchorage. Not disease prone but very sensitive to any decline in water quality. sometimes difficult to feed, as tubular mouths accommodate only tiny foods, like brine shrimp, bloodworms, glassworms, livebearer fry.

260) DACTYLOPTERIDAE
Flying gurnards. Benthic; generally peaceful. Large pectoral fins easily damaged—transport with care. Eat mostly bottom invertebrates, small fishes. Hardy.

268) TRIGLIDAE
Searobins. Active benthic predators. Like a fairly open substrate area on which to "crawl" around. Spiny but nonvenomous. Will eat anything that can be swallowed. Sometimes aggressive with conspecifics. Very hardy.

269) PLATYCEPHALIDAE
Flatheads. Similar in care to searobins.

271) ANOPLOPOMATIDAE,
272) HEXAGRAMMIDAE
Sablefishes and greenlings. Large, midwater or benthic. Often coldwater species that require large tanks. Feed mostly on crustaceans, molluscs, fishes; good captive foods include chopped fish or clam, shrimp, crab, squid. Rocky tanks with plants.

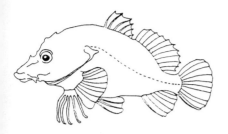

283) PERCICHTHYIDAE
Temperate basses. Very large, generally suitable only for large public aquaria. Very active fish eaters. Sensitive to water quality fluctuations.

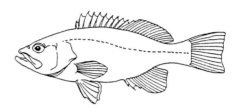

276) COTTIDAE
Sculpins. Generally small (but there are exceptions), very territorial with conspecifics but relatively peaceful with other species. Spiny; handle with care. Usually very tolerant of environmental variations, and many are very adaptable to salinity changes. Like plenty of hiding places and a cool, dim tank. Will eat small fishes and invertebrates; good foods include guppies or goldfish, brine shrimp, bloodworms, glassworms, chopped clam, prawn.

280) AGONIDAE
Poachers. Coldwater, benthic. Peaceful. Sensitive to variations in water quality. Generally eat small invertebrates: try brine shrimp, bloodworms, glassworms, small prepared foods.

284) SERRANIDAE
Groupers and sea basses. Large-mouthed predators, often aggressive with conspecifics but peaceful with other fishes if they cannot be swallowed. Will eat crustaceans and some molluscs but leave most other invertebrates alone. Like to dig in bottom substrate. Coral aquascape with shadowy grottos preferred. Many very large when full-grown. Foods: feeder fishes, chopped clam, squid, shrimp, crab. Very hardy.

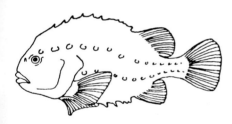

281) CYCLOPTERIDAE
Lumpfishes and snailfishes. Benthic, coldwater. Need water that is heavily aerated. Some rather large. Very hardy; few disease problems. Peaceful with anything too large to swallow. Foods: chunks of clam, squid, shrimp, crab, also brine shrimp and bloodworms for smaller species.

282) CENTROPOMIDAE
Snooks. Large shallow water, estuarine predators; too big for most tanks. Require plenty of open swimming room. Eat mostly live fishes but may accept chunks of fish, mollusc, crustacean meats. Fairly tolerant of salinity variations. Fairly hardy. Related family (Ambassidae—often considered subfamily of Centropomidae) includes the small, planktivorous glassfishes, which are peaceful and harmless.

285) GRAMMISTIDAE
Soapfishes. Peaceful but will eat small fishes. Hardy, disease resistant. However, when stressed or attacked may release a toxic mucus that can kill tankmates. Shy; coral caves preferred. Often tame easily. All foods accepted.

286) PSEUDOCHROMIDAE,
287) GRAMMIDAE
Dottybacks and basslets respectively. Mostly tropical. Small, peaceful grouper-like fishes, generally do not bother fish or invertebrate tankmates. May be aggressive with conspecifics, however. Prefer heavily aquascaped tanks with

lots of hiding places. Good in reef aquaria. Most are hardy fishes. Small foods: brine shrimp, bloodworms, glassworms, small slivers of chopped clam or shrimp. Most will learn to take flake foods and other prepared diets.

288) PLESIOPIDAE
Devilfishes or roundheads. Shy at first, but very hardy. Like shadowy grottos. May be aggressive with conspecifics. Usually slow to feed at first—start with live guppies and wean over to brine shrimp, prawn, chopped clam, prepared foods. Very disease resistant.

289) ACANTHOCLINIDAE
Habits and care as for dottybacks (Pseudochromidae).

290) GLAUCOSOMATIDAE
Similar to serranids in appearance, habits, and care.

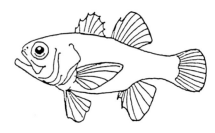

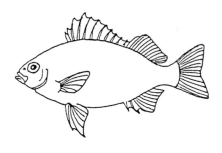

297) APOGONIDAE
Cardinalfishes. Most small and peaceful. Good for reef tanks. Somewhat light-shy. Many are schooling fishes. Have spawned in aquaria. Mostly tropical temperatures. Hardy and resistant to disease and water quality variations. Foods: brine shrimp, bloodworms, glassworms, chopped clam, shrimp. Large species will accept small guppies.

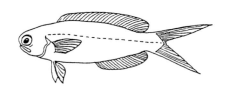

291) TERAPONIDAE, 293) KUHLIIDAE
Grunters and aholeholes. Schools preferred. Large, spacious aquaria. Very hardy; not prone to disease; very tolerant of water quality fluctuations. Wide salinity tolerance. Foods: brine shrimp, bloodworms, prawn, all prepared foods, feeder fishes.

299) SILLAGINIDAE,
300) MALACANTHIDAE
Smelt-whitings and tilefishes. Most aquarium species are small, colorful, and shy. Feed mostly on small invertebrates. Good for reef aquaria. Often feed poorly, but good foods would include live brine shrimp, bloodworms, glassworms, chopped clam. They are very sensitive to any fluctuation in water quality. Prone to transport shock; acclimate with care.

301) LABRACOGLOSSIDAE,
302) LACTARIIDAE
False trevallies and relatives. Midwater schooling fishes; large spacious tank needed. Feed mostly on small fishes; incompatible with smaller fishes.

303) POMATOMIDAE
Bluefishes. Ravenous schooling predators; incompatible with all other species. Very large tanks needed. Fairly hardy. Wide salinity tolerance.

304) RACHYCENTRIDAE
Cobia. Active, very large. Feed on fishes and crustaceans, so incompatible with most other species.

296) PRIACANTHIDAE
Bigeyes. Nocturnal predators, peaceful and shy with all they cannot swallow. Prefer dark overhangs and grottos in their tanks. Hardy, but some too large for average home aquaria.

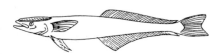

305) ECHENEIDIDAE

Remoras. Very hardy in aquaria. May attach to larger fishes with sucker disc (for transport only; no injury is involved). However, this does annoy some fishes. Best hosts are sharks. Live, frozen, prepared foods accepted with gusto.

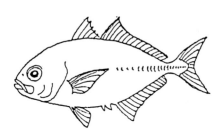

306) CARANGIDAE

Jacks. Fast-moving schooling fishes. Fairly hardy but require large tanks. Somewhat prone to transport shock. Many species have wide salinity tolerance. Will eat small fishes but are usually peaceful with those of like size. Foods: fish, chopped clam, prawn, squid.

307) NEMATISTIIDAE

Roosterfish. Care similar to Carangidae.

308) CORYPHAENIDAE

Dolphinfishes. Young specimens sometimes kept but are delicate; feed brine shrimp, bloodworms. Adults are swift, pelagic—not suitable for any but large public aquaria.

309) APOLECTIDAE

Habits and care similar to Carangidae.

310) MENIDAE,
311) LEIOGNATHIDAE,
312) BRAMIDAE

Moonfish, slipmouths, pomfrets, respectively. Mostly oceanic schooling fishes. Peaceful but active; spacious tanks necessary. Sensitive to fluctuations in water quality. Foods: small fishes, brine shrimp, bloodworms, squid, chopped clam.

314) ARRIPIDAE

Australian salmon. Midwater schoolers feeding on small fishes. Cool water, large tanks. Hardy but seldom seen.

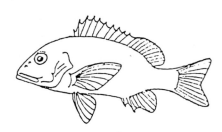

315) EMMELICHTHYIDAE,
316) LUTJANIDAE,
317) CAESIONIDAE

Rovers, snappers, fusiliers, respectively. Bottom to midwater schooling fishes. Very hardy, not prone to disease. Not sensitive to water quality fluctuations, and many species are euryhaline. Peaceful with fishes too large to swallow. Many species grow large and need big tanks. Coral/rock aquascape. All foods accepted.

318) LOBOTIDAE

Tripletails. Very hardy, euryhaline. Young specimens need planted tanks for camouflage. Peaceful with large fishes, will eat small ones. Most grow large but are fairly sedentary; provide caves for shelter.

319) GERREIDAE

Mojarras. Small, active schooling fishes. Peaceful, relatively hardy, euryhaline. Need plenty of open swimming room. Feed on planktonic crustaceans-live brine shrimp is a good substitute.

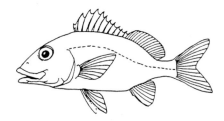

320) HAEMULIDAE,
321) INERMIIDAE,
322) SPARIDAE

Grunts, bonnetmouths, porgies, respectively.

Generally, snapper-like fishes. Midwater- to bottom-schoolers. Very hardy; disease resistant; tolerant of variations in water quality. Coral/rock/plant aquascapes. Often aggressive. All foods taken, but many species also graze on algae.

324) LETHRINIDAE, 325) NEMIPTERIDAE
Emperors and threadfin breams. Similar in appearance, behavior, and requirements to snappers.

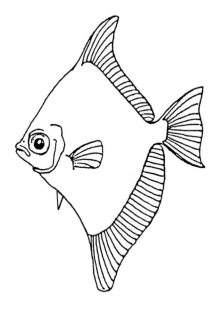

326) SCIAENIDAE
Drums. Bottom dwellers feeding on benthic invertebrates. Fairly hardy, but tropical species are prone to *Cryptocaryon* infection. Most species (except the very largest) are peaceful and shy. Coral rock aquascape. Prefer groups of same species. Foods: brine shrimp, bloodworms, glassworms, chopped clam, frozen prepared foods.

328) MONODACTYLIDAE
Monos. Brackish to marine. Planted aquaria, roots and rocks as decorations. Prefer groups of same species, but tend to be scrappy. Generally peaceful with other species. Not overly sensitive to water quality, but prone to ich. Foods: bloodworms, brine shrimp, glassworms, prepared foods.

329) PEMPHERIDIDAE
Sweepers. Small fishes, school in massive shoals in open water adjoining reefs; large tanks necessary. Very delicate and sensitive; not often seen in the hobby. Feed on planktonic invertebrates—try brine shrimp and bloodworms.

327) MULLIDAE
Goatfishes. Use barbels to find benthic invertebrates. Excellent scavengers. Hardy when acclimated, but very prone to transport shock. Peaceful. Soft substrate preferable. Foods: chopped clam, prawn, crab, brine shrimp, bloodworms, all prepared foods.

334) KYPHOSIDAE
Sea chubs. Open-water schoolers; large tanks. Often reach a large size. Very hardy. Peaceful but will eat small fishes. Algae necessary in diet. Other foods include clam, squid, crab, prawn.

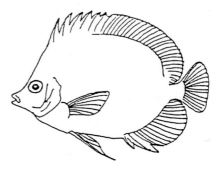

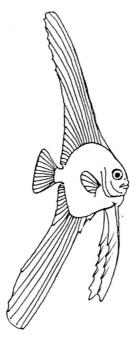

338) CHAETODONTIDAE
Butterflyfishes. Very beautiful reef fishes. Of variable hardiness. Some feed only on live corals and are almost impossible to keep alive in captivity. Most are sensitive to any change in water quality, and some are fairly disease-prone. Butterflyfishes are very peaceful with other species but may be aggressive with conspecifics, except for mated pairs collected together. Foods: brine shrimp, bloodworms, glassworms, chopped clam, frozen prepared foods.

335) EPHIPPIDIDAE
Spadefishes and batfishes. Very hardy (with the exception of *Platax pinnatus*) but grow quite large. Very peaceful. Young prefer plant thickets for camouflage. Need very deep tanks. Coral/rock aquascape for mature specimens. All foods taken vigorously.

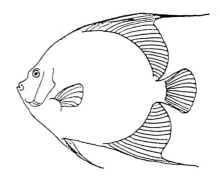

339) POMACANTHIDAE
Angelfishes. Hardy with relatively few exceptions. Not sensitive to most variations in water quality, but some suffer in the presence of high nitrates. Most are fairly disease-resistant but some species are prone to lymphocystis. Generally peaceful, but large specimens are often dominant fishes. Often aggressive with conspecifics. Coral/rock aquascape with lots of nooks will partition territories. Algae very necessary in diet. Other foods: brine shrimp, bloodworms, glassworms, chopped clam, prawn, boiled spinach.

340) ENOPLOSIDAE
Oldwife. Hardy, accepts all foods.

343) OPLEGNATHIDAE
Knifejaws. Peaceful but may eat small fishes. Will consume most crustaceans and many mol-

336) SCATOPHAGIDAE
Scats. Brackish to marine; very hardy but prone to lymphocystis. Not overly aggressive but sometimes nip at fins. Prefer schools but are often scrappy with conspecifics. Algae necessary in diet. Other foods: brine shrimp, bloodworms, prepared foods.

luscs. Coral/rock aquascape. Foods: shrimp, crab, chopped clam.

345) EMBIOTOCIDAE
Surfperches. Most need cool aquaria, but are otherwise hardy. Some reach a large size. Livebearers. Peaceful but will eat small fishes. Algae necessary in diet. Rock/plant aquascape.

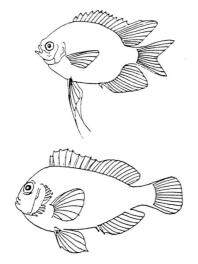

346) POMACENTRIDAE
Damselfishes (including anemonefishes). Very hardy and resistant to disease. Very scrappy with conspecifics, and very territorial in relations with other species. Lots of coral decorations will help to break up territories. Among the easiest of marine fishes to breed; they are parental-guarding substrate spawners. Anything edible will be accepted. Algae is also welcome.

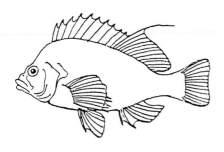

348) CIRRHITIDAE
Hawkfishes. Benthic predators. Like high coral heads as "perches" and vantage points. Some species grow rather large. Peaceful, extremely hardy. Coral/rock aquascape. Foods: guppies, bloodworms, brine shrimp, chopped clam, prepared foods.

350) APLODACTYLIDAE,
351) CHILODACTYLIDAE,
352) LATRIDIDAE
Morwongs and relatives. Generally similar to hawkfishes in habits and care, but most are somewhat larger.

353) OWSTONIIDAE,
354) CEPOLIDAE
Owstoniids and bandfishes. Somewhat delicate, not generally available. Feed on small invertebrates.

355) MUGILIDAE
Mullets. Fast-swimming schoolers. Euryhaline. Excellent jumpers tank must be well-covered. Feed on small invertebrates in substrate, but will accept most aquarium foods. Sensitive to fungal infections. Prone to transport shock; acclimate carefully. Large spacious tanks necessary.

356) SPHYRAENIDAE
Barracudas. Large midwater predators. Some are schooling fishes. Very large tanks needed. Toothy; handle with care. Usually accept only live fishes, but acclimated specimens will sometimes accept chunks of fish. Hardy and long-lived, but aggressive and difficult to house with other fishes.

357) POLYNEMIDAE
Threadfins. Benthic fishes. Peaceful, active. Feed on small fishes and crustaceans.

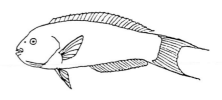

358) LABRIDAE
Wrasses. Most very hardy and peaceful. Most are at least territorial with conspecifics—provide plenty of room. Most invertebrates will not be molested except for small crustaceans. Many burrow in sand; soft substrate needed. Coral aquascape. With few exceptions, all foods will be taken. Good for reef aquaria.

359) ODACIDAE
Weed whitings. Wrasselike, care similar. Cool water and plainrock aquascape. Seldom seen.

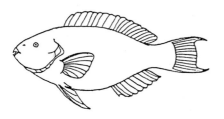

360) SCARIDAE
Parrotfishes. Moderately hardy, but some species are delicate. Often difficult to feed. Coral heads heavily encrusted with algae necessary in diet. Other foods: brine shrimp, chopped clam, bloodworms. Peaceful with other fishes but sometimes aggressive with conspecifics. Some species very large and need big tanks.

361) BATHYMASTERIDAE,
362) ZOARCIDAE,
363) STICHAEIDAE, 365) PHOLIDIDAE,
366) ANARHICHADIDAE
Eelpouts, wolffishes, and relatives. Bottom dwellers, some very large. Cool to cold water. Most eat benthic invertebrates, but fishes that can be caught will also be eaten. Rock aquascape with caves preferred. Very hardy. Foods: whole or chopped fish, clam, squid, also shrimp and crab.

375) OPISTOGNATHIDAE
Jawfishes. Secretive. Excavate burrows in sand and gravel bottoms. Very territorial with conspecifics, but conflicts are largely bluff. Peaceful with other species. Good with most invertebrates; excellent in reef aquaria. Hardy. Foods: brine shrimp, bloodworms, glassworms, prepared foods.

376) CONGROGADIDAE
Eelblennies. Very hardy. Some are euryhaline. Aggressive; will eat many tankmates (both fish and invertebrate). Bred in captivity. Foods: goldfish, chopped clam, prawn. Rocks and caves needed for shelter.

379) NOTOGRAPTIDAE
Similar to eelblennies in habits and care.

380) PHOLIDICHTHYIDAE
Convict blenny. Hardy. Prefers groups of conspecifics. Coral/rock aquascape with shady hollows. Peaceful. All foods accepted.

381) TRICHODONTIDAE
Sandfishes. Ambush predators. Need soft sand for burying. Incompatible with fishes small enough to swallow. Somewhat delicate and disease prone. Food: small feeder fishes.

382) TRACHINIDAE
Weeverfishes. Venomous, some lethally so. Ambush predators. Seldom kept.

383) URANOSCOPIDAE
Stargazers. Capable of delivering startling but sublethal electrical shock. Ambush predators; soft sand needed for burrowing. Will eat any fishes and crustaceans small enough to swallow. Some quite large and need big tanks. Foods: small live fishes as well as shrimp and crab meat. Very hardy.

384) TRICHONOTIDAE, 385) CREEDIIDAE,
386) LEPTOSCOPIDAE, 388) MUGILOIDIDAE
Sanddivers, sandperches, and relatives. Similar to sandfishes and lizardfishes in habits and care.

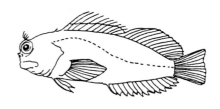

390) TRIPTERYGIIDAE,
391) DACTYLOSCOPIDAE,
392) LABRISOMIDAE, 393) CLINIDAE,
394) CHAENOPSIDAE, 395) BLENNIIDAE
Blennies and their relatives. Most are small, benthic fishes. Often very territorial with both conspecifics and other species; provide plenty of room. Extremely hardy, disease resistant, and tolerant of variations in water quality. Provide rocky caves and plenty of plants. Most eat small invertebrates, and algae is an important dietary item to many species. Other foods: brine shrimp, bloodworms, glassworms, chopped clam, most prepared foods including flake food.

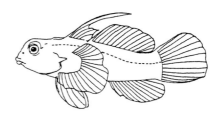

399) CALLIONYMIDAE
Dragonets. Benthic micropredators. Some hardy, others very delicate. Excellent for reef

tanks. Very peaceful with other fishes, but may fight among themselves. Very slow-moving, deliberate feeders that cannot compete with voracious feeders like damselfishes. Foods: brine shrimp, bloodworms, glassworms.

402) ELEOTRIDIDAE
Sleeper gobies. Ambush predators. Euryhaline. Lethargic; need shadowy holes for shelter. Very hardy and disease resistant. Many grow large; big tanks needed. Rocks, plants, driftwood. Foods: small fishes, chunks of beef heart, clam, squid, shrimp.

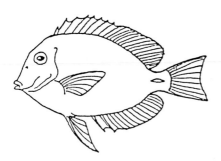

403) GOBIIDAE
Gobies. Benthic micropredators. Most small and peaceful, but territorial with their own species. Most fairly hardy. Some commonly bred in captivity (especially *Gobiosoma* Sp.). Coral aquascape including empty mollusc shells. Excellent in reef tanks. Foods: brine shrimp, bloodworms, glassworms, prepared foods.

404) GOBIOIDIDAE,
405) TRYPAUCHENIDAE,
407) MICRODESMIDAE
Eel-like gobies. Some are burrowers; soft substrate needed. Feed on small benthic invertebrates. Delicate and seldom seen.

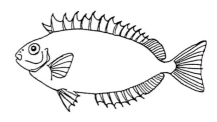

410) SIGANIDAE
Rabbitfishes. Dorsal spines venomous; handle with care. Habits and care as for Acanthuridae.

412) GEMPYLIDAE, 413) TRICHIURIDAE
Snake mackerels and cutlassfishes. Elongate, pelagic, often deep water. Piscivorous. Delicate; never kept.

414) SCOMBRIDAE
Tunas and mackerels. Extremely active pelagic schooling fishes. Too large except for public aquaria. Delicate and prone to shock. Many very large. Food almost exclusively small fishes.

415) XIPHIIDAE, 417) ISTIOPHORIDAE
Swordfishes. Pelagic piscivores. Do not adapt well to captivity—bills easily damaged.

419) CENTROLOPHIDAE, 420) NOMEIDAE
Medusafishes and driftfishes. Often commensal with jellyfishes or other pelagic cnidarians. Usually small and peaceful, but other fishes will be in danger from cnidarian host. Somewhat delicate and sensitive to water quality variations. Feed on planktonic invertebrates; good substitutes are brine shrimp and bloodworms.

423) STROMATEIDAE
Butterfishes. Open-water schooling fishes. Delicate; prone to transport injury and shock. Foods: brine shrimp, bloodworms.

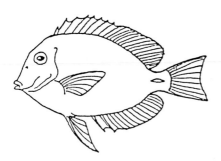

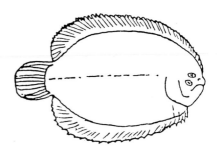

409) ACANTHURIDAE
Surgeonfishes. Algal grazers. Peaceful with other species; may be aggressive with conspecifics. Generally hardy but somewhat prone to *Oodinium* and *Cryptocaryon*. Coral aquascape. Foods: boiled spinach and vegetable-based prepared foods, brine shrimp, bloodworms, chopped clam.

432) PSETTODIDAE, 434) BOTHIDAE,
435) PLEURONECTIDAE,
436) CYNOGLOSSIDAE, 437) SOLEIDAE
Flatfishes. Benthic; need soft sand for burying. Some very large. Often euryhaline. Tropical as well as coldwater species. Generally peaceful with other fishes but will eat most benthic invertebrates. Fairly hardy. Foods: chopped fish, shrimp, crab, squid.

438) TRIACANTHODIDAE,
439) TRIACANTHIDAE
Spikefishes and triplespines. Relatively peaceful. Feed on small benthic invertebrates. Seldom kept.

442) TRIODONTIDAE
Three-toothed puffer. Habits and care as for Tetraodontidae.

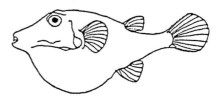

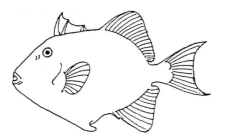

443) TETRAODONTIDAE
Puffers. Very hardy. May be nippy. May inflate when disturbed. If inflated with air, deflation often troublesome. All foods taken.

440) BALISTIDAE
Triggerfishes and filefishes. Triggers are aggressive, sometimes downright vicious. Files are a bit more peaceful but still may be nippy. Teeth and jaws powerful; large specimens may bite. Given to rearranging aquascape. Large triggers may attack heaters, filters, etc. Incompatible with most other species, and usually aggressive with conspecifics. Will eat anything remotely edible and may sample things that are not

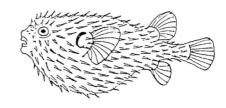

444) DIODONTIDAE
Porcupinefishes. Habits and care as for Tetraodontidae.
445) MOLIDAE
Molas. Huge, never kept. Feed mostly on jelly-fishes.

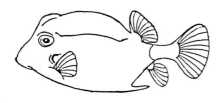

441) OSTRACIIDAE
Boxfishes. Very peaceful and shy. Slow moving; do not compete well with more active fishes. Some species will exude toxins when stressed, so acclimate with care. Fairly hardy but may be slow to begin feeding. Foods: brine shrimp, bloodworms, chopped clam.

CAPTIONS

The captions, where possible, identify the fishes, the family they belong to, feeding habits, aquarium lighting, temperament, aquarium decor, and swimming habits. Additionally, information about the greatest size each species normally reaches is included; this information should be useful in determining the tank size that should be used.

SYMBOLS:

Feeding:

℗ represents prepared foods, usually frozen. Very few marine fishes will do well on flake food although some types are quite nutritious.

∿ represents invertebrates. Although the symbol is obviously a worm, most marine fishes prefer crustaceans and/or molluscs. Many fishes have specific diets and must be supplied certain invertebrates or they will decline (ex. some butterflyfishes need live coral polyps). Please refer to the family write-ups in this book as well as other TFH marine fish books for more detailed information.

➤ represents live fishes.

✲ represents plant matter. This is usually in the form of algae and can be provided rather easily by growing your own in the tank or buying prepared algal foods from your dealer.

Light: These symbols represent the amount of light recommended for the tank. Remember, however, that some fishes are nocturnal and will be seen only when light levels are low. Others are at their best during the days, showing their colors off in direct sunlight.

◐ Bright with occasional sunlight

◑ Bright, no sunlight   ◕ As dark as possible as long as fishes are visible

Aggressiveness/Compatibility: Here again there is commonly no cut-and-dried distinction. A fish may be quite peaceful when small, but its appetite grows with its size and it may soon be foraging on the other tankmates. Others may be aggressive only during spawning seasons.

♥ Peaceful community fish   ☠ Not recommended for beginners

Tank Decoration:

🐚 represents a tank that should be supplied with plant life (normally algae). In some instances algae must be supplied because the inhabitants of the tank feed on it; in others it is decorative.

🪨 represents coral for the tropical tanks and rocks for the more temperate aquaria.

🏖 represents a sand or gravel bottom. Sand is required by some fishes (burrowers, etc.), while marine aquarists use dolomite or similar material in order to maintain the proper pH balance.

🐠 represents a balance between the two where corals (or rocks) are used along with plants for decoration. The new reef tanks would come under this designation.

Tank Level: Fish Locality: Many species are tied to one area of the tank. There are bottom species (like flatfishes) and surface fishes (like needlefishes) and some that will stay just about in the middle range all the time. Most species of marine fishes are wanderers and will swim pretty much around the tank investigating every corner.

⊡ Bottom swimmer   ⊡ Top swimmer

⊡ No special swimming level   ⊟ Swims in middle of water

cm: the length of a fish (standard length). This is the maximum length a fish is said to attain. Most specimens seen are normally much smaller, and aquarium specimens are generally smaller yet. Many species grow so large only juveniles are kept by aquarists. (An inch equals 2.54 cm.)

Family Number:
Family numbers follow the species name. The list of families is given beginning on page 25.

Systematic List of the Families of Fishes of the World
(after Nelson, 1984)

Myxiniformes
1 Myxinidae
Petromyzontiformes
2 Petromyzontidae
Chimaeriformes
3 Callorhynchidae
4 Chimaeridae
5 Rhinochimaeridae
Hexanchiformes
6 Chlamydoselachidae
7 Hexanchidae
Heterodontiformes
8 Heterodontidae
Lamniformes
9 Rhincodontidae
10 Orectolobidae
11 Odontaspididae
12 Lamnidae
13 Scyliorhinidae
14 Carcharhinidae
15 Sphyrnidae
Squaliformes
16 Squalidae
17 Pristiophoridae
18 Squatinidae
Rajiformes
19 Pristidae
20 Torpedinidae
21 Rhinobatidae
22 Rajidae
23 Dasyatidae
24 Potamotrygonidae
25 Hexatrygonidae
26 Myliobatididae
27 Mobulidae
Ceratodontiformes
28 Ceratodontidae
Lepidosireniformes
29 Lepidosirenidae
30 Protopteridae
Coelacanthiformes
31 Latimeriidae
Polypteriformes
32 Polypteridae
Acipenseriformes
33 Acipenseridae
34 Polyodontidae
Lepisosteiformes
35 Lepisosteidae
Amiiformes
36 Amiidae

Osteoglossiformes
37 Osteoglossidae
38 Pantodontidae
39 Hiodontidae
40 Notopteridae
41 Mormyridae
42 Gymnarchidae
Elopiformes
43 Elopidae
44 Megalopidae
45 Albulidae
46 Halosauridae
47 Notacanthidae
48 Lipogenyidae
Anguilliformes
49 Anguillidae
50 Heterenchelyidae
51 Moringuidae
52 Xenocongridae
53 Myrocongridae
54 Muraenidae
55 Nemichthyidae
56 Cyematidae
57 Synaphobranchidae
58 Ophichthidae
59 Nettastomatidae
60 Colocongridae
61 Macrocephenchelyidae
62 Congridae
63 Derichthyidae
64 Serrivomeridae
65 Saccopharyngidae
66 Eurypharyngidae
67 Monognathidae
Clupeiformes
68 Denticipitidae
69 Clupeidae
70 Engraulididae
71 Chirocentridae
Gonorynchiformes
72 Chanidae
73 Gonorynchidae
74 Kneriidae
75 Phractolaemidae
Cypriniformes
76 Cyprinidae
77 Psilorhynchidae
78 Homalopteridae
79 Cobitididae
80 Gyrinocheilidae
81 Catostomidae

Characiformes
82 Citharinidae
83 Hemiodontidae
84 Curimatidae
85 Anostomidae
86 Erythrinidae
87 Lebiasinidae
88 Gasteropelecidae
89 Ctenoluciidae
90 Hepsetidae
91 Characidae
Siluriformes
92 Diplomystidae
93 Ictaluridae
94 Bagridae
95 Cranoglanididae
96 Siluridae
97 Schilbidae
98 Pangasiidae
99 Amblycipitidae
100 Amphiliidae
101 Akysidae
102 Sisoridae
103 Clariidae
104 Heteropneustidae
105 Chacidae
106 Olyridae
107 Malapteruridae
108 Ariidae
109 Plotosidae
110 Mochokidae
111 Doradidae
112 Auchenipteridae
113 Pimelodidae
114 Ageneiosidae
115 Helogenidae
116 Cetopsidae
117 Hypophthalmidae
118 Aspredinidae
119 Trichomycteridae
120 Callichthyidae
121 Loricariidae
122 Astroblepidae
Gymnotiformes
123 Sternopygidae
124 Rhamphichthyidae
125 Hypopomidae
126 Apteronotidae
127 Gymnotidae
128 Electrophoridae

Perciformes

282 Centropomidae
283 Percichthyidae
284 Serranidae
285 Grammistidae
286 Pseudochromidae
287 Grammidae
288 Plesiopidae
289 Acanthoclinidae
290 Glaucosomatidae
291 Teraponidae
292 Banjosidae
293 Kuhliidae
294 Centrarchidae
295 Percidae
296 Priacanthidae
297 Apogonidae
298 Dinolestidae
299 Sillaginidae
300 Malacanthidae
301 Labracoglossidae
302 Lactariidae
303 Pomatomidae
304 Rachycentridae
305 Echeneididae
306 Carangidae
307 Nematistiidae
308 Coryphaenidae
309 Apolectidae
310 Menidae
311 Leiognathidae
312 Bramidae
313 Caristiidae
314 Arripidae
315 Emmelichthyidae
316 Lutjanidae
317 Caesionidae
318 Lobotidae
319 Gerreidae
320 Haemulidae
321 Inermiidae
322 Sparidae
323 Centracanthidae
324 Lethrinidae
325 Nemipteridae
326 Sciaenidae
327 Mullidae
328 Monodactylidae
329 Pempherididae
330 Leptobramidae
331 Bathyclupeidae
332 Toxotidae
333 Coracinidae
334 Kyphosidae
335 Ephippididae
336 Scatophagidae
337 Rhinoprenidae
338 Chaetodontidae

339 Pomacanthidae
340 Enoplosidae
341 Pentacerotidae
342 Nandidae
343 Oplegnathidae
344 Cichlidae
345 Embiotocidae
346 Pomacentridae
347 Gadopsidae
348 Cirrhitidae
349 Chironemidae
350 Aplodactylidae
351 Cheilodactylidae
352 Latrididae
353 Owstoniidae
354 Cepolidae
355 Mugilidae
356 Sphyraenidae
357 Polynemidae
358 Labridae
359 Odacidae
360 Scaridae
361 Bathymasteridae
362 Zoarcidae
363 Stichaeidae
364 Cryptacanthodidae
365 Pholididae
366 Anarhichadidae
367 Ptilichthyidae
368 Zaproridae
369 Scytalinidae
370 Bovichthyidae
371 Nototheniidae
372 Harpagiferidae
373 Bathydraconidae
374 Channichthyidae
375 Opistognathidae
376 Congrogadidae
377 Chiasmodontidae
378 Champsodontidae
379 Notograptidae
380 Pholidichthyidae
381 Trichodontidae
382 Trachinidae
383 Uranoscopidae
384 Trichonotidae
385 Creediidae
386 Leptoscopidae
387 Percophidae
388 Mugiloididae
389 Cheimarrhichthyidae
390 Tripterygiidae
391 Dactyloscopidae
392 Labrisomidae
393 Clinidae
394 Chaenopsidae
395 Blenniidae
396 Icosteidae

397 Schindleriidae
398 Ammodytidae
399 Callionymidae
400 Draconettidae
401 Rhyacichthyidae
402 Eleotrididae
403 Gobiidae
404 Gobioididae
405 Trypauchenidae
406 Kraemeriidae
407 Microdesmidae
408 Kurtidae
409 Acanthuridae
410 Siganidae
411 Scombrolabracidae
412 Gempylidae
413 Trichiuridae
414 Scombridae
415 Xiphiidae
416 Luvaridae
417 Istiophoridae
418 Amarsipidae
419 Centrolophidae
420 Nomeidae
421 Ariommatidae
422 Tetragonuridae
423 Stromateidae
424 Anabantidae
425 Belontiidae
426 Helostomatidae
427 Osphronemidae
428 Luciocephalidae
429 Channidae
430 Mastacembelidae
431 Chaudhuriidae

Pleuronectiformes

432 Psettodidae
433 Citharidae
434 Bothidae
435 Pleuronectidae
436 Cynoglossidae
437 Soleidae

Tetraodontiformes

438 Triacanthodidae
439 Triacanthidae
440 Balistidae
441 Ostraciidae
442 Triodontidae
443 Tetraodontidae
444 Diodontidae
445 Molidae

Photographers
Pictorial Identification Section

We are indebted to the following photographers, many of whom are certainly of world-class caliber and have taken some of the finest fish portraits that have ever graced the pages of any book. We sincerely apologize to anybody who has been inadvertently omitted from this list.

Robert Abrams
Ray Allard
Dr. Gerald R. Allen
Paul Allen
Neil Armstrong
Charles Arneson
Glen S. Axelrod
Dr. Herbert R. Axelrod
Wayne Baldwin
Heiko Bleher
Guy van den Bossche
Dr. Martin R. Brittan
Dr. Warren E. Burgess
Taylor Cafferey
Bruce Carlson
Kok-Hang Choo
Neville Coleman
Dr. Patrick L. Colin
Walter Deas
Helmut Debelius
Dr. Guido Dingerkus
Wade Doak
Douglas Faulkner
Stanislav Frank
U. Erich Friese
Karl Frogner
Michio Goto (Marine Life Documents)
Daniel W. Gotshall
Hilmar Hansen (Aquarium Berlin)
Edmund Hobson
Scott Johnson
Earl Kennedy
Alex Kerstitch

Karl Knaack
Alexandr Kochetov
Sergei Kochetov
Rudie Kuiter
Pierre Laboute
Dr. Roger Lubbock
Ken Lucas (Steinhart Aquarium)
Gerhard Marcuse
George Miller
Robert F. Myers
Arend van den Nieuwenhuizen
Aaron Norman
James H. O'Neill
Brian J. Parkinson
Klaus Paysan
Nicholas Polunin
Allan Power
Dr. John E. Randall
Andre Roth
Barry C. Russell
Dr. Shih-Chieh Shen
Dr. Victor G. Springer
Dr. Walter A. Starck II
Roger Steene
William Stephens
Katsumi Suzuki
Yoshio Takemura
Dr. Denis Terver
Dr. R. E. Thresher
Gene Wolfsheimer
Dr. Loren P. Woods
Dr. Fujio Yasuda

In addition, the following photographers provided photos used in the Marine Aquarium Set-up and Maintenance Section beginning on page 673: Aqua Module; David Axelrod; John Burleson; Bernd Degen; Dr. Mark P. Dulin; Dr. C. W. Emmens; W. Frickhinger; Dr. R. Geisler; Michael Gilroy; Peter T. Jam; Burkhard Kahl; Dr. Don E. McAllister; Midori Shobo, Fish Magazine, Japan; W. Paccagnella; Steve Robinson; Fred Rosenzweig; Fritz Seidel; Dr. Kenneth Simpson; George Smit; Gunther Spies; W. Tomey; Dieter Untergasser; Peter Wilkens; Steve Wright.

The line drawings of family representatives were accomplished by Mr. John Quinn.

Acknowledgments

We wish to thank the many people who have helped with every aspect of the formation of this book. We particularly acknowledge Lourdes A. Burgess, Mary Ellen Sweeney, and Ray Weigand for their contributions in researching the material used in the writing of the original captions, and to Jerry Walls for reading most of the material and making relevant suggestions for its improvement.

Many thanks to those who have helped in other ways such as continuing to provide reprints of their papers or by identifying the color proofs sent to them. Unfortunately, lack of time prevented others from completing this task and they must accept our identifications based on the literature at hand. Hopefully future editions will be able to carry any corrections made in the interim.

WEIGHTS & MEASURES

CUSTOMARY U.S. MEASURES AND EQUIVALENTS

METRIC MEASURES AND EQUIVALENTS

LENGTH

1 inch (in)		= 2.54 cm
1 foot (ft)	= 12 in	= .3048 m
1 yard (yd)	= 3 ft	= .9144 m
1 mile (mi)	= 1760 yd	= 1.6093 km
1 nautical mile	= 1.152 mi	= 1.853 km

1 millimeter (mm)		= .0394 in
1 centimeter (cm)	= 10 mm	= .3937 in
1 meter (m)	= 1000 mm	= 1.0936 yd
1 kilometer (km)	= 1000 m	= .6214 mi

AREA

1 square inch (in^2)		= 6.4516 cm^2
1 square foot (ft^2)	= 144 in^2	= .093 m^2
1 square yard (yd^2)	= 9 ft^2	= .8361 m^2
1 acre	= 4840 yd^2	= 4046.86 m^2
1 square mile (mi^2)	= 640 acre	= 2.59 km^2

1 sq centimeter (cm^2)	= 100 mm^2	= .155 in^2
1 sq meter (m^2)	= 10,000 cm^2	= 1.196 yd^2
1 hectare (ha)	= 10,000 m^2	= 2.4711 acres
1 sq kilometer (km^2)	= 100 ha	= .3861 mi^2

WEIGHT

1 ounce (oz)	= 437.5 grains	= 28.35 g
1 pound (lb)	= 16 oz	= .4536 kg
1 short ton	= 2000 lb	= .9072 t
1 long ton	= 2240 lb	= 1.0161 t

1 milligram (mg)		= .0154 grain
1 gram (g)	= 1000 mg	= .0353 oz
1 kilogram (kg)	= 1000 g	= 2.2046 lb
1 tonne (t)	= 1000 kg	= 1.1023 short tons
1 tonne		= .9842 long ton

VOLUME

1 cubic inch (in^3)		= 16.387 cm^3
1 cubic foot (ft^3)	= 1728 in^3	= .028 m^3
1 cubic yard (yd^3)	= 27 ft^3	= .7646 m^3
1 fluid ounce (fl oz)		= 2.957 cl
1 liquid pint (pt)	= 16 fl oz	= .4732 l
1 liquid quart (qt)	= 2 pt	= .946 l
1 gallon (gal)	= 4 qt	= 3.7853 l
1 dry pint		= .5506 l
1 bushel (bu)	= 64 dry pt	= 35.2381 l

1 cubic centimeter (cm^3)		= .061 in^3
1 cubic decimeter (dm^3)	= 1000 cm^3	= .353 ft^3
1 cubic meter (m^3)	= 1000 dm^3	= 1.3079 yd^3
1 liter (l)	= 1 dm^3	= .2642 gal
1 hectoliter (hl)	= 100 l	= 2.8378 bu

TEMPERATURE

C°	25°	-18°	-10°	0°	10°	20°	30°	40°	50°	60°	70°	80°	90°	100°
F°	-13°	0°	10°	20°	32° 40°	50°	60°	70°	80°	90°	100° 110°	120° 130°	140° 150°	160° 170° 180° 190° 200° 212°

$$CELSIUS° = 5/9 \ (F° - 32°) \qquad FAHRENHEIT° = 9/5 \ C° + 32°$$

PICTORIAL IDENTIFICATION SECTION

The following pictorial identification section includes thousands of photographs of marine fishes from around the world. The families are in systematic sequence according to Nelson (1984), or as close to that as practical. (See page 25.) The individual photos are coded with a family number for ready reference. In some cases some photos are out of sequence due to reasons beyond our control, but the family numbers should make it easy to place these "orphans" in their proper sequence.

Aquarists can scan through these pages until they find the correct family for the fish they are trying to identify. They can then look more carefully at the species depicted until they find the photo that most closely resembles their fish. One has to remember that differences in size, sex, geography, and even temperament may cause the fish to look different. However, in most cases an identification can be made with reasonable certainty. Once a correct name is arrived at the species can be researched in other references or the caption may provide all the information needed.

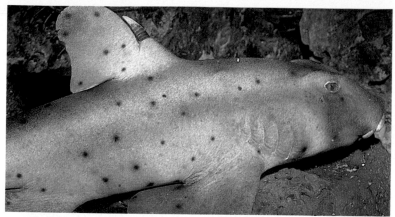

Heterodontus francisci 8 ∿ ○ ✕ ▣ ▭96 cm
Heterodontus mexicanus 8 ∿ ○ ✕ ▣ ▭ 96 cm

Stegostoma varium 10 ↘ ◑ ♥ ▣ ▭ 230 cm

Hydrolagus collei 4 ∿ ↖ ◑ ♥ 🖻 ⊡ 96 cm
Heterodontus zebra 8 ∿ ○ ✕ 🖻 ⊡ 100 cm

Rhincodon typus 9 ∿ ○ ♥ ⊡ 1000-1200 cm

Ginglymostoma cirratum 10 ⌇ ⤸ ◐ ✻ 🔲 ⬜ 430 cm
Hemiscyllium trispeculare 10 ⌇ ◐ ✻ 🔲 ⬜ 62 cm

Orectolobus maculatum 10 ⌇ ◐ ✻ 🔲 ⬜ 320 cm

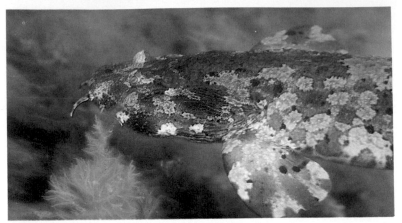

Orectolobus ornatus 10 ⌇ ↘ ◐ ✖ ▱ ▱ 215 cm
Eucrossorhinus dasypogon 10 ⌇ ↘ ◐ ✖ ▱ ▱ 120 cm

Chiloscyllium confusum 10 ⌇ ↘ ◐ ✖ ▰ ▱ 220 cm

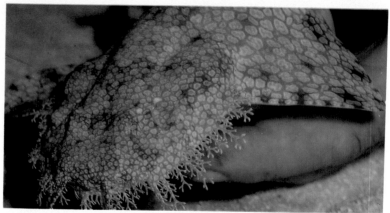

Eucrossorhinus dasypogon 10 〜 ❯ ◖ ✕ ▱ ▱ 120 cm
Chiloscyllium griseum 10 〜 ❯ ◖ ✕ ▣ ▣ 240 cm

Chiloscyllium punctatum 10 〜 ❯ ◖ ✕ ▣ ▣ 110 cm

Scyliorhinus canicula 13 ❟ ◐ ✖ ▦ ▭ 80 cm
Cephaloscyllium ventriosum 13 ↷ ❟ ◐ ✖ ▣ ▭ 100 cm

Mustelus henlei 14 ❟ ◐ ✖ ▦ ▭ 94 cm

Triakis scyllia 14 〜 ↘ ◑ ✴ 🔲 🗄 100 cm
Galeocerdo cuvier 14 〜 ↘ ◐ ✴ 🗄 730 cm

Carcharhinus melanopterus 14 ↘ ◐ ✴ 🗄 200 cm

Triakis semifasciatus 14 ⤻ ◐ ✗ 🖻 ⊟ 200 cm
Carcharhinus amblyrhynchos 14 ⤻ ◐ ✗ 🖻 ⊟ 250 cm

Carcharhinus plumbeus 14 ⤻ ◐ ✗ ⊟ 220 cm

Hypnos monopterygium 20 ∿ ◑ ✕ ▱ ▱ 70 cm
Narcine brunneus 20 ∿ ↘ ◑ ✕ ▱ ▱ 20 cm

Narcine brasiliensis 20 ∿ ↘ ◑ ✕ ▱ ▱ 45 cm

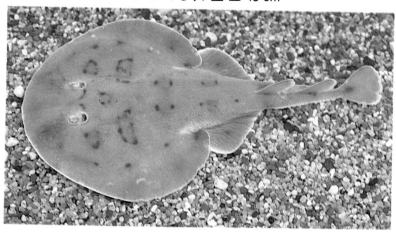

Torpedo sinuspersici 20 ᭜ ↘ ◖ ✂ ▱ ▱ 130 cm
Trygonorhina fasciata 21 ᭜ ↘ ◖ ✂ ▱ ▱ 100 cm

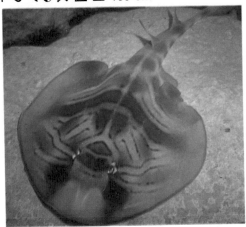

Platyrhinoidis triseriata 21 ᭜ ↘ ◖ ✂ ▱ ▱ 91 cm

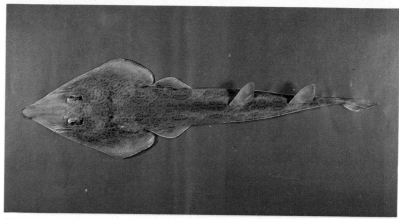

Rhinobatos hynnicephalus 21 ∿ ◑ ♥ ▱ ▱ 100 cm
Rhinobatos vincentiana 21 ∿ ◑ ♥ ▱ ▱ 100 cm

Taeniura lymma 23 ∿ ◑ ♥ ▱ ▱ 25 cm

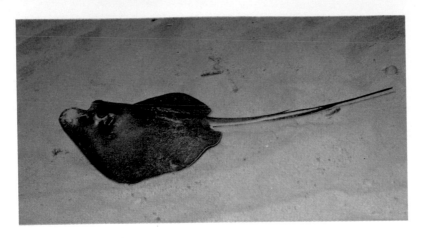

Dasyatis americana 22 ⌇ ↘ ◑ ✕ ▱ ▱ 90 cm
Urolophus aurantiacus 22 ⌇ ↘ ◑ ✕ ▱ ▱ 40 cm

Urolophus halleri 22 ⌇ ↘ ◑ ✕ ▱ ▱ 56 cm

Urolophus mucosus 22 〜 ↘ ◑ ✕ ▭ ▭ 35 cm
Urolophus jamaicensis 22 〜 ↘ ◑ ✕ ▭ ▭ 62 cm

Urolophus concentricus 22 〜 ↘ ◑ ✕ ▭ ▭ 45 cm

Latimeria chalumnae 31 ↘ ◑ ✕ 🔒 ⬚ 135 cm
Elops machnata 43 ↘ ◑ ♥ ⬚ ⊟ 80 cm

Megalops cyprinoides 44 ↘ ◑ ✕ ⬚ ⊟ 150 cm

Anguilla japonica 49 ⤳ ↖ ◑ ✂ ▣ ☐ 100 cm
Moringa microchir 51 ⤳ ◑ ♥ ▣ ☐ 39 cm

Echidna catenata 54 ⤳ ↖ ◑ ✂ ▣ ☐ 70 cm

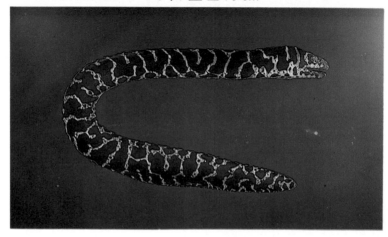

Gymnomuraena zebra 54 ∿ ↘ ◑ ✖ 🔳 ⬜ 90 cm
Gymnothorax fimbriatus 54 ∿ ↘ ◑ ✖ 🔳 ⬜ 80 cm

Gymnothorax rueppelliae 54 ∿ ↘ ◑ ✖ 🔳 ⬜ 55 cm

Uropterygius concolor 54 〜 ＼ ◑ ✕ 🔳 ▭ 32 cm
Gymnothorax steindachneri 54 〜 ＼ ◑ ✕ 🔳 ▭ 91 cm

Gymnothorax woodwardi 54 〜 ＼ ◑ ✕ 🔳 ▭ 75 cm

Gymnothorax flavimarginatus 54 ⤳ ↘ ◐ ✻ 📷 🖵 100 cm
Gymnothorax moringa 54 ⤳ ↘ ◐ ✻ 📷 🖵 70 cm

Gymnothorax castaneus 54 ⤳ ↘ ◐ ✻ 📷 🖵 120 cm

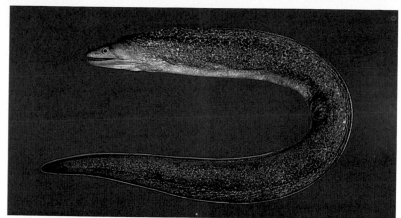

Gymnothorax vicinus 54 〜 ↖ ◑ ✕ 🎦 ▭ 120 cm
Gymnothorax funebris 54 〜 ↖ ◑ ✕ 🎦 ▭ 240 cm

Gymnothorax mordax 54 〜 ↖ ◑ ✕ 🎦 ▭ 200 cm

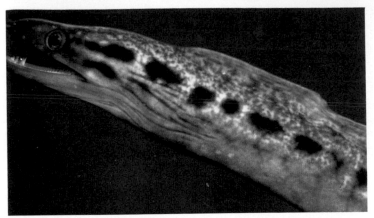

Gymnothorax margaritophorus 54 〰 ↖ ◐ ✕ ▣ ▢ 36 cm
Gymnothorax meleagris 54 〰 ↖ ◐ ✕ ▣ ▢ 90 cm

Gymnothorax permistus 54 〰 ↖ ◐ ✕ ▣ ▢ 75 cm

Gymnothorax nubilis 54 ～ ┑ ◑ ✕ 🎞 🖵 50 cm
Gymnothorax obesus 54 ～ ┑ ◑ ✕ 🎞 🖵 178 cm

Gymnothorax ramosus 54 ～ ┑ ◑ ✕ 🎞 🖵 80 cm

Siderea picta 54 ᔑ ᖰ ◐ ✖ ▣ ▢ 68 cm
Gymnothorax undulatus 54 ᔑ ᖰ ◐ ✖ ▣ ▢ 150 cm

Gymnothorax zonipectus 54 ᔑ ᖰ ◐ ✖ ▣ ▢ 31 cm

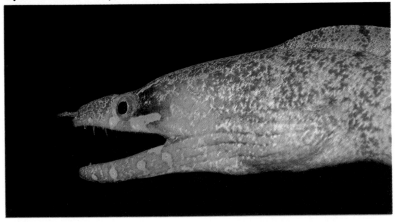

Gymnothorax flavimarginatus? 54 〰 ➘ ◐ ☆ 📷 🖼 80 cm
Gymnothorax prasinus 54 〰 ➘ ◐ ☆ 📷 🖼 30 cm

Muraena pardalis 54 〰 ➘ ◐ ☆ 📷 🖼 100 cm

Echidna nebulosa 54 〜 ⌇ ◗ ✕ 🖼 ⬒ 70 cm
Gymnothorax breedeni 54 〜 ⌇ ◗ ✕ 🖼 ⬒ 65 cm

Echidna nebulosa 54 ～ヽ ● ✕ 🔲 🔲 70 cm
Enchelycore bayeri 54 ～ヽ ● ✕ 🔲 🔲 70 cm

Gymnothorax favagineus 54 ～ヽ ● ✕ 🔲 🔲 250 cm

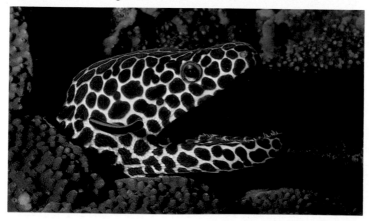

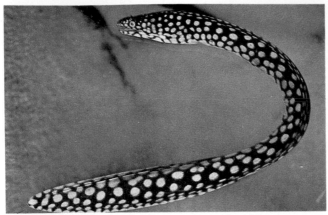

Muraena melanotis 54 ～ 丶 ◑ ✄ 🎦 ▭ 100 cm
Muraena lentiginosa 54 ～ 丶 ◑ ✄ 🎦 ▭ 60 cm

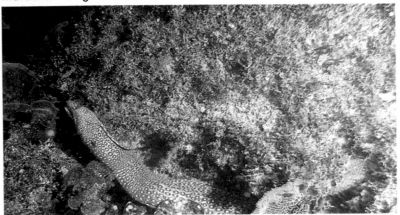

Gymnothorax miliaris 54 ～ 丶 ◑ ✄ 🎦 ▭ 60 cm

Muraenichthys tasmaniensis 58 〜 ↘ ◑ ☆ 📷 🖥 35 cm
Myrichthys acuminatus 58 〜 ↘ ◑ ☆ 📷 🖥 102 cm

Myrichthys maculosus 58 〜 ↘ ◑ ☆ 📷 🖥 100 cm

Leiuranus semicinctus 58 〰 ↘ ◑ ✕ 🎥 ☐ 60 cm
Myrichthys colubrinus 58 〰 ↘ ◑ ✕ 🎥 ☐ 75 cm

Myrichthys ocellatus 58 〰 ↘ ◑ ✕ 🎥 ☐ 100 cm

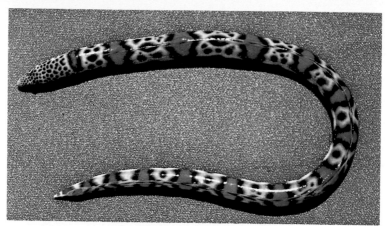

Quassiremus notochir 58 〜 ↘ ◑ ✖ ▱ ▱ 85 cm
Sphagebranchus flavicauda 58 〜 ◑ ♥ ▱ ▱ 30 cm

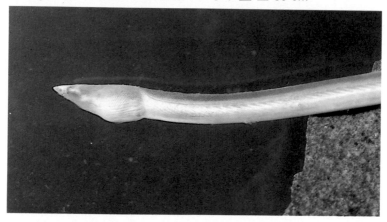

Aprognathodon platyventris 58 〜 ↘ ◑ ✖ ▱ ▱ 23 cm

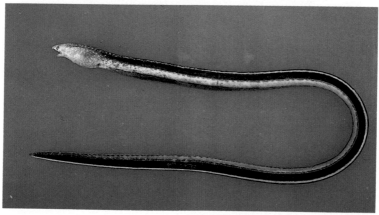

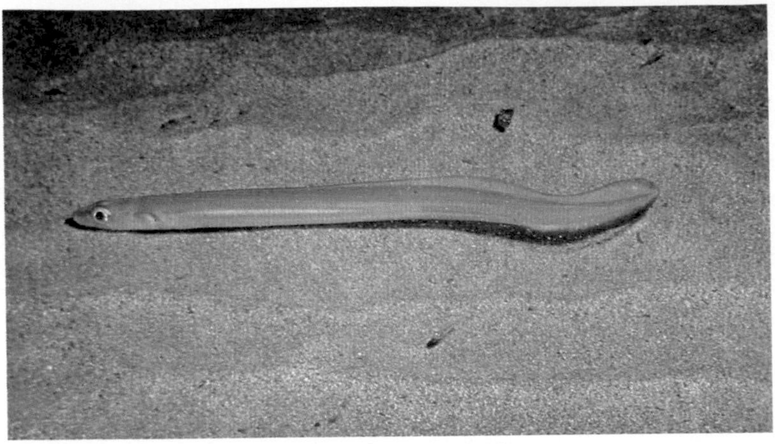

Ariosoma impressa 62 18 cm

Conger wilsoni 62 150 cm

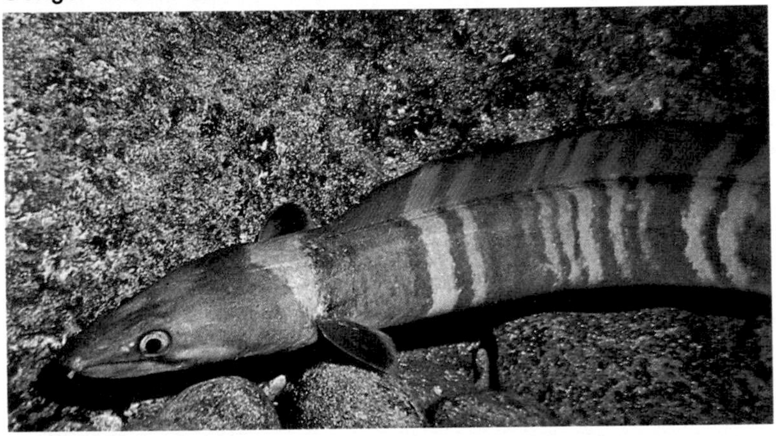

Gorgasia preclara 62 40 cm

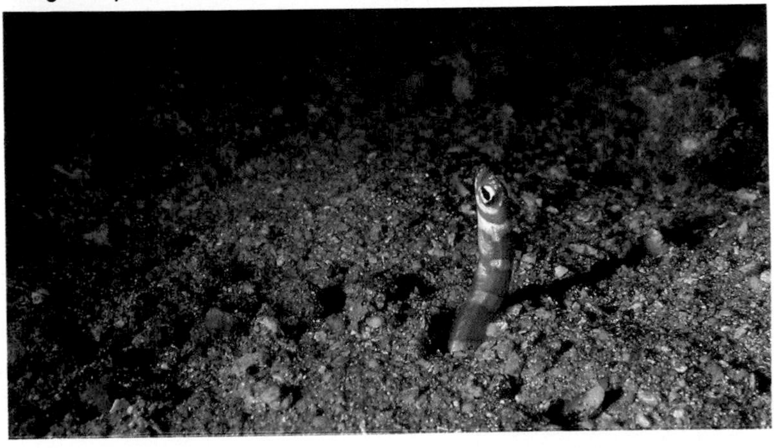

Rhinomuraena quaesita 54 〜 ↘ ◐ ✕ 📷 🖥 120 cm
Myrichthys maculosus 58 〜 ↘ ◐ ✕ 📷 🖥 100 cm

Gorgasia preclara 62 〜 ↘ ◐ ✕ 📷 🖥 40 cm

Brachysomophis crocodilinus 58 〜 ﹨ ◑ ✕ ▣ ▭ 110 cm
Taenioconger hassi 62 〜 ﹨ ◑ ✕ ▭ ▭ 36 cm

Anodontostoma chacunda 69 ∿ ◑ ♥ 🔲 🔄 17 cm
Konosirus punctatus 69 ∿ ◑ ♥ 🔲 🔄 30 cm

ARITA

Opisthonema libertate 69 ∿ ◑ ♥ 🔲 🔄 23 cm

Platycephalus haackei 269 ᴠᴦ ↘ ◐ ✕ ▱ ▱ 40 cm
Aulopus purpurissatus 153 ᴠᴦ ↘ ◐ ✕ ▱ ▱ 60 cm

Synodus jaculum 157 ᴠᴦ ↘ ◐ ✕ ▱ ▱ 15 cm

Synodus rubromarmoratus 157 ᴠ ᴠ ◐ ⚒ ▦ ▱ 7.5 cm
Synodus ulae 157 ᴠ ᴠ ◐ ⚒ ▦ ▱ 25 cm

Synodus synodus 157 ᴠ ᴠ ◐ ⚒ ▦ ▱ 30 cm

Synodus synodus 157 ∿ ↖ ◑ ✕ ▱ ▱ 30 cm
Synodus lacertinus 157 ∿ ↖ ◑ ✕ ▱ ▱ 10 cm

Synodus sp. 157 ∿ ↖ ◑ ✕ ▱ ▱ 15 cm

? Lotella sp. 171 ∿ ⌇ ◑ ✕ 🔳 ▱ 50 cm
Lotella rhacinus 171 ∿ ⌇ ◑ ✕ 🔳 ▱ 45 cm

Pseudophycis breviusculus 171 ∿ ◑ ♥ 🔳 ▱ 7 cm

Gadus morhua 174 ⌇ ↘ ◑ ✕ ▣ ▭ 150 cm
Urophycis chuss 174 ⌇ ↘ ◑ ✕ ▣ ▭ 52 cm

Coryphaenoides asper 176 ⌇ ↘ ◑ ✕ ▣ ▭ 10 cm

Lepophidium prorates 177 ∿ ◑ ✕ ▨ ▢ 27 cm
Brotula multibarbata 177 ↖ ◑ ✕ ▣ ▢ 100 cm

Onuxodon margaritiferae 178 ∿ ◑ ♥ ▨ ▢ 9 cm

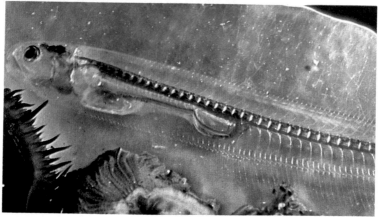

Stygnobrotula latibricola 179 〜 ◐ ♥ 📷 🖥 8 cm
Brotulina fusca 179 〜 ◐ ♥ 📷 🖥 12 cm

Ogilbia cayorum 179 〜 ◐ ♥ 🖼 🖥 10 cm

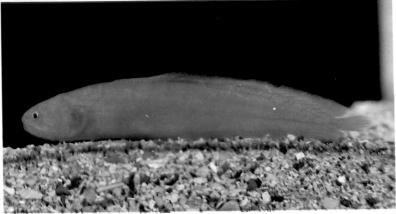

Sanopus splendidus 181 ∿ ↘ ◑ ✕ 🎥 ▭ 20 cm
Porichthys notatus 181 ∿ ↘ ◑ ✕ 🎥 ▭ 38 cm

Batrachomoeus trispinosus 181 ∿ ↘ ◑ ✕ 🎥 ▭ 20 cm

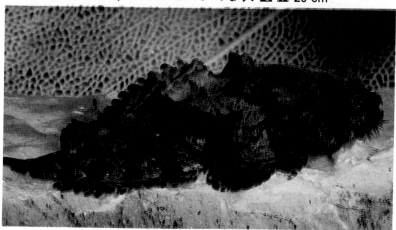

Halophryne diemensis 181 ⤳ ↘ ◑ ✳ 🖼 ⬛ 30 cm
Sanopus splendidus 181 ⤳ ↘ ◑ ✳ 📷 ⬛ 20 cm

Porichthys margaritatus 181 ⤳ ↘ ◑ ♥ 📷 ⬛ 11.3 cm

Lophiocharon trisignatus 183 ↘ ◑ ✕ 🐟 ☐ 15 cm
Tathicarpus butleri 183 ↘ ◑ ✕ 🐟 ☐ 10 cm

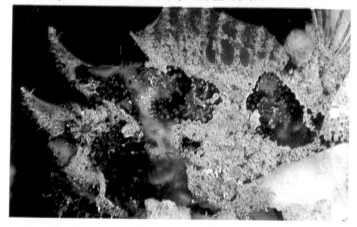

Antennarius maculatus 183 ↘ ◑ ✕ 🐟 ☐ 8.5 cm

Tathicarpus butleri 183 ↘ ◑ ✕ 🐟 ▭ 10 cm
Histrio histrio 183 ↘ ○ ✕ 🐟 ▭ 14 cm

Antennarius biocellatus 183 ↘ ◑ ✕ 🐟 ▭ 12 cm

Antennarius strigatus 183 ↘ ◑ ✕ 🖼 ▭ 8 cm
Antennarius multiocellatus 183 ↘ ◑ ✕ 🖼 ▭ 11 cm

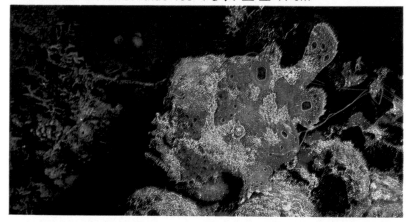

Antennarius ocellatus 183 ↘ ◑ ✕ 🖼 ▭ 32 cm

Antennarius avalonis 183 ↘ ◐ ✕ 🖾 ☐ 33 cm
Antennarius pauciradiatus 183 ↘ ◐ ✕ 🖾 ☐ 4 cm

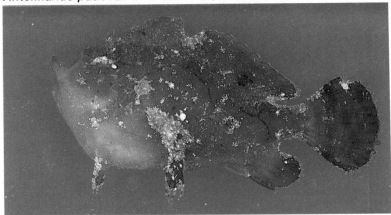

Antennarius tuberosus 183 ↘ ◐ ✕ 🖾 ☐ 7 cm

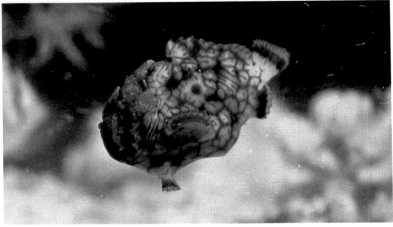

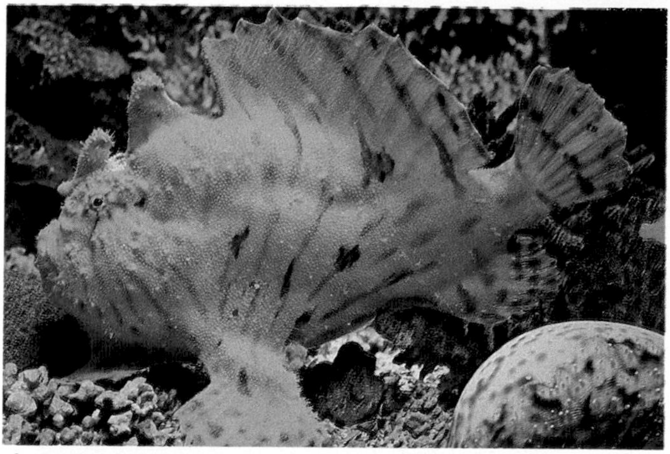

Antennarius hispidus 183 ↘ ◐ ☠ ▨ ◻ 15 cm

Antennarius commersoni 183 ↘ ◐ ☠ ▨ ◻ 29 cm

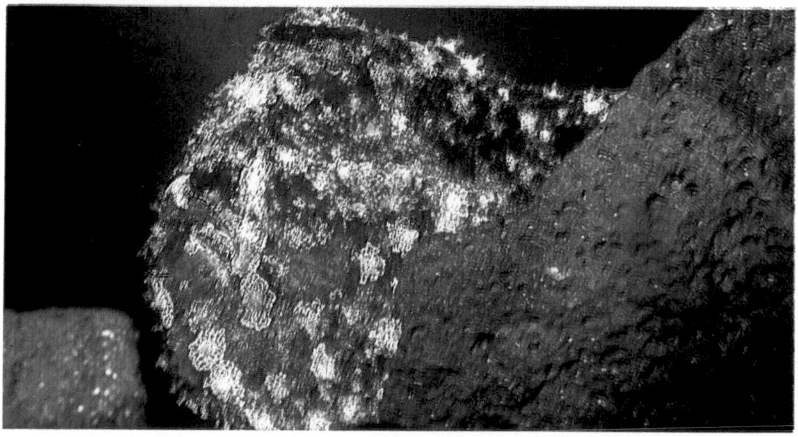

Antennarius nummifer 183 ↘ ◐ ☠ ▨ ◻ 29 cm

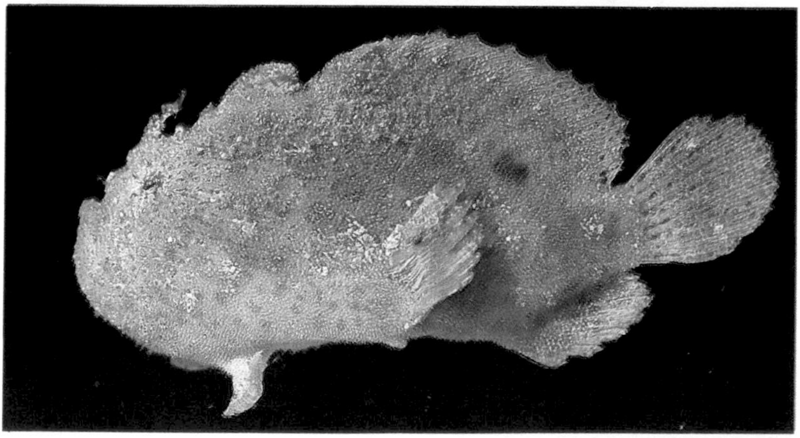

Antennarius striatus 183 ↖ ◑ ✕ 🖼 ▭ 15.5 cm
Antennarius coccineus 183 ↖ ◑ ✕ 🖼 ▭ 9 cm

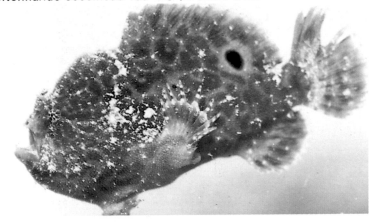

Antennarius indicus 183 ↖ ◑ ✕ 🖼 ▭ 19 cm

Halieutichthys aculeatus 186 ～ ◐ ♥ ▱ ▱ 10 cm
Halieutaea stellata 186 ～ ◐ ♥ ▱ ▱ 30 cm

Ogcocephalus nasutus 186 ～ ◐ ♥ ▱ ▱ 38 cm

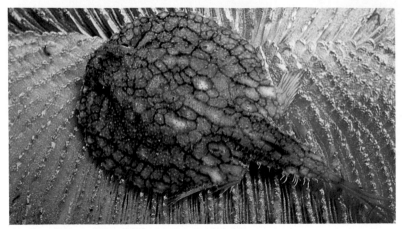

Halieutaea retifera 186 ↰ ◑ ♥ 🖼 ⬛ 10 cm
Zalieutes elator 186 ↰ ◑ ♥ 🖼 ⬛ 15 cm

Ogcocephalus corniger 186 ↰ ◑ ♥ 🖼 ⬛ 23 cm

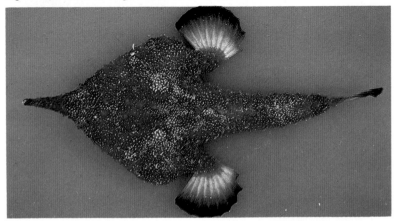

Gobiesocid 198 ∿ ◑ ♥ 🎞 ▭ 2.5 cm
Cochleoceps spatula 198 ∿ ◑ ♥ 🎞 ▭ 2.5 cm

Sicyases sanguineus 198 ∿ ↘ ◑ ✖ ▭ 22 cm

Gobiesocid 198 🜚 ◐ ♥ 📷 ▭ 2.5 cm
Aspasmogaster tasmaniensis 198 🜚 ◐ ♥ 📷 ▭ 1.5 cm

Sicyases sanguineus 198 🜚 ↘ ◐ ✄ 📷 ▭ 22 cm

Lepadogaster candollei 198 〰 ◑ ♥ 🖼 🖵 12 cm
Discotrema crinophila 198 〰 ◑ ♥ 🖼 🖵 5 cm

Tomicodon humeralis 198 〰 ◑ ♥ 🖼 🖵 8.3 cm

Lepadogaster lepadogaster 198 ⌇ ◑ ♥ 📷 ⬒ 10 cm
Lepadichthys frenatus 198 ⌇ ◑ ♥ 📷 ⬒ 3 cm

Gobiesox maeandricus 198 ⌇ ◑ ♥ 📷 ⬒ 16 cm

Hirundichthys sp. 200 ∿ ↘ ◐ ♥ ▱ 26 cm
Cheilopogon pinnatibarbatus japonicus 200 ∿ ↘ ◐ ♥ ▱ 30 cm

Cypselurus hiraii 200 ∿ ↘ ◐ ♥ ▱ 30 cm

Monocentrus japonicus 229 ↝ ◑ ♥ 🖾 ⬚ 16 cm
Cleidopus gloriamaris 229 ↝ ◑ ♥ 🖾 ⬚ 28 cm

Hoplostethus elongatus 230 ↝ ◑ ♥ 🖾 ⬚ 12 cm

Anomalops katoptron 231 ∿ ↘ ◑ ✕ 🔳 ⊟ 30 cm
Beryx decadactylus 234 ∿ ↘ ◑ ✕ 🔳 ⊟ 60 cm

Trachichthodes affinis 230 ∿ ↘ ◑ ✕ 🔳 ⊟ 45 cm

Sargocentron diadema 235 ♀ ∿ ⌐ ◐ ✕ ▣ ▭ 23 cm
Sargocentron xantherythrus 235 ∿ ⌐ ◐ ✕ ▣ ▭ 18 cm

Sargocentron violaceum 235 ♀ ∿ ⌐ ◐ ✕ ▣ ▭ 25 cm

Sargocentron tieroides 235 ♀ ∿ ⤙ ◑ ✖ 🖻 ▭ 30 cm
Sargocentron bullisi 235 ♀ ∿ ⤙ ◑ ✖ 🖻 ▭ 13 cm

Sargocentron vexillarium 235 ♀ ∿ ⤙ ◑ ✖ 🖻 ▭ 12.6 cm

Sargocentron diadema 235 ♀ ∿ ⤙ ◑ ✕ 📷 🖵 23 cm
Sargocentron microstoma 235 ♀ ∿ ⤙ ◑ ✕ 📷 🖵 18 cm

Sargocentron rubrum 235 ♀ ∿ ⤙ ◑ ✕ 📷 🖵 18 cm

Sargocentron caudimaculatum 235 ♀ ⤳ ⤸ ◐ ✻ 🖿 ⬓ 25 cm
Sargocentron spiniferum 235 ♀ ⤳ ⤸ ◐ ✻ 🖿 ⬓ 45 cm

Sargocentron tiere 235 ♀ ⤳ ⤸ ◐ 🖿 ⬓ 30 cm

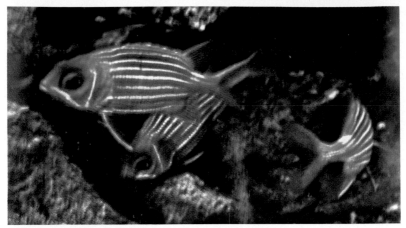

Holocentrus hastatus 235 ♀ ∿ ⌇ ◑ ✄ 📷 ⬚ 10 cm
Holocentrus rufus 235 ♀ ∿ ⌇ ◑ ✄ 📷 ⬚ 26 cm

Sargocentron rubrum 235 ♀ ∿ ⌇ ◑ ✄ 📷 ⬚ 36 cm

Neoniphon opercularis 235 ♀ ∿ ↘ ◐ ✕ 📷 🖼 35 cm
Sargocentron ensifer 235 ♀ ∿ ↘ ◐ ✕ 📷 🖼 22.5 cm

Sargocentron melanospilos 235 ♀ ∿ ↘ ◐ ✕ 📷 🖼 27 cm

Neoniphon marianus 235 ♀ ∿ ↘ ◐ ✕ 🔳 ▭ 17 cm
Neoniphon sammara 235 ♀ ∿ ↘ ◐ ✕ 🔳 ▭ 30 cm

Neoniphon scythrops 235 ♀ ∿ ↘ ◐ ✕ 🔳 ▭ 25 cm

Plectrypops retrospinis 235 ∿ ↘ ◑ ✕ 🔲 ▭ 10 cm
Pristilepis oligolepis 235 ∿ ↘ ◑ ✕ 🔲 ▭ 30 cm

Myripristis leiognathus 235 ∿ ↘ ◑ ✕ 🔲 ▭ 17.5 cm

Plectrypops lima 235 〜 ↘ ◐ ✖ 🖼 ▭ 17.5 cm
Ostichthys japonicus 235 〜 ↘ ◐ ✖ 📷 ▭ 45 cm

Myripristis jacobus 235 〜 ↘ ◐ ✖ 📷 ▭ 20 cm

Myripristis adustus 235 〜 ↘ ◑ ✕ 🎥 ▭ 32 cm
Myripristis kuntee 235 〜 ↘ ◑ ✕ 🎥 ▭ 15 cm

Myripristis murdjan 235 〜 ↘ ◑ ✕ 🎥 ▭ 24 cm

Myripristis hexagonatus 235 〜 ↘ ◐ ✸ 🔲 ⬜ 20 cm
Myripristis melanostictus 235 〜 ↘ ◐ ✸ 🔲 ⬜ 30 cm

Myripristis violacea 235 〜 ↘ ◐ ✸ 🔲 ⬜ 20 cm

Myripristis axillaris 235 〜 ↘ ◑ ✂ 🖻 ▭ 27 cm
Myripristis xanthacrus 235 〜 ↘ ◑ ✂ 🖻 ▭ 17 cm

Myripristis pralinia 235 〜 ↘ ◑ ✂ 🖻 ▭ 20 cm

Pegasus volitans 253 ∿ ◑ ♥ 🎞 ▭ 17.5 cm
Aulostomus maculatus 254 ∿ ⌐ ◑ ✕ 🖼 ⊟ 91 cm

Aulostomus chinensis 254 ∿ ⌐ ◑ ♥ 🖼 ⊟ 100 cm

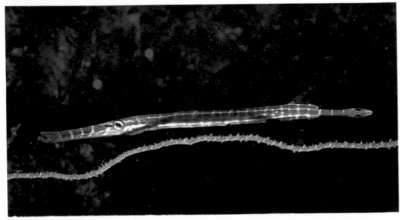

Macrorhamphosus scolopax 256 〜 ◑ ♥ 🖼 🔲 20 cm
Solenostomus paegnius 258 〜 ◑ ♥ 🖼 🔲 10 cm

Centriscus scutatus 257 ⤳ ◐ ♥ 🖻 ▭ 15 cm
Solenostomus cyanopterus 258 ⤳ ◐ ♥ 🖻 ▭ 17 cm

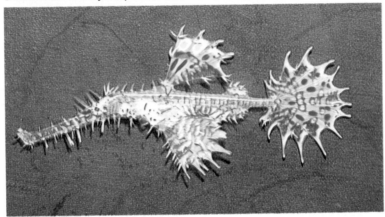

Solenostomus paradoxus 258 ⤳ ◐ ♥ 🖻 ▭ 16.5 cm

Phycodurus eques 259 ∿ ◑ ♥ 💹 ⬜ 35 cm
Phycodurus eques 259 ∿ ◑ ♥ 💹 ⬜ 35 cm

Phyllopteryx taeniolatus 259 ∿ ◐ ♥ 🆆 ☐ 23 cm
Phyllopteryx taeniolatus 259 ∿ ◐ ♥ 🆆 ☐ 23 cm

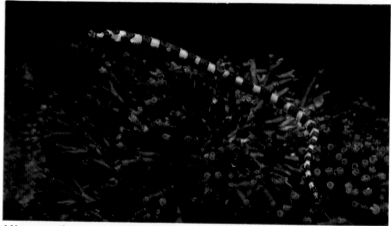

Micrognathus ensenadae 259 ∿ ◑ ♥ 🔲 ☐ 12.5 cm
Stigmatopora nigra 259 ∿ ◑ ♥ 🔲 ☐ 17.3 cm

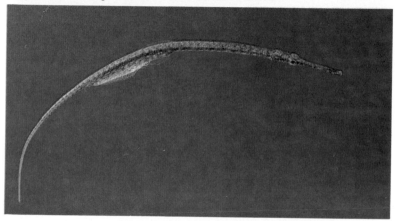

Syngnathoides biaculeatus 259 ∿ ◑ ♥ 🔲 ☐ 30 cm

Hippichthys penicillus 259 ∿ ◐ ♥ 🖼 ☐ 8.0 cm
Halicampus grayi 259 ∿ ◐ ♥ 🖼 ☐ 20 cm

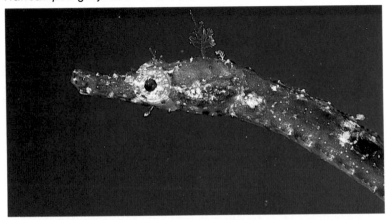

Maroubra perserrata 259 ∿ ◐ ♥ 🖼 ☐ 7 cm

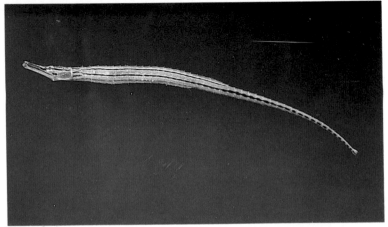

Doryrhamphus dactyliophorus 259 〜 ❶ ♥ 🔛 ⬛ 18 cm
Doryrhamphus japonicus 259 〜 ❶ ♥ 🔛 ⬛ 8 cm

Corythoichthys intestinalis 259 〜 ❶ ♥ 🔛 ⬛ 16 cm

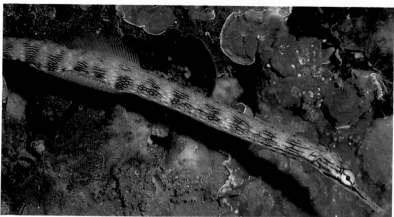

Doryrhamphus e. excisus 259 〜 ◐ ♥ 🔽 ▭ 7 cm
Corythoichthys paxtoni 259 〜 ◐ ♥ 🔽 ▭ 14 cm

Corythoichthys amplexus 259 〜 ◐ ♥ 🔽 ▭ 9 cm

Corythoichthys amplexus 259 〜 ◑ ♥ 🔳 ▭ 9 cm
Corythoichthys schultzi 259 〜 ◑ ♥ 🔳 ▭ 15 cm

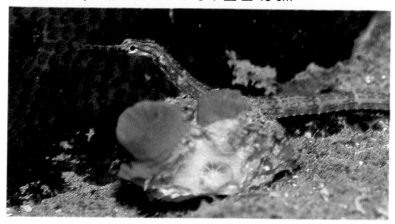

Halicampus spinirostris 259 〜 ◑ ♥ 🔳 ▭ 12 cm

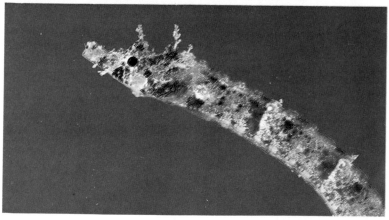

Doryrhamphus multiannulatus 259 ⌢ ◑ ♥ ⊠ ⊡ 18 cm
Hippocampus angustus 259 ⌢ ◑ ♥ ⊠ ⊡ 10 cm

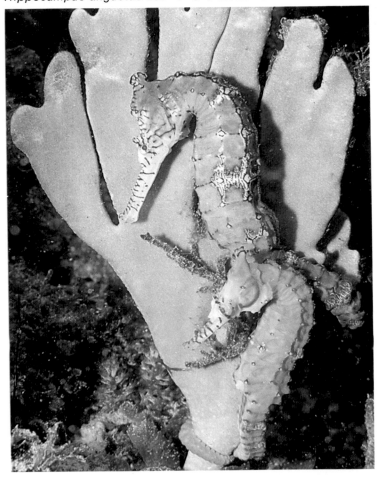

Hippocampus japonicus 259 〜 ◑ ♥ ▨
Hippocampus ingens 259 〜 ◑ ♥ ▧ ▭

Hippocampus kuda 259 〜 ◑ ♥ ▧ ▭
Hippocampus hippocampus 259 〜 ◑ ♥

Dactyloptena peterseni 260 〜 ↘ ◐ ✕ 🖾 ☐ 30 cm
Ablabys taenianotus 262 〜 ↘ ◐ ✕ 🖾 ☐ 9 cm

Neosebastes pandus 262 〜 ↘ ◐ ✕ 🖾 ☐ 30 cm

Dactylopterus volitans 260 〰 ↘ ◑ ✳ 🖼 ☐ 45 cm
Apistus carinatus 262 〰 ↘ ◑ ✳ 🖼 ☐ 15 cm

Peristrominous dolosus 265 〰 ↘ ◑ ✳ 🖼 ☐ 8 cm

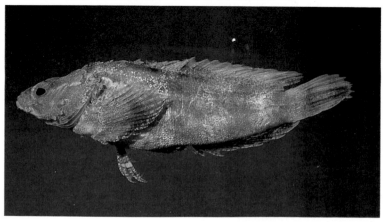

Pterois antennata 262 ∿ ↘ ◑ ✶ ▨ ⊟ 30 cm
Pterois volitans 262 ∿ ↘ ◑ ✶ ▨ ⊟ 35 cm

Pterois miles 262 ∿ ↘ ◑ ✶ ▨ ⊟ 31 cm

Dendrochirus zebra 262 ᴖ ᴝ ◐ ✕ ⛭ ⬚ 30 cm
Dendrochirus biocellatus 262 ᴖ ᴝ ◑ ✕ ⛭ ⬚ 12 cm

Taenianotus triacanthus 262 ᴖ ᴝ ◑ ✕ ⛭ ⬚ 10 cm

Dendrochirus brachypterus 262 ⌇ ⌇ ◑ ✻ 🖼 ⬚ 18 cm
Taenianotus triacanthus 262 ⌇ ⌇ ◑ ✻ 🖼 ⬚ 10 cm

Scorpaena scrofa 262 30 cm
Scorpaena porcus 262 30 cm

Scorpaena mystes 262 30 cm

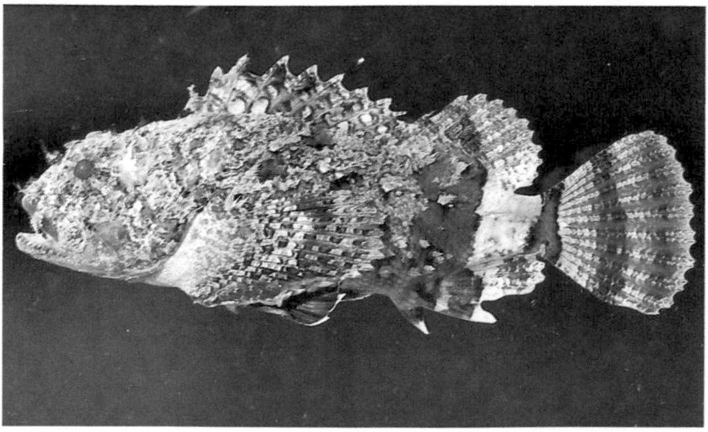

Scorpaena coniorta 262 ⌇ ⤵ ◑ ✕ ▨ ☐ 7.5 cm
Sebastapistes cyanostigma 262 ⌇ ⤵ ◑ ✕ ▨ ☐ 8 cm

Scorpaena sp. 262 ⌇ ⤵ ◑ ✕ ▨ ☐ 10 cm

Iracundus signifier 262 🗦 ↘ ◑ ✕ 🗊 ▭ 10 cm
Sebastapistes cyanostigma 262 🗦 ↘ ◑ ✕ 🖼 ▭ 7 cm

Scorpaenopsis sp. 262 🗦 ↘ ◑ ✕ 🖼 ▭ 8 cm

Scorpaena picta 262 ∿ ↘ ◐ ✕ 📷 ▱ 12 cm
Scorpaena sp. 262 ∿ ↘ ◐ ✕ 📷 ▱ 22 cm

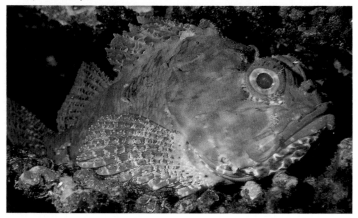

Scorpaenopsis sp. 262 ∿ ↘ ◐ ✕ 📷 ▱ 20 cm

Sebastes carnatus 262 ᭙ ᭡ ◑ ✕ ▣ ▱ 40 cm
Sebastes chrysomelas 262 ᭙ ᭡ ◑ ✕ ▣ ▱ 39 cm

Sebastes dallii 262 ᭙ ᭡ ◑ ✕ ▣ ▱ 20 cm

Sebastes caurinus 262 ⌇ ⟍ ◑ ✗ 📷 🖵 40 cm
Sebastes constellatus 262 ⌇ ⟍ ◑ ✗ 📷 🖵 30 cm

Sebastes maliger 262 ⌇ ⟍ ◑ ✗ 📷 🖵 60 cm

Sebastes nigrocinctus 262 ∿ ↘ ◑ ✕ ▣ ☐ 61 cm
Sebastes paucispinis 262 ↘ ◑ ✕ ▣ ▦ ☐ 90 cm

Sebastes rosaceus 262 ∿ ↘ ◑ ✕ ▣ ☐ 36 cm

Sebastes melanops 262 ∿ ↘ ◑ ✕ 🖻 ⬜ 60 cm
Sebastes nebulosus 262 ∿ ↘ ◑ ✕ 🖻 ⬜ 43 cm

Sebastes pinniger 262 ∿ ↘ ◑ ✕ 🖻 ⬜ 76 cm

Sebastes serriceps 262 ⌇ ⌇ ◑ ✕ ▨ ▭ 41 cm
Sebastes miniatus 262 ⌇ ⌇ ◑ ✕ ▣ ▭ 76 cm

Sebastes serranoides 262 ⌇ ⌇ ◑ ✕ ▣ ▭ 61 cm

Sebastes rubrivinctus 262 ∿ ↘ ◑ ✕ 🔲 ⬜ 51 cm
Sebastes umbrosus 262 ∿ ↘ ◑ ✕ 🔲 ⬜ 27 cm

Sebastes marinus 262 ∿ ↘ ◐ ✕ 🔲 ⬜ 51 cm

Sebastes trivittatus 262 〜 ↘ ◑ ✕ 🎞 ▭ 40 cm
Sebastes nivosus 262 〜 ↘ ◑ ✕ 📷 ▭ 40 cm

Sebastes baramenuke 262 〜 ↘ ◑ ✕ 🎞 ▭ 45 cm

Scorpaenopsis diabolus 262 〜 ↘ ◑ ✕ 🎞 ⬜ 30 cm
Scorpaena ballieui 262 〜 ↘ ◑ ✕ 📷 ⬜ 10 cm

Scorpionfish 262 〜 ↘ ◑ ✕ 📷 ⬜ 15 cm

Erosa erosa 263 ⤳ ↘ ◑ ✕ 🖻 ☐ 15 cm
Inimicus filamentosus 263 ⤳ ↘ ◑ ✕ 🖻 ☐ 18 cm

Inimicus caledonicus 263 ⤳ ↘ ◑ ✕ 🖻 ☐ 15 cm

Synanceia verrucosa 263 🐟 ➘ ◐ 🍴 🎞 🖵 35 cm
Glyptauchen panduratus 262 🐟 ➘ ◐ 🍴 🎞 🖵 10 cm

Ablabys taenionotus 267 🐟 ➘ ◐ 🍴 🎞 🖵 12 cm

Inimicus sinensis 263 ∿ ⤙ ◑ ☒ ▨ ▭ 25 cm
Perryena leucometopon 267 ∿ ⤙ ◑ ☒ ▨ ▭ 16 cm

Caracanthus maculatus 264 ∿ ⤙ ◑ ♥ ▨ ▭ 5.5 cm

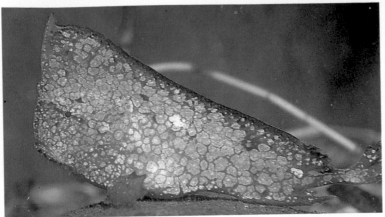

Neopataecus waterhausi 266 ～ ＼ ◑ ✕ ▱ ⊡ 6.0 cm
Richardsonichthys leucogaster 262 ～ ＼ ◑ ✕ ▱ ⊡ 11 cm

Paracentropogon vespa 262 ～ ＼ ◑ ✕ ▱ ⊡ 9 cm

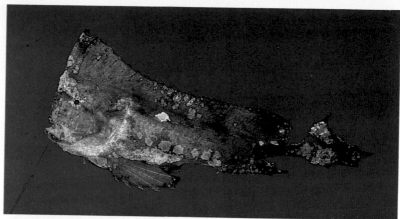

Neopataecus waterhausi 266 ✧ ↘ ◑ ✖ 🎦 ⬚ 6.0 cm
Richardsonichthys leucogaster 262 ✧ ↘ ◑ ✖ 🎦 ⬚ 5 cm

Hypodytes sp. 262 ✧ ↘ ◑ ✖ 🎦 ⬚ 5 cm

Chelidonichthys spinosus 268 ♀ ∿ ↘ ◑ ✕ 📷 ⬛ 40 cm
Satyrichthys laticephalus 268 ♀ ∿ ↘ ◑ ✕ 📷 ⬛ 30 cm

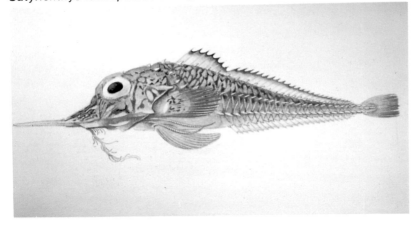

Pterygotrigla multiocellatus 268 ♀ ∿ ↘ ◑ ✕ 📷 ⬛ 35 cm

Prionotus scitulus 268 ♀ ◡ ⟍ ◑ ☆ 📷 🖵 25 cm
Prionotus rubio 268 ♀ ◡ ⟍ ◑☆ 📷 🖵 23 cm

Cociella crocodila 269 ♀ ◡ ⟍ ◑☆ 📷 🖵 50 cm

Oxylebias pictus 272 ∿ ↘ ◐ ✕ ▨ ⬚ 15 cm
Pleurogrammus azonus 272 ∿ ↘ ◐ ✕ ▨ ⬚ 70 cm

Zaniolepis latipinnis 273 ∿ ↘ ◐ ✕ ▨ ⬚ 30 cm

Hemilepidotus gilberti 276 ♀ ∿ ↘ ◑ ✕ 🖼 ⬛ 36 cm
Hemitripteris americanus 276 ♀ ∿ ↘ ◑ ✕ 🖼 ⬛ 40 cm

Jordania zonope 276 ♀ ∿ ↘ ◑ ✕ 🖼 ⬛ 15 cm

Hemitripteris villosus 276 ♀ ∿ ↘ ◑ ✻ ▨ ▭ 40 cm
Leiocottus hirundo 276 ♀ ∿ ↘ ◑ ✻ ▨ ▭ 25 cm

Myoxocephalus scorpius 276 ♀ ∿ ↘ ◑ ✻ ▨ ▭ 60 cm

Pseudoblennius cottoides 276 ♀ ∿ ◑ ♥ 🖼 ⬚ 7 cm
Taurulus bubalis 276 ♀ ∿ ↘ ◑ ✴ 🖼 ⬚ 17 cm

Psychrolutes paradoxus 279 ♀ ∿ ◑ ♥ 🖼 ⬚ 6.4 cm

Podothecus acipenserinus 280 ∿ ↘ ◑ ✕ 🎞 ▭ 30 cm
Podothecus sachi 280 ∿ ↘ ◑ ✕ 🎞 ▭ 50 cm

Liparis pulchellus 281 ∿ ↘ ◑ ✕ 🎞 ▭ 30 cm

Agonomalus proboscidalis 280 ∿ ↘ ◑ ✕ ▓ ▭ 20 cm
Occella verrucosa 280 ∿ ↘ ◑ ✕ ▓ ▭ 30 cm

Cyclopterus lumpus 281 ∿ ↘ ◑ ✕ ▓ ▭ 56 cm

Centropomus nigrescens 282 ⌇ ⌇ ◑ ✕ ▣ ⊟ 100 cm
Lates calcarifer 282 ⌇ ◑ ✕ ▨ ⊡ 150 cm

Chanda ranga 282 ⌇ ⌇ ◐ ♥ ▨ ⊡ 8 cm

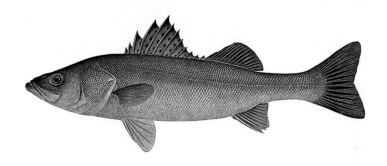

Lateolabrax japonicus 283 ⌇ ⟍ ◑ ✖ 🖼 ⊟ 150 cm
Niphon spinosus 283 ⌇ ⟍ ◑ ✖ 🖼 ⊟ 100 cm

Stereolepis gigas 283 ⌇ ⟍ ◑ ✖ 🖼 ⊟ 220 cm

Morone saxatilis 283 〜 ↘ ◑ ✕ 📷 ⊟ 183 cm
Polyprion oxygeneiosus 283 〜 ↘ ◑ ✕ 📷 ⊟ 220 cm

Stereolepis ischinagi 283 〜 ↘ ◑ ✕ 🖼 ⊟ 200 cm

Mirolabrichthys bicolor 284 ～ ＼ ◑ ♥ 🖾 ⊟ 13 cm
Pseudanthias sp. 284 ～ ＼ ◑ ♥ 🖾 ⊟ 10 cm

Pseudanthias luzonensis 284 ～ ＼ ◑ ♥ 🖾 ⊟ 10 cm

Pseudanthias kashiwae 284 ∿ ↘ ◑ ♥ 🖼 ⊟ 11 cm
Mirolabrichthys ignitus 284 ∿ ↘ ◑ ♥ 🖼 ⊟ 10 cm

Pseudanthias truncatus 284 ∿ ↘ ◑ ♥ 🖼 ⊟ 10 cm

Anthias anthias 284 〜 ↘ ◐ ♥ 🐟 ⊟ 25 cm
Pseudanthias pleurotaenia 284 〜 ↘ ◐ ♥ 🐟 ⊟ 10 cm

Mirolabrichthys thompsoni 284 〜 ↘ ◐ ♥ 🐟 ⊟ 19 cm

Pseudanthias rubrizonatus 284 〜 ↘ ◑ ♥ 🖼 ▤ 12 cm
Pseudanthias ventralis hawaiiensis 284 〜 ↘ ◑ ♥ 🖼 ▤ 8 cm

Pseudanthias squamipinnis 284 〜 ↘ ◑ ♥ 🖼 ▤ 10.5 cm

Pseudanthias hutchi 284 ∿ ↘ ◐ ♥ 🖼 ⊟ 10 cm
Pseudanthias sp. 284 ∿ ↘ ◐ ♥ 🖼 ⊟ 10 cm

Pseudanthias taira 284 ∿ ↘ ◐ ♥ 🖼 ⊟ 10 cm

Pseudanthias fasciatus 284 ∿ ⟍ ◑ ♥ 🖼 ⊟ 9 cm
Pseudanthias engelhardi 284 ∿ ⟍ ◑ ♥ 🖼 ⊟ 4 cm

Pseudanthias pictilis 284 ∿ ⟍ ◑ ♥ 🖼 ⊟ 9 cm

Mirolabrichthys bartletti 284 ∿ ➘ ◑ ♥ 🎞 ▤ 17 cm
Mirolabrichthys tuka 284 ∿ ➘ ◑ ♥ 🎞 ▤ 13 cm

Mirolabrichthys tuka (juv.) 284 ∿ ➘ ◑ ♥ 🎞 ▤ 13 cm

Mirolabrichthys pascalus 284 ∿ ↘ ◐ ♥ 🖼 ▭ 17 cm
Mirolabrichthys imeldae 284 ∿ ↘ ◐ ♥ 🖼 ▭ 9 cm

Paranthias furcifer 284 ∿ ↘ ◐ ♥ 🖼 ▭ 38 cm

Pseudanthias squamipinnis 284 〜 ↘ ◑ ♥ 🖼 ⊟ 10.5 cm
Acanthistius pardalotus 284 〜 ↘ ◑ ♥ 🖼 ⊟ 20 cm

Anyperodon leucogrammicus 284 〜 ↘ ◑ ✖ 🖼 ▭ 50 cm

Mirolabrichthys dispar 284 〜 ↘ ◑ ♥ 🎞 ⊟ 10 cm
Pseudanthias squamipinnis 284 〜 ↘ ◑ ♥ 🎞 ⊟ 10.5 cm

Aethaloperca rogaa 284 〜 ↘ ◑ ♥ 🎞 ⊟ 60 cm

Serranocirrhitus latus 284 ♀ ∿ ◑ ♥ 🖼 ▭ 8 cm
Sacura margaritacea 284 ♀ ∿ ◑ ♥ 🖼 ▭ 13 cm

Odontanthias elizabethae 284 ♀ ∿ ◑ ♥ 🖼 ▭ 17 cm

Caesioperca rasor 284 ∿ ⬎ ◑ ♥ 🖼 ⬜ 18 cm
Caprodon schlegeli 284 ∿ ⬎ ◑ ✖ 🖼 ⬜ 40 cm

Caprodon longimanus 284 ∿ ⬎ ◑ ♥ 🖼 ⬜ 3.5 cm

Gracila albomarginata 284 ♀ ∿ ↘ ◑ ✳ 📷 🖵 50 cm
Cephalopholis argus 284 ♀ ∿ ↘ ◑ ✳ 📷 🖵 50 cm

Cephalopholis boenak 284 ♀ ∿ ↘ ◑ ✳ 📷 🖵 30 cm

Cephalopholis polleni 284 ♀ ∿ ⟍ ◑ ✳ 📷 🖵 35 cm
Cephalopholis formosa 284 ♀ ∿ ⟍ ◑ ✳ 📷 🖵 35 cm

Cephalopholis leopardus 284 ♀ ∿ ⟍ ◑ ✳ 📷 🖵 20 cm

Cephalopholis miniata 284 ♀ ∿ ↘ ◐ ✕ ▣ ▭ 45 cm
Cephalopholis sonnerati 284 ♀ ∿ ↘ ◐ ✕ ▣ ▭ 57 cm

Cephalopholis miniata 284 ♀ ∿ ↘ ◑ ✳ 📷 ⬜ 45 cm
Cephalopholis sexmaculata 284 ♀ ∿ ↘ ◑ ✳ 📷 ⬜ 50 cm

Cephalopholis sonnerati 284 ♀ ∿ ↘ ◑ ✳ 📷 ⬜ 57 cm

Cephalopholis fulva 284 ♀ ⌇ ↘ ◑ ✕ 📷 🖵 30 cm
Cephalopholis analis 284 ♀ ⌇ ↘ ◑ ✕ 📷 🖵 60 cm

Cephalopholis taeniops 284 ♀ ⌇ ↘ ◑ ✕ 📷 🖵 30 cm

Cephalopholis fulva 284 ♀ ⌇ ↘ ◐ ✳ ▦ ▭ 30 cm
Cephalopholis hemistiktos 284 ♀ ⌇ ↘ ◐ ✳ ▦ ▭ 25 cm

Epinephelus fasciatus 284 ♀ ⌇ ↘ ◐ ✳ ▦ ▭ 35 cm

Epinephelus awoara 284 ♀ ⌇ ↘ ◑ ✶ 🖻 ☐ 45 cm
Callanthias japonica 284 ♀ ⌇ ◑ ♥ 🖻 ☐ 20 cm

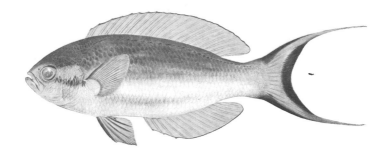

Callanthias allporti 284 ♀ ⌇ ◑ ✶ 🖻 ☐ 20 cm

Epinephelus corallicola 284 ♀ ⤳ ⤙ ◑ ✕ 📷 🖵 45 cm
Epinephelus merra 284 ♀ ⤳ ⤙ ◑ ✕ 📷 🖵 45 cm

Epinephelus merra 284 ♀ ⤳ ⤙ ◑ ✕ 📷 🖵 45 cm

Epinephelus caeruleopunctatus 284 ♀ ∿ ⤙ ◑ ✕ 📷 ⬜ 75 cm
Epinephelus fasciatus 284 ♀ ∿ ⤙ ◑ ✕ 📷 ⬜ 35 cm

Epinephelus rivulatus 284 ♀ ∿ ⤙ ◑ ✕ 📷 ⬜ 35 cm

Epinephelus bontoides 284 ♀ ⌇ ⬚ ◑ ✳ 📷 ⬚ 40 cm
Epinephelus malabaricus 284 ♀ ⌇ ⬚ ◑ ✳ 📷 ⬚ 150 cm

Epinephelus multinotatus 284 ♀ ⌇ ⬚ ◑ ✳ 📷 ⬚ 100 cm

Epinephelus maculatus 284 ♀ ∿ ↘ ◑ ✕ 📷 🖵 70 cm
Epinephelus tukula 284 ♀ ∿ ↘ ◑ ✕ 📷 🖵 200 cm

Epinephelus undulosus 284 ♀ ∿ ↘ ◑ ✕ 📷 🖵 75 cm

Epinephelus maculatus 284 ♀ ～ ↘ ◑ ✕ 📷 ▭ 60 cm
Epinephelus septemfasciatus 284 ♀ ～ ↘ ◑ ✕ 📷 ▭ 100 cm

Epinephelus moara 284 ♀ ～ ↘ ◑ ✕ 📷 ▭ 100 cm

Epinephelus diacanthus 284 ♀ ∿ ↘ ◑ ✖ ▣ ⊒ 55 cm
Epinephelus polyphekadion 284 ♀ ∿ ↘ ◑ ✖ ⊒ 90 cm

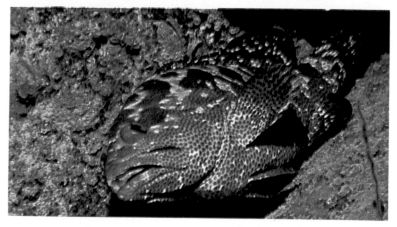

Epinephelus quoyanus 284 ♀ ∿ ↘ ◑ ✖ ▣ ⊒ 55 cm

Epinephelus akaara 284 ♀ ⌇ ⤸ ◐ ♥ 📷 ⬜ 40 cm
Epinephelus flavocoeruleus 284 ♀ ⌇ ⤸ ◐ ✖ 📷 ⬜ 70 cm

Epinephelus cyanopodus 284 ♀ ⌇ ⤸ ◐ ✖ 📷 ⬜ 80 cm

Epinephelus niveatus 284 ♀ ⌇ ↖ ◑ ✕ 📷 ⌗ 90 cm
Epinephelus dermatolepis 284 ♀ ⌇ ↖ ◑ ✕ 📷 ⌗ 110 cm

Epinephelus acanthistius 284 ♀ ⌇ ↖ ◑ ✕ 📷 ⌗ 71 cm

Epinephelus guttatus 284 ♀ ⌇ ↘ ◑ ✖ 📷 🖵 38 cm
Epinephelus adscensionis 284 ♀ ⌇ ↘ ◑ ♥ 📷 🖵 60 cm

Epinephelus striatus 284 ♀ ⌇ ↘ ◑ ✖ 📷 🖵 30 cm

Epinephelus guttatus 284 ♀ ∿ ↘ ◑ ✕ 🔲 ▭ 60 cm
Epinephelus mystacinus 284 ♀ ∿ ↘ ◑ ✕ 🔲 ▭ 150 cm

Epinephelus striatus 284 ♀ ∿ ↘ ◑ ✕ 🔲 ▭ 30 cm

Mycteroperca interstitialis 284 ♀ ⌒ ↘ ◑ ✻ 🎣 ⬚ 76 cm
Mycteroperca rosacea 284 ♀ ⌒ ↘ ◑ ✻ 🎣 ⬚ 100 cm

Mycteroperca rubra 284 ♀ ⌒ ↘ ◑ ✻ 🎣 ⬚ 68 cm

Mycteroperca prionura 284 ♀ ⤳ ⤸ ◑ ✄ 📷 ▭ 80 cm
Mycteroperca interstitialis 284 ♀ ⤳ ⤸ ◑ ✄ 📷 ▭ 76 cm

Mycteroperca rosacea 284 ♀ ⤳ ⤸ ◑ ✄ 📷 ▭ 100 cm

Mycteroperca rosacea 284 ♀ ∿ ↘ ◑ ✂ 🔒 🖵 100 cm
Mycteroperca rosacea 284 ♀ ∿ ↘ ◑ ✂ 🔒 🖵 100 cm

Cromileptes altivelis 284 ♀ ∿ ↘ ◑ ✂ 🔒 🖼 65 cm

Mycteroperca tigris (juv.) 284 ♀ ∿ ↘ ◐ ✕ 📷 🖵 100 cm
Mycteroperca microlepis 284 ♀ ∿ ↘ ◐ ✕ 📷 🖵 60 cm

Epinephelus lanceolatus 284 ♀ ∿ ↘ ◐ ✕ 📷 🖵 270 cm

Serranus tortugarum 284 ♀ ↝ ↖ ◑ ♥ 🐟 ☐ 80 cm
Serranus cabrilla 284 ♀ ↝ ↖ ◑ ✳ 🐟 ☐ 25 cm

Serranus scriba 284 ♀ ↝ ↖ ✳ ◑ 🐟 ☐ 25 cm

Serranus fasciatus 284 ♀ ∿ ⟍ ◑ ✴ 🎞 ☐ 18 cm
Serranus annularis 284 ♀ ∿ ⟍ ◑ ♥ 🎞 ☐ 8 cm

Serranus flaviventris 284 ♀ ∿ ⟍ ◑ ♥ 🎞 ☐ 7.5 cm

Serranus baldwini 284 ♀ ∿ ↘ ◑ ✗ ▣ ☲ 5 cm
Serranus tigrinus 284 ♀ ∿ ↘ ◑ ✗ ▣ ☲ 10 cm

Serranus subligarius 284 ♀ ∿ ↘ ◑ ✗ ▣ ☲ 15 cm

Liopropoma carmabi 284 ∿ ↖ ◐ ♥ 🖼 ⬚ 5 cm
Liopropoma fasciatum 284 ∿ ↖ ◐ ♥ 🖼 ⬚ 15 cm

Liopropoma mowbrayi 284 ∿ ↖ ◐ ♥ 🖼 ⬚ 8 cm

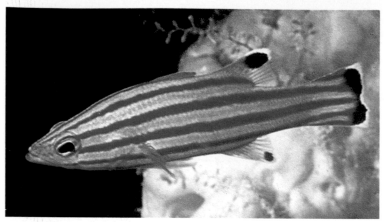

Liopropoma rubre 284 ⤳ ↘ ◑ ♥ 🖼 ▭ 8 cm
Liopropoma eukrines 284 ⤳ ↘ ◑ ♥ 🖼 ▭ 5 cm

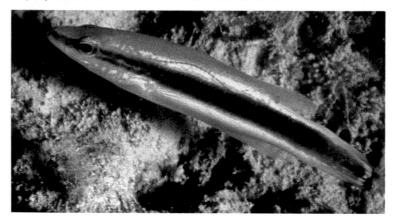

Liopropoma swalesi 284 ⤳ ↘ ◑ ♥ 🖼 ▭ 6 cm

Hypoplectrodes nigrorubrum 284 〜 ↘ ◐ ♥ 📼 ☐ 9 cm
Ellerkeldia hunti 284 〜 ↘ ◐ ♥ 📷 ☐ 38 cm

Ellerkeldia rubra 284 〜 ↘ ◐ ♥ 📼 ☐ 9 cm

Ellerkeldia annulata 284 ∿ ↘ ◑ ♥ 🎞 ⬜ 20 cm
Ellerkeldia maccullochi 284 ∿ ↘ ◑ ♥ 🎞 ⬜ 20 cm

Ellerkeldia wilsoni 284 ∿ ↘ ◑ ✕ 🎞 ⬜ 20 cm

Hypoplectrus nigricans 284 ♀ ∿ ↘ ◐ ✄ 🖭 ⬜ 15 cm
Hypoplectrus chlorurus 284 ♀ ∿ ↘ ◐ ✄ 🖭 ⬜ 12.5 cm

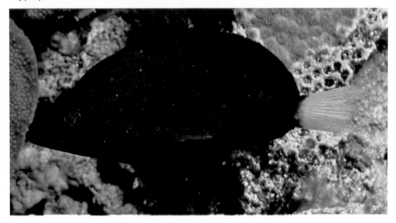

Hypoplectrus gemma 284 ♀ ∿ ↘ ◐ ✄ 🖭 ⬜ 11 cm

Hypoplectrus gummigutta 284 ♀ ∿ ↘ ◐ ✳ 📷 🖵 13 cm
Hypoplectrus unicolor 284 ♀ ∿ ↘ ◐ ✳ 📷 🖵 10 cm

Hypoplectrus guttavarius 284 ♀ ∿ ↘ ◐ ✳ 📷 🖵 11.3 cm

Acanthistius serratus 284 ⤳ ↘ ◑ ✕ 🖻 🖵 45 cm
Trachypoma macracantha 284 ⤳ ↘ ◑ ✕ 🖻 🖵 22 cm

Paralabrax clathratus 284 ⤳ ↘ ◑ ✕ 🂠 🖵 45 cm

Trisotropis dermopterus 284 〜 ⌁ ◑ ✕ 🖻 ▭ 45 cm
Paralabrax nebulifer 284 〜 ⌁ ◑ ✕ 🖾 ▭ 50 cm

Paralabrax maculatofasciatus 284 〜 ⌁ ◑ ✕ 🖻 ▭ 56 cm

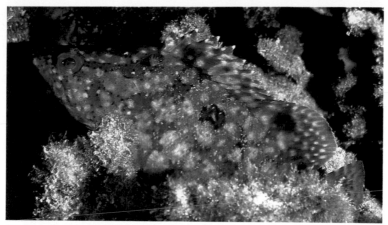

Alphestes afer 284 ♀ ⤳ ↘ ◑ ✗ 📷 ▭ 27 cm
Variola albomarginata 284 ♀ ⤳ ↘ ◑ ✗ 📷 ▭ 65 cm

Plectropomus maculatus 284 ♀ ⤳ ↘ ◑ ✗ 📷 ▭ 60 cm

Alphestes multiguttatus 284 ♀ 〜 ↘ ◑ ✕ 📷 ⬜ 20 cm
Variola louti 284 ♀ 〜 ↘ ◑ ✕ 📷 ⬜ 80 cm

Plectropomus leopardus 284 ♀ 〜 ↘ ◑ ✕ 📷 ⬜ 65 cm

Variola louti 284 ♀ ⌇ ↘ ◑ ✕ 📷 ⬜ 80 cm
Plectropomus laevis 284 ♀ ⌇ ↘ ◑ ✕ 📷 ⬜ 100 cm

Plectropomus leopardus 284 ♀ ⌇ ↘ ◑ ✕ 📷 ⬜ 66 cm

Plectropomus laevis 284 ♀ ∿ ↘ ◑ ✕ 🔳 ⬜ 100 cm
Pogonoperca punctata 285 ♀ ∿ ↘ ◑ ✕ 🔳 ⬜ 35 cm

Grammistes sexlineatus 285 ♀ ∿ ↘ ◑ ✕ 🔳 ⬜ 27 cm

Cromileptes altivelis 284 70 cm
Ellerkeldia wilsoni 284 20 cm

Othos dentex 284 70 cm

Dinoperca petersi 284 ♀ ∿ ↘ ◑ ✕ 🎦 ⊡ 62 cm
Luzonichthys microlepis 284 ♀ ∿ ↘ ◑ ✕ 🎦 ⊡ 6 cm

Belonoperca chabanaudi 285 ♀ ∿ ↘ ◑ ✕ 🎦 ⊡ 14.5 cm

Diploprion bifasciatus 285 ♀ ∿ ↘ ◑ ✖ ▣ ▭ 25 cm
Diploprion bifasciatus (var.) 285 ♀ ∿ ↘ ◑ ✖ ▣ ▭ 25 cm

Grammistes sexlineatus 285 ♀ ∿ ↘ ◑ ✖ ▣ ▭ 27 cm

Diploprion bifasciatus (juv.) 285 ♀ ∿ ↘ ◑ ✕ ▣ ☐ 25 cm
Diploprion drachi 285 ♀ ∿ ↘ ◑ ✕ ▣ ☐ 14 cm

Grammistops ocellatus 285 ♀ ∿ ↘ ◑ ✕ ▣ ☐ 10 cm

Suttonia lineata 285 ♀ ∿ ↘ ◐ ♥ 🎥 ▭ 12.5 cm
Rypticus subbifrenatus 285 ♀ ∿ ↘ ◐ ♥ 🎥 ▭ 18 cm

Rypticus bicolor (?) 285 ♀ ∿ ↘ ◑ ✕ 🎥 ▭ 33 cm

Rypticus bistrispinus 285 ♀ ∿ ↘ ◐ ♥ 🖼 ▭ 15 cm
Rypticus saponaceus 285 ♀ ∿ ↘ ◐ ✚ 🖼 ▭ 33 cm

Labracinus cyclophthalmus 286 ♀ ∿ ↘ ◐ ♥ 🖼 ▭ 14 cm

Pseudochromis wilsoni 286 ♀ ∿ ○ ♥ 🖾 ▭ 7 cm
Pseudochromis marshallensis 286 ♀ ∿ ○ ♥ 🖾 ▭ 7 cm

Ogilbyina velifera 286 ♀ ∿ ○ ♥ 🖾 ▭ 7 cm

Pseudochromis bitaeniatus 286 ♀ ∿ ○ ♥ 🖼 ⬚ 6 cm
Ogilbyina novaehollandiae 286 ♀ ∿ ○ ♥ 🖼 ⬚ 19 cm

Pseudochromis aureus 286 ♀ ∿ ○ ♥ 🖼 ⬚ 10 cm

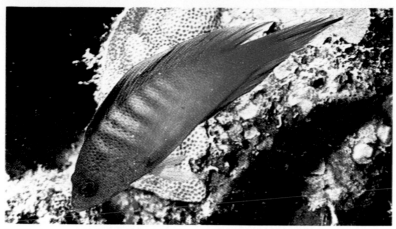

Ogilbyina queenslandiae 286 ♀ ∿ ☾ ♥ 🖼 ▭ 9.6 cm
Cypho purpurascens 286 ♀ ∿ ☾ ♥ 🖼 ▭ 10 cm

Pseudochromis moorei 286 ♀ ∿ ☾ ♥ 🖼 ▭ 10 cm

Pseudochromis springeri 286 ♀ ∿ ◐ ♥ 🎞 ⬛ 5.5 cm
Pseudochromis pesi 286 ♀ ∿ ◐ ♥ 🎞 ⬛ 8 cm

Pseudochromis dixurus 286 ♀ ∿ ◐ ♥ 🎞 ⬛ 9 cm

Pseudochromis porphyreus 286 ♀ ∿ ☉ ♥ 🖼 ☐ 5.5 cm
Pseudochromis diadema 286 ♀ ∿ ↘ ☉ ♥ 🖼 ☐ 27.4 cm

Pseudochromis cyanotaenia 286 ♀ ∿ ☉ ✕ 🖼 ☐ 5 cm

Pseudochromis paccagnellae 286 ♀ ∿ ◐ ♥ 🖼 ▭ 5 cm
Pseudochromis perspicillatus 286 ♀ ∿ ◐ ♥ 🖼 ▭ 12 cm

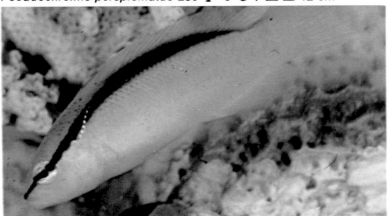

Pseudochromis flammicauda 286 ♀ ∿ ◐ ♥ 🖼 ▭ 5.5 cm

Pseudochromis dutoiti 286 ♀ ∿ ○ ♥ 🖻 ⬜ 8.8 cm
Pseudochromis fridmani 286 ♀ ∿ ○ ♥ 🖻 ⬜ 6 cm

Labracinus lineatus 286 ♀ ∿ ○ ♥ 🖻 ⊟ 14 cm

Pseudochromis flavivertex 286 ♀ ∿ ◑ ✳ 🖼 ⊟ 7 cm
Pseudochromis melas 286 ♀ ∿ ◐ ♥ 🖼 ⊟ 9 cm

Glaucosoma hebraicum 290 ∿ ↘ ◐ ✳ 📷 ⊟ 60 cm

Lipogramma trilineata 287 ♀ ⤴ ◐ ♥ 🎞 ▭ 3.5 cm
Gramma linki 287 ♀ ⤴ ◐ 🎞 ▭ 6.5 cm

Assessor flavissimus 288 ♀ ⤴ ◐ ♥ 🎞 ▭ 8 cm

Lipogramma klayi 287 ♀ ∿ ◐ ✕ 🖼 ☐ 4 cm
Gramma loreto 287 ♀ ∿ ◐ ✕ 🖼 ☐ 7.5 cm

Gramma melacara 287 ♀ ∿ ◐ ✕ 🖼 ☐ 10 cm

Calloplesiops altivelis 287 ♀ ∿ ↘ ◑ ♥ 🖼 ⊟ 16 cm
Plesiops oxycephalus 287 ♀ ∿ ↘ ◐ ♥ 🖼 ⊟ 10 cm

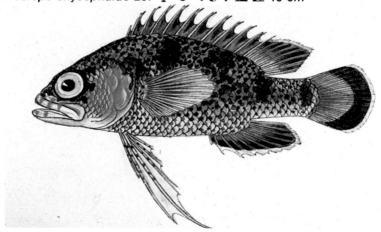

Plesiops cephalotaenia 287 ◌ ∿ ↘ ◑ ♥ 🖼 ⊟ 3.5 cm

Calloplesiops argus 288 ♀ ⌇ ↘ ◑ ♥ 📷 🔲 16 cm
Plesiops corallicola 288 ♀ ⌇ ↘ ◑ ♥ 📷 🔲 15 cm

Paraplesiops poweri 288 ♀ ⌇ ↘ ◑ ♥ 📷 🔲 20 cm

Terapon jarbua 291 ∿ ↘ ◑ ✕ 🖼 ⊟ 23 cm
Pelates sexlineatus 291 ♀ ↘ ◑ ✕ 🖼 ⊟ 25 cm

Mesopristes argenteus 291 ∿ ↘ ◑ ✕ 🖼 ⊟ 25 cm

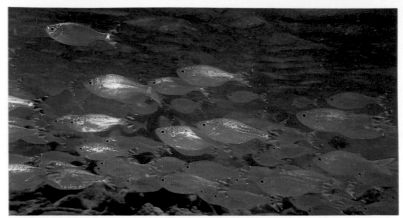

Kuhlia mugil 293 〜 ↘ ◐ ✕ ▨ ⊟ 20 cm
Pristigenys multifasciata 296 〜 ↘ ◐ ✕ ▨ ⊡ 27 cm

Pristigenys serrula 296 〜 ↘ ◐ ✕ ▨ ⊡ 15 cm

Kuhlia sandvicensis 293 〜 ＼ ◑ ✕ 🐟 ⊟ 30 cm
Pristigenys alta 296 〜 ＼ ◑ ✕ 🐟 ▭ 30 cm

Pristigenys niphonia 296 〜 ＼ ◑ ✕ 🐟 ▭ 26 cm

Apogon angustatus 297 ♀ ∿ ◐ ♥ 🖼 ⬜ 8 cm
Apogon robustus 297 ♀ ∿ ◐ ♥ 🖼 ⬜ 6 cm

Apogon cookii 297 ♀ ∿ ◐ ♥ 🖼 ⬜ 10 cm

Apogon robustus 297 ♀ ∿ ◐ ♥ 🖼 ▭ 6 cm
Apogon victoriae 297 ♀ ∿ ◐ ♥ 🖼 ▭ 8 cm

Apogon nitidus 297 ♀ ∿ ◐ ♥ 🖼 ▭ 8 cm

Apogon nigrofasciatus 297 ♀ ∿ ↘.◑ ✕ 📷 ⬜ 10 cm
Apogon novemfasciatus 297 ♀ ∿ ↘ ◐ ✕ 📷 ⬜ 8 cm

Apogon limenus 297 ♀ ∿ ↘ ◐ ✕ 📷 ⬜ 11.5 cm

Apogon compressus 297 ♀ ∿ ◑ ♥ 🖼 🖵 11.5 cm
Apogon doederleini 297 ♀ ∿ ◑ ♥ 🖼 🖵 7 cm

Apogon cookii 297 ♀ ∿ ◑ ♥ 🖼 🖵 10 cm

Apogon chrysotaenia 297 ♀ ∿ ◑ ♥ 🖼 ⬜ 12 cm
Apogon novaeguinea (? = *cyanosoma*) 297 ♀ ∿ ◑ ♥ 🖼 ⬜ 10 cm

Apogon doederleini 297 ♀ ∿ ◑ ♥ 🖼 ⬜ 7 cm

Apogon exostigma 297 ♀ ∿ ◐ ♥ 🖼 ▭ 10 cm
Apogon kallopterus 297 ♀ ∿ ◐ ♥ 🖼 ▭ 15 cm

Apogon notatus 297 ♀ ∿ ◐ ♥ 🖼 ▭ 10 cm

Apogon kallopterus 297 ♀ ∿ ◐ ♥ 🖼 ▱ 15 cm
Apogon fragilis 297 ♀ ∿ ◐ ♥ 🖼 ▱ 4 cm

Apogon hoeveni 297 ♀ ∿ ◐ ♥ 🖼 ▱ 60 cm

Apogon sangiensis 297 ♀ ∿ ◑ ♥ 🖼 ▭ 9 cm
Apogon ceramensis 297 ♀ ∿ ◑ ♥ 🖼 ▭ 10 cm

Apogon hyalosoma 297 ♀ ∿ ◑ ♥ 🖼 ▭ 17 cm

Apogon trimaculatum 297 ♀ ∿ ◑ ♥ ▨ ▭ 15 cm
Apogon leptacanthus 297 ♀ ∿ ◑ ♥ ▨ ▭ 6 cm

Apogon dispar 297 ♀ ∿ ◑ ♥ ▨ ▭ 4.5 cm

Apogon menesemus 297 ♀ ∿ ◐ ♥ 🎞 ▭ 10 cm
Apogon guamensis 297 ♀ ∿ ◐ ♥ 🎞 ▭ 10 cm

Apogon maculifera 297 ♀ ∿ ◐ ♥ 🎞 ▭ 15 cm

Apogon pseudomaculatus 297 ♀ ∿ ◑ ♥ 🖼 ▭ 8 cm
Apogon quadrisquamatus 297 ♀ ∿ ◑ ♥ 🖼 ▭ 8 cm

Apogon leptocaulus 297 ♀ ∿ ◑ ♥ 🖼 ▭ 6 cm

Apogon maculatus 297 ♀ ∿ ◑ ♥ 🖼 ⌷ 15 cm
Apogon aurolineatus 297 ♀ ∿ ◑ ♥ 🖼 ⌷ 5 cm

Apogon retrosella 297 ♀ ∿ ◑ ♥ 🖼 ⌷ 10 cm

Apogon planifrons 297 ♀ ∿ ◐ ♥ 🌿 ▭ 11 cm
Apogon townsendi 297 ♀ ∿ ◐ ♥ 🌿 ▭ 7.5 cm

Apogon phenax 297 ♀ ∿ ◐ ♥ 🌿 ▭ 8 cm

Apogon sp. 297 ♀ ∿ ◑ ♥ 🎞 ☐ 8 cm
Apogon nigripes 297 ♀ ∿ ◑ ♥ 🎞 ☐ 7 cm

Apogon elliotti 297 ♀ ∿ ◑ ♥ 🎞 ☐ 12 cm

Apogon evermanni 297 ♀ ∿ ◑ ♥ 🎞 ▭ 10 cm
Apogon selas 297 ♀ ∿ ◑ ♥ 🎞 ▭ 7.5 cm

Sphaeramia nematoptera 297 ♀ ∿ ◑ ♥ 🎞 ▭ 10 cm

Apogon apogonides 297 ♀ ∿ ◑ ♥ 🎞 ⬛ 10 cm
Apogon kallopterus 297 ♀ ∿ ◑ ♥ 🎞 ⬛ 15 cm

Apogon sp. 297 ♀ ∿ ◑ ♥ 🎞 ⬛ 6 cm

Apogon aureus 297 ♀ ∿ ◑ ♥ 🎞 ⬚ 12 cm
Apogon taeniatus 297 ♀ ∿ ◑ ♥ 🎞 ⬚ 10 cm

Cheilodipterus quinquelineatus 297 ♀ ∿ ◑ ♥ 🎞 ⬚ 12 cm

Cheilodipterus macrodon 297 ♀ ∿ ↘ ◑ ✗ 🎞 ▭ 24 cm
Cheilodipterus sp. 297 ♀ ∿ ◑ ♥ 🎞 ▭ 8 cm

Cheilodipterus zonatus 297 ♀ ∿ ◑ ♥ 🎞 ▭ 9 cm

Cheilodipterus lachneri 297 ♀ ∿ ◑ ♥ 🖼 ▭ 12 cm
Cheilodipterus sp. 297 ♀ ∿ ◑ ♥ 🖼 ▭ 12 cm

Apogon affinis 297 ♀ ∿ ◑ ♥ 🖼 ▭ 10 cm

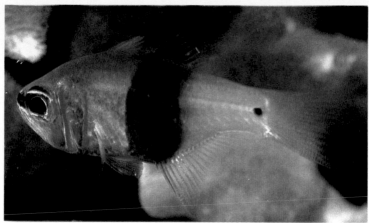

Archamia zosterophora 297 ♀ ᘰ ◐ ♥ 🖼 ⬚ 6 cm
Archamia fucata 297 ♀ ᘰ ◐ ♥ 🖼 ⬚ 8 cm

Archamia biguttata 297 ♀ ᘰ ◐ ♥ 🖼 ⬚ 7 cm

Astrapogon stellatus 297 ♀ ∿ ◑ ♥ 🖼 ⬒ 6.5 cm
Phaeoptyx xenus 297 ♀ ∿ ◑ ♥ 🖼 ⬒ 7.5 cm

Pterapogon mirifica 297 ♀ ∿ ◑ ♥ 🖼 ⬒ 8 cm

Astrapogon punticulatus 297 ♀ ∿ ◐ ♥ 📺 ⬒ 6.5 cm
Phaeoptyx pigmentaria 297 ♀ ∿ ◐ ♥ 📺 ⬒ 6.5 cm

Phaeoptyx conklini 297 ♀ ∿ ◐ ♥ 📺 ⬒ 6.5 cm

Siphamia fuscolineata 297 ♀ ∿ ◑ ♥ 📷 ▭ 4 cm
Siphamia sp. 297 ♀ ∿ ◑ ♥ 📷 ▭ 4 cm

Siphamia mossambica (?) 297 ♀ ∿ ◑ ♥ 📷 ▭ 4 cm

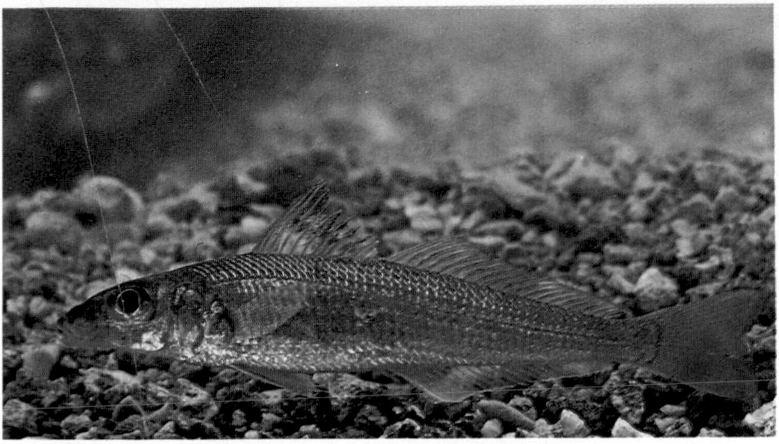

Sillago macrolepis 299 20 cm

Sillago maculata 299 30 cm

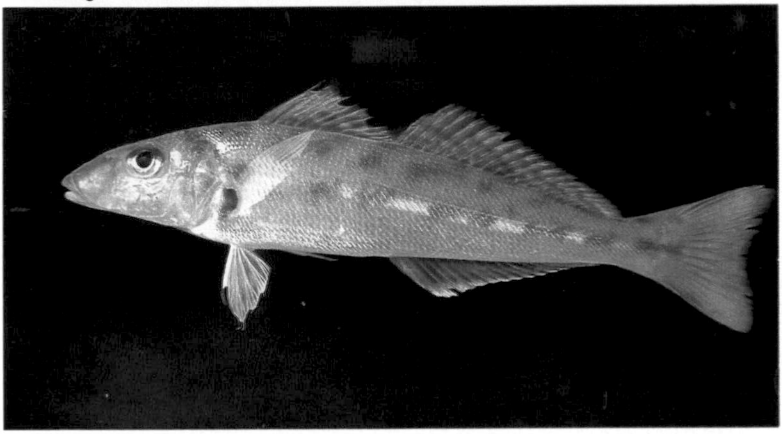

Lopholatilus chameleonticeps 300 42 cm

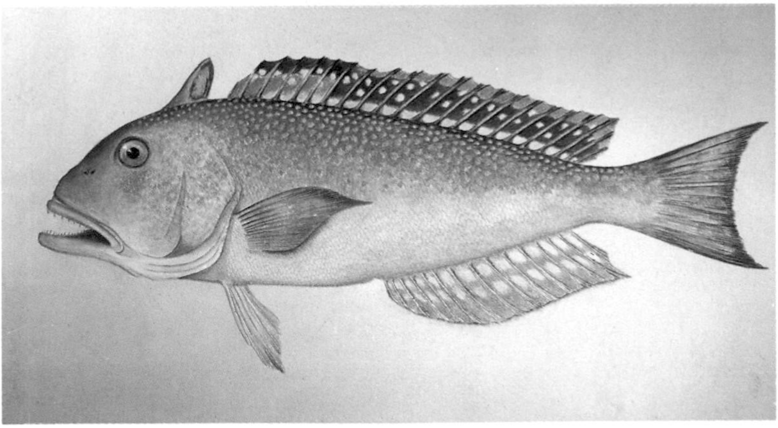

Hoplolatilus cuniculus 300 〰 ◑ ♥ 🔲 🔲 9 cm
Hoplolatilus starcki 300 〰 ◑ ♥ 🔲 🔲 11 cm

Hoplolatilus fourmanoiri 300 〰 ◑ ♥ 🔲 🔲 20 cm

Hoplolatilus marcosi 300 ∿ ◑ ♥ 🖻 ⊟ 20 cm
Hoplolatilus starcki 300 ∿ ◑ ♥ 🖻 ⊟ 11 cm

Hoplolatilus purpureus 300 ∿ ◑ ♥ 🖻 ⊟ 15 cm

Malacanthus latovittatus 300 〜 ↘ ◑ ✕ 🎞 🖼 🖵 45 cm
Malacanthus brevirostris 300 〜 ↘ ◑ ✕ 🎞 🖼 🖵 60 cm

Malacanthus brevirostris 300 〜 ↘ ◑ ✕ 🎞 🖼 🖵 60 cm

Rachycentron canadum 304 〜 〵 ◯ ✚ ▣ ▭ 180 cm
Echeneis naucrates 305 〵 ◯ ✚ ▣ ▭ 100 cm

Echeneis sp. 305 〵 ◯ ✚ ▣ ▭ 100 cm

Gnathanodon speciosus 306 ↘ ◑ ✕ 🏞 ▱ 90 cm
Carangoides orthogrammus 306 ∿ ↘ ◑ ✕ 📷 ⊟ 40 cm

Carangoides dinema (juv.) 306 ∿ ↘ ○ ✕ 📷 ⊟

Caranx melampygus 306 ⬧ ◑ ✘ ▭ ⊟ 80 cm
Caranx lugubris 306 ⬦ ⬧ ◑ ✘ ▥ ⊟ 93 cm

Caranx crysos 306 ⬧ ◑ ✘ ▭ ⊟ 58 cm

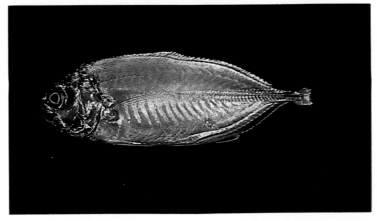

Alectis indicus 306 ↭ ↘ ◑ ✳ 🖼 ⊟ 50 cm
Selene brevoortii 306 ↭ ↘ ◑ ♥ 🖼 ⊟ 30 cm

Selene vomer 306 ↭ ↘ ◑ ♥ 🖼 ⊟ 30 cm

Seriola lalandi dorsalis 306 ∿ ↘ ◐ ✖ ▨ ⊟ 100 cm
Seriola rivoliana 306 ↘ ◐ ✖ ▨ ⊟ 100 cm

Seriola quinqueradiata 306 ↘ ◐ ✖ ▨ ⊟ 100 cm

Naucrates ductor 306 60 cm
Seriolina nigrofasciata 306 90 cm

Chloroscombrus chrysurus 306 30 cm

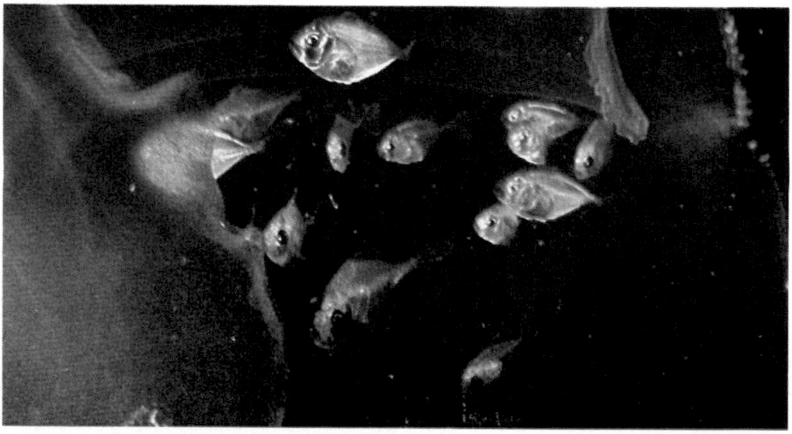

Seriolina nigrofasciata (juv.) 306 ↘ ◖ ✕ 🔦 ⊟ 90 cm
Pantolabus radiatus 306 ↘ ◖ ✕ ▦ ⊟ 25 cm

Pseudocaranx dentex 306 ↘ ↘ ◖ ✕ ▦ ⊟ 95 cm

Trachinotus goodei 306 〜 ↘ ◐ ✕ ▨ ⊟ 32 cm
Trachinotus baillonii 306 ↘ ◐ ✕ ▨ ⊟ 60 cm

Trachinotus kennedyi 306 〜 ↘ ◐ ✕ ▨ ⊟ 61 cm

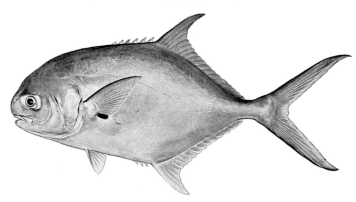

Gazza minuta 311 〰 ⤙ ◑ ♥ ▱ ▱ 15 cm
Leiognathus nuchalis 311 〰 ⤙ ◑ ♥ ▱ ▱ 17 cm

Leiognathus equulus 311 〰 ⤙ ◑ ♥ ▱ ▣ 30 cm

Leiognathus fasciata 311 〜 〜 ◑ ♥ ▱ ▱ 20 cm
Leiognathus nuchalis 311 〜 〜 ◑ ♥ ▱ ▱ 17 cm

Leiognathus rivulatus 311 〜 ◑ ♥ ▱ ▱ 10 cm

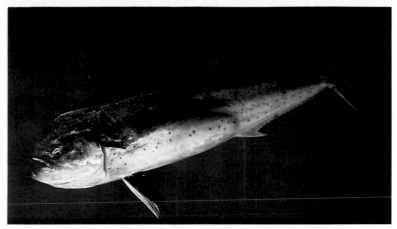

Coryphaena hippurus 308 ↘ ◐ ✄ ▨ ▭ 150 cm
Brama japonicus 312 ↝ ↘ ◐ ✄ ▨ ▭ 122 cm

Pterycombus petersii 312 ↝ ↘ ◐ ✄ ▨ ▣ 31 cm

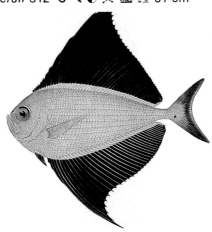

Paracaesio xanthura 316 ⤻ ⬎ ◑ ✖ 🎴 ⊡ 38 cm
Etelis oculatus 316 ⤻ ⬎ ◑ ✖ 🎴 ⊡ 91 cm

Rhomboplites aurorubens 316 ⬎ ◑ ✖ 🎴 ⬜ 60 cm

Lutjanus bengalensis 316 ♀ ⌇ ↖ ◐ ✕ 🖼 ⊟ 13 cm
Lutjanus notatus 316 ♀ ⌇ ↖ ◐ ✕ 🖼 ⊟ 22 cm

Lutjanus argentimaculatus 316 ♀ ⌇ ↖ ◐ ✕ 🖼 ⊟ 60 cm

Lutjanus decussatus 316 ♀ ∿ ↖ ◑ ✳ 🖻 ⊡ 30 cm
Lutjanus kasmira 316 ♀ ∿ ↖ ◑ ✳ 🖻 ⊡ 40 cm

Lutjanus ehrenbergii 316 ♀ ∿ ↖ ◑ ✳ 🖻 ⊡ 30 cm

Lutjanus carponotatus 316 ♀ ⌇ ⤙ ◐ ✕ 🐟 ▭ 40 cm
Lutjanus vitta 316 ♀ ⌇ ◐ ✕ 🐟 ▭ 37.5 cm

Lutjanus quinquelineatus 316 ♀ ⌇ ⤙ ◐ ✕ 🐟 ▭ 40 cm

Lutjanus monostigma 316 ᔓ ↘ ◑ ✕ 🖼 ⊟ 60 cm
Lutjanus madras 316 ♀ ᔓ ↘ ◑ ✕ 🖼 ⊟ 20 cm

Lutjanus gibbus 316 ᔓ ↘ ◑ ✕ 🖼 ⊟ 40 cm

Lutjanus fulvus 316 ♀ ⤳ ↘ ◐ ✳ 🖼 ⊟ 60 cm
Lutjanus lunulatus 316 ♀ ⤳ ↘ ◐ ✳ 🖼 ⊟ 25 cm

Lutjanus biguttatus 316 ♀ ⤳ ↘ ◐ ✳ 🖼 ⊟ 28 cm

Lutjanus jocu 316 ♀ ∿ ↘ ◑ ✳ 🖼 ⊟ 100 cm
Lutjanus mahogoni 316 ♀ ∿ ↘ ◐ ✳ 🖼 ⊟ 36 cm

Lutjanus viridis 316 ♀ ∿ ↘ ◐ ✳ 🖼 ⊟ 30 cm

Lutjanus analis 316 ♀ 〰 ⌇ ◑ ✕ 🐟 ⊟ 65 cm
Lutjanus buccanella (juv.) 316 ♀ 〰 ⌇ ◑ ✕ 🐟 ⊟ 75 cm

Lutjanus apodus 316 ♀ 〰 ⌇ ◑ ✕ 🐟 ⊟ 60 cm

Lutjanus argentimaculatus 316 ♀ ∿ ↘ ◑ ✕ 🖻 🖾 ⊟ 60 cm
Lutjanus erythropterus (juv.) 316 ♀ ∿ ↘ ◑ ✕ 🖾 ⊟ 60 cm

Lutjanus sp. *(lemniscatus?)* 316 ♀ ∿ ↘ ◑ ✕ 🖾 ⊟ 45 cm

Symphorichthys spilurus 316 ♀ 〜 ↘ ◑ ✕ 🐟 ⊟ 50 cm
Symphorus nematophorus 316 ♀ 〜 ↘ ◑ ✕ 🐟 ⊟ 80 cm

Lutjanus colorado 316 ♀ 〜 ↘ ◑ ✕ 🐟 ⊟ 75 cm

Lutjanus sebae 316 ♀ ∿ ↘ ◑ ✕ ▨ ▣ 90 cm
Macolor niger (juv.) 316 ♀ ∿ ↘ ◑ ✕ ▨ ▣ 75 cm

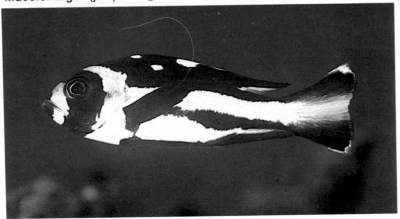

Macolor niger 316 ♀ ∿ ↘ ◑ ✕ ▨ ▣ 75 cm

Caesio teres 317 ♀ ∿ ↘ ◑ ✗ 🖼 ⊟ 30 cm
Pterocaesio chrysozona 317 ∿ ◑ ♥ 📷 ⊟ 18 cm

Pterocaesio tile 317 ∿ ◑ ♥ 📷 ⊟ 30 cm

Pterocaesio lativittata 317 🗝 ◑ ♥ 📷 ⊟ 18 cm
Pterocaesio tile 317 🗝 ◑ ♥ 📷 ⊟ 30 cm

Caesio caerulaurea 317 🗝 ◑ ♥ 📷 ⊟ 18 cm

Pterocaesio digramma 317 〜 ◑ ♥ 📷 ⊟ 25 cm
Caesio teres 317 ♀ 〜 ↘ ◑ ✕ 🖼 ⊟ 30 cm

Caesio pisang 317 〜 ◑ ♥ 🖼 ⊟ 20 cm

Caesio lunaris 317 〜 ↘ ◑ ✕ 🔲 ☒ 30 cm
Caesio caerulaurea 317 〜 ◑ ♥ 🔲 ☒ 18 cm

Lobotes surinamensis 317 〜 ↘ ◑ ✕ 🔲 ☒ 96 cm

Parequula melbournenses 319 ∿ ❶ ♥ 📷 ⊟ 10 cm
Gerres acinaces 319 ∿ ❶ ♥ 📷 ⊟ 25 cm

Gerres cinereus 319 ∿ ❶ ♥ 📷 ⊟ 20 cm

Haemulon parrai 320 ♀ ∿ ↘ ◐ ✕ 🔲🔳 40 cm
Haemulon sciurus 320 ♀ ∿ ↘ ◐ ✕ 🔲 🔳 18 cm

Haemulon flavolineatum 320 ♀ ∿ ↘ ◐ ✕ 🔲 🔳 30 cm

Haemulon melanurum 320 ♀ ～ 丶 ◑ ♥ 🔛 ⊟ 33 cm
Haemulon plumieri 320 ♀ ～ 丶 ◑ ✂ 🔛 ⊟ 45 cm

Haemulon flavolineatum 320 ♀ ～ 丶 ◑ ✂ 🔛 ⊟ 30 cm

Haemulon album 320 60 cm

Haemulon aurolineatum 320 25 cm

Haemulon sciurus 320 30 cm

Anisotremus surinamensis 320 ♀ ∿ ↘ ◑ ✕ 📷 ⊟ 60 cm
Anisotremus taeniatus 320 ♀ ∿ ↘ ◑ ✕ 📷 ⊟ 25 cm

Anisotremus virginicus 320 ♀ ∿ ↘ ◑ ✕ 📷 ⊟ 38 cm

Scolopsis ghanam 325 ♀ ∿ ↘ ◑ ♥ 🎞 ▭ 25 cm
Plectorhinchus chaetodonoides (juv.) 320 ♀ ∿ ↘ ◑ ♥ 🎞 ▭ 60 cm

Plectorhinchus chaetodonoides 320 ♀ ∿ ↘ ◑ ♥ 🎞 ▭ 60 cm

Scolopsis ghanam 325 ♀ ∿ ↘ ◑ ♥ 🖼 ⬜ 25 cm
Plectorhinchus chaetodonoides 320 ♀ ∿ ↘ ◑ ♥ 🖼 ⬜ 60 cm

Plectorhinchus flavomaculatus 320 ♀ ∿ ↘ ◑ ♥ 🖼 ⬜ 60 cm

Plectorhinchus flavomaculatus 320 ♀ ⌇ ↘ ◑ ♥ 🐟 ⬜ 60 cm
Plectorhinchus picus 320 ⌇ ↘ ◑ ♥ 🐟 ⬜ 60 cm

Hapalogenys nigripinnis 320 ♀ ⌇ ↘ ◑ ✖ 🐟 ⬜ 60 cm

Plectorhinchus albovittatus (juv.) 320 〜 ➤ ◑ ♥ 🖼 ⬒ 20 cm
Plectorhinchus multivittatus 320 ♀ 〜 ➤ ◑ ♥ 🖼 ⬒ 60 cm

Plectorhinchus chaetodonoides 320 ♀ 〜 ➤ ◑ ♥ 🖼 ⬒ 60 cm

Plectorhinchus goldmanni 320 ♀ ∿ ↘ ◑ ♥ 🖼 ⬜ 60 cm
Plectorhinchus diagrammus 320 ∿ ↘ ◑ ♥ 🖼 ⬜ 50 cm

Plectorhinchus nigrus 320 ∿ ↘ ◑ ♥ 🖼 ⬜ 60 cm

Plectorhinchus gaterinus 320 ∿ ↘ ◑ ♥ 🎞 ▭ 60 cm
Plectorhinchus lineatus 320 ∿ ↘ ◑ ♥ 🎞 ▭ 35 cm

Plectorhinchus picus 320 ∿ ↘ ◑ ♥ 🎞 ▭ 60 cm

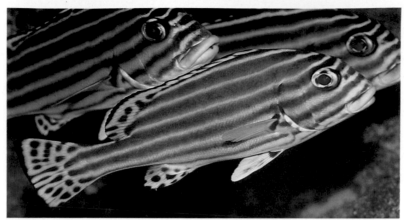

Plectorhinchus lineatus 320 〰 ↘ ◑ ✖ 🖼 ▭ 35 cm
Plectorhinchus picus 320 〰 ↘ ◑ ♥ 🖼 ▭ 60 cm

Plectorhinchus pictus 320 〰 ↘ ◑ ♥ 🖼 ▭ 90 cm

Inermia vittata 321 〜 ↘ ⚲ ◑ ✖ 🖼 ⊟ 13 cm
Spicara maena flexuosa 323 〜 ↘ ⚲ ◑ ✖ 🖼 ⊟ 21 cm

Boops salpa 322 〜 ↘ ⚲ ◑ ✖ 🖼 ⊟ 45 cm

Diplodus puntazzo 322 45 cm

Lithognathus mormyrus 322 55 cm

Diplodus vulgaris 322 30 cm

Pagrus major 322 ⤳ ↘ ◑ ✕ 🐟 ⊟ 70 cm
Pagrus major 322 ⤳ ↘ ◑ ✕ 🐟 ⊟ 70 cm

Gymnocranius audleyi 322 ⤳ ↘ ◑ ✕ 🐟 ⊟ 41 cm

Calamus pennatula 322 ♀ ∿ ↘ ◑ ✵ ⌦ ⊟ 35 cm
Calamus nodosus 322 ♀ ∿ ↘ ◑ ✵ ⌦ ⊟ 46 cm

Pagrus pagrus 322 ♀ ∿ ↘ ◑ ✵ ⌦ ⊟ 90 cm

Archosargus probatocephalus 322 91 cm

Lagodon rhomboides 322 35 cm

Acanthopagrus bifasciatus 322 50 cm

Diplodus sargus 322 〜 〜 ↑ ◑ ✕ 🖾 ⊟ 40 cm
Rhabdosargus thorpei 322 〜 〜 ◑ ✕ 🖾 ⊟ 40 cm

Monotaxis grandoculus 324 〜 〜 ◑ ✕ 🖾 ⊟ 45 cm

Diplodus cervinus 322 55 cm

Porcostoma dentata 322 30 cm

Gymnocranius griseus 324 50 cm

Lethrinus obsoletus 324 〜 ↘ ◑ ✕ 🖼 ▭ 60 c.
Lethrinus atkinsoni 324 〜 ↘ ◑ ✕ 🖼 ▭ 50 cm

Lethrinus rubrioperculatus 324 〜 ↘ ◑ ✕ 🖼 ▭ 60 cm

Lethrinus harak 324 〜 丶 ◐ 火 🖼 ⊟ 60 cm
Lethrinus variegatus 324 〜 丶 ◐ 火 🖼 ⊟ 60 cm

Gnathodentex aurolineatus 324 〜 丶 ◐ 火 🖼 ⊟ 45 cm

Lethrinus enigmaticus 324 ⌇ ↘ ◑ ✵ ▨ ▭ 40 cm
Lethrinus erythracanthus 324 ↘ ⌇ ◑ ✵ ▨ ▭ 80 cm

Lethrinus variegatus 324 ⌇ ↘ ◑ ✵ ▨ ▭ 60 cm

Pentapodus nemurus 325 ～ ↘ ◑ ✕ 🐟 ☐ 46 cm
Pentapodus paradiseus 325 ～ ↘ ◑ ✕ 🐟 ☐ 35 cm

Pentapodus paradiseus 325 ～ ↘ ◑ ✕ 🐟 ☐ 35 cm

Scolopsis bimaculatus 325 〜 ↘ ◑ ✕ 🖼 🖾 30 cm
Scolopsis frenatus 325 〜 ↘ ◑ ✕ 🖼 🖾 31 cm

Scolopsis ciliatus 325 〜 ↘ ◑ ✕ 🖼 🖾 20 cm

Scolopsis bimaculatus? ⤳ ⬎ ◐ ✕ 🎞 ⊟ 30 cm
Scolopsis frenatus ⤳ ⬎ ◐ ✕ 🎞 ⊟ 31 cm

Scolopsis vosmeri 325 ⤳ ⬎ ◐ ✕ 🎞 ⊟ 25 cm

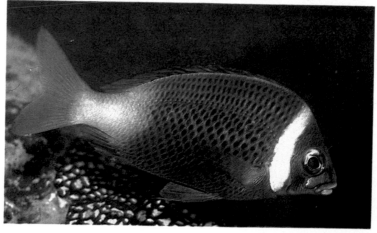

Scolopsis dubiosus 325 〜 ◑ ♥ ▣ ⊡ 40 cm
Scolopsis bilineatus 325 〜 ◑ ♥ ▣ ⊡ 25 cm

Scolopsis xenochrous 325 〜 ◑ ♥ ▣ ⊡ 11 cm

Scolopsis bilineatus 325 〜 ◐ ♥ 🔳 ⊟ 25 cm

Scolopsis bleekeri 325 〜 ◐ ♥ 🔳 ⊟ 25 cm

Scolopsis lineatus 325 〜 ◐ ♥ 🔳 ⊟ 25 cm

Nemipterus virgatus 325 〜 ◐ ♥ 🔲 ⊟ 30 cm
Equetus acuminatus 326 ♀ 〜 ◐ ♥ 🖾 ⊡ 23 cm

Equetus acuminatus 326 ♀ 〜 ◐ ♥ 🖾 ⊡ 23 cm

Equetus lanceolatus 326 ♀ ∿ ◑ ♥ 🖼 ⬚ 25 cm
Equetus punctatus 326 ♀ ∿ ◑ ♥ 🖼 ⬚ 25 cm

Equetus viola 326 ♀ ∿ ◑ ♥ 🖼 ⬚ 25 cm

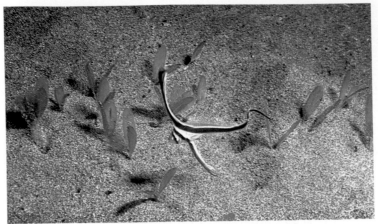

Equetus lanceolatus (juv.) 326 ♀ ∿ ◑ ♥ 🎞 ▭ 25 cm
Equetus punctatus 326 ♀ ∿ ◑ ♥ 🎞 ▭ 25 cm

Equetus viola 326 ♀ ∿ ◑ ♥ 🎞 ▭ 25 cm

Umbrina roncador 326 ∿ ◑ ♥ ▣ ▢ 51 cm
Genyonemus lineatus 326 ∿ ◑ ♥ ▣ ▢ 41 cm

Odontoscion dentex 326 ∿ ◑ ♥ ▣ ▢ 20 cm

Parupeneus rubescens 327 〜 ◐ ♥ ▱ ▱ 42 cm
Upeneus tragula 327 〜 ◐ ♥ ▱ ▱ 25 cm

Parupeneus macronema 327 〜 ◐ ♥ ▱ ▱ 35 cm

Upeneichthys lineatus 327 ∿ ◑ ♥ ▦ ▭ 35 cm
Upeneus bensasi 327 ∿ ◐ ♥ ▦ ▭ 20 cm

Upeneus tragula 327 ∿ ◐ ♥ ▦ ▭ 25 cm

Upeneus moluccensis 327 ∿ ◑ ♥ ▭ ▭ 23 cm
Upeneus vittatus 327 ∿ ◑ ♥ ▭ ▭ 30 cm

Pseudupeneus maculatus 327 ∿ ◑ ♥ ▭ ▭ 30 cm

Mulloidichthys flavolineatus 327 ♀ ∿ ◑ ♥ ▦ ▭ 40 cm
Parupeneus atrocingulatus 327 ♀ ∿ ◑ ♥ ▦ ▭ 25 cm

Parupeneus pleurostigma 327 ♀ ∿ ◑ ♥ ▦ ▭ 25 cm

Parupeneus heptacanthus 327 ♀ ∿ ◑ ♥ ▱ ▭ 28 cm
Parupeneus forsskalii 327 ♀ ∿ ◑ ♥ ▱ ▭ 28 cm

Parupeneus indicus 327 ♀ ∿ ◑ ♥ ▱ ▭ 40 cm

Mulloidichthys vanicolensis 327 ♀ ∿ ◑ ♥ ▱ ▱ 38 cm
Parupeneus barberinus 327 ♀ ∿ ◑ ♥ ▱ ▱ 50 cm

Parupeneus barberinoides 327 ♀ ∽ ● ♥ ▭ ▭ 25 cm
Parupeneus bifasciatus 327 ♀ ∽ ● ♥ ▭ ▭ 40 cm

Parupeneus cyclostomus 327 ♀ ∽ ● ♥ ▭ ▭ 52 cm

Monodactylus sebae 328 ♀ ⤳ ⚲ ◐ ♥ 🖼 🖻 20 cm
Parapriacanthus dispar 329 ♀ ⤳ ◐ ♥ 🖼 🖻 8 cm

Pempheris adspersa 329 ♀ ∿ ◑ ♥ 🖻 ⊡ 16 cm
Pempheris cf *japonicus* 329 ♀ ∿ ◑ ♥ 🖻 ⊡ 18 cm

Pempheris klunzinger 329 ♀ ∿ ◑ ♥ 🖻 ⊡ 16 cm

Parapriacanthus ransonneti 329 ♀ ∿ ⟍ ◑ ✖ 🖼 ⊟ 75 cm
Pempheris vanicolensis 329 ♀ ∿ ⟍ ◑ ✖ 🖼 ⊟ 20 cm

Pempheris schwenki 329 ♀ ∿ ⟍ ◑ ✖ 🖼 ⊟ 30 cm

Atypichthys latus 334 ♀ ∿ ◐ ♥ 🖼 ▱ 13 cm
Girella nigricans 334 ♀ ∿ ◐ ♥ 🖼 ▱ 66 cm

Girella simplicidens 334 ♀ ∿ ◐ ♥ 🖼 ▱ 46 cm

Kyphosus cornelii 334 ♀ ∿ ⚲ ◑ ♥ 📷 ⊟ 38 cm
Girella zebra 334 ♀ ∿ ⚲ ◑ ♥ 📷 ⊟ 30 cm

Neatypus obliquus 334 ♀ ∿ ⚲ ◑ ♥ 📷 ⊟ 14 cm

Tilodon sexfasciatum 334 ♀ ∿ ⚘ ◐ ♥ 🖼 ⊟ 30 cm
Kyphosus sydneyanus 334 ♀ ∿ ⚘ ◑ ♥ 🖼 ⊟ 80 cm

Microcanthus strigatus 334 ♀ ∿ ⚘ ◐ ♥ 🖼 ⊟ 20 cm

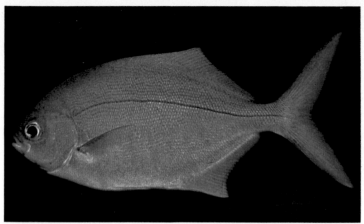

Scorpis aequipinnis 334 〰 ⟍ ◑ ✕ 🖼 ▱ 40 cm
Kyphosus elegans 334 ♀ 〰 ⟋ ◑ ✕ 🖼 ▱ 38 cm

Kyphosus cinerascens (golden form) 334 ♀ 〰 ⟋ ◑ ✕ 🖼 ▱ 50 cm

Kyphosus fuscus 334 ♀ ᔆ ⚓ ◑ ✕ 🎦 ⊟ 60 cm
Scatophagus argus 336 ♀ ᔆ ⚓ ◑ ✕ 🎦 ⊟ 35 cm

Selenotoca multifasciata 336 ♀ ᔆ ⚓ ◑ ✕ 🎦 ⊟ 40 cm

314

Kyphosus sydneyanus 334 ♀ ∿ ⚓ ◑ ♥ 🎥 ▣ 80 cm
Scatophagus tetracanthus 336 ♀ ∿ ⚓ ◑ ♥ 🎦 ▣ 40 cm

Scatophagus argus 336 ♀ ∿ ⚓ ◑ ♥ 🎦 ▣ 35 cm

Chaetodipterus faber 335 ♀ ⌇ ⚹ ◑ ♥ 🖼 ⊟ 91 cm
Chaetodipterus zonatus 335 ♀ ⌇ ⚹ ◑ ♥ 🖼 ⊟ 65 cm

Platax orbicularis 335 ♀ ⌇ ⚹ ◑ ♥ 🖼 ⊟ 50 cm

Drepane punctata 335 ♀ ∿ ⚲ ◑ ♥ 🎞 ⊟ 40 cm
Zabidius novemaculeatus 335 ♀ ∿ ⚲ ◑ ♥ 🎞 ⊟ 20 cm

Platax pinnatus 335 ♀ ∿ ⚲ ◑ ♥ 🎞 ⊟ 40 cm

Drepane punctata 335 ♀ ∿ ⚡ ◑ ♥ 🖼 ▭ 40 cm
Platax teira 335 ♀ ∿ ⚡ ◑ ♥ 🖼 ▭ 50 cm

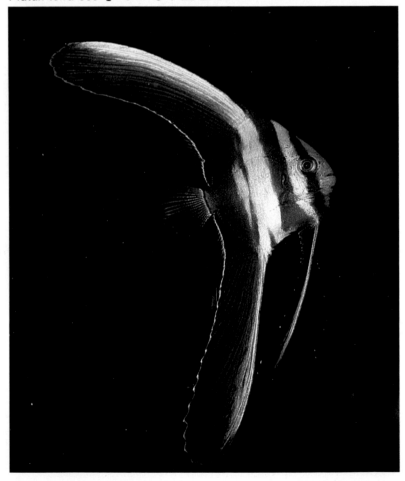

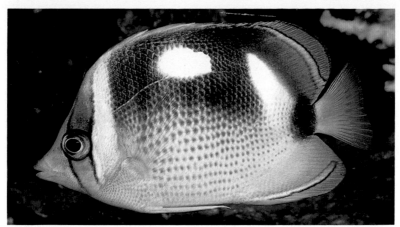

Chaetodon quadrimaculatus 338 〜 ◐ ♥ 🐟 ▣ 15 cm
Chaetodon robustus 338 〜 ◐ ♥ 🐟 ▣ 15 cm

Chaetodon ocellatus 338 〜 ◐ ♥ 🐟 ▣ 19 cm

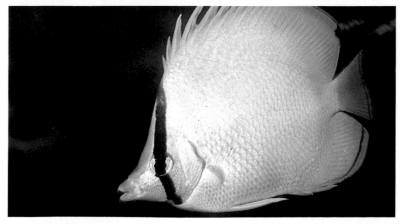

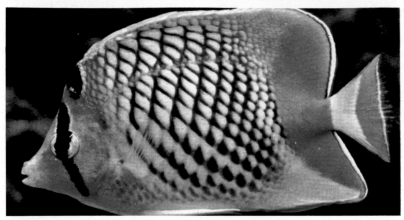

Chaetodon xanthurus 338 ∿ ◑ ♥ ▨ ▱ 15 cm
Chaetodon trifascialis 338 ∿ ◑ ♥ ▨ ▱ 15 cm

Chaetodon trifasciatus 338 ∿ ◑ ♥ ▨ ▱ 17 cm

Chaetodon mertensii 338 〜 ◑ ♥ 🖼 ⊟ 14 cm
Chaetodon baronessa 338 〜 ◑ ♥ 🖼 ⊟ 15 cm

Chaetodon reticulatus 338 〜 ◑ ♥ 🖼 ⊟ 18 cm

Chaetodon leucopleura 338 ᔑ ◐ ♥ 🖼 ⊟ 19 cm
Chaetodon paucifasciatus 338 ᔑ ◐ ♥ 🖼 ⊟ 15 cm

Chaetodon melannotus 338 ᔑ ◐ ♥ 🖼 ⊟ 17 cm

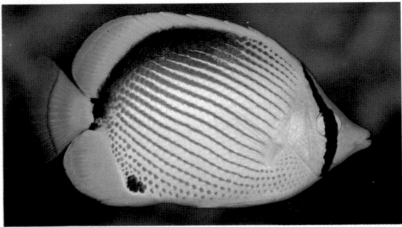

Chaetodon larvatus 338 〜 ◐ ♥ 🖼 ☒ 14 cm
Chaetodon trifascialis 338 〜 ◐ ♥ 🖼 ☒ 15 cm

Chaetodon madagascariensis 338 〜 ◐ ♥ 🖼 ☒ 15 cm

Chaetodon auripes 338 ∿ ○ ♥ 🖼 ⊟ 19 cm
Chaetodon adiergastos 338 ∿ ○ ♥ 🖼 ⊟ 19 cm

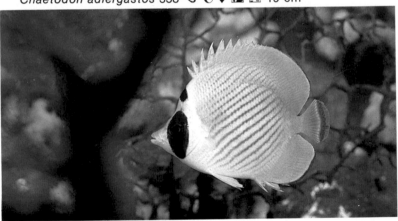

Chaetodon mesoleucos 338 ∿ ○ ♥ 🖼 ⊟ 15 cm

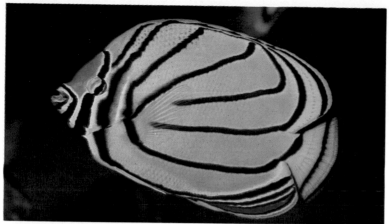

Chaetodon meyeri 338 ↘ ◑ ♥ 🖼 ⊟ 18 cm
Chaetodon nigropunctatus 338 ↘ ◑ ♥ 🖼 ⊟ 16 cm

Chaetodon lineolatus 338 ↘ ◑ ♥ 🖼 ⊟ 29 cm

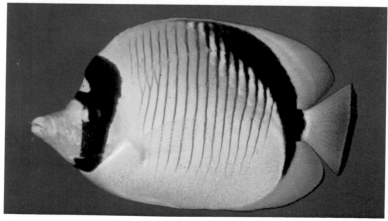

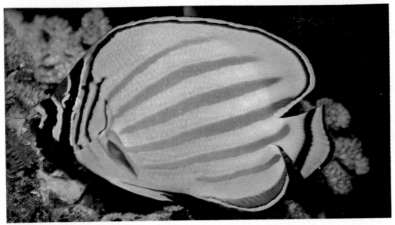

Chaetodon ornatissimus 338 ∿ ◐ ♥ 🖼 ⊟ 19 cm
Chaetodon octofasciatus 338 ∿ ◐ ♥ 🖼 ⊟ 13 cm

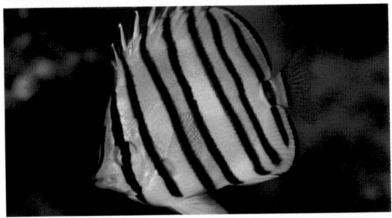

Chaetodon plebeius 338 ∿ ◐ ♥ 🖼 ⊟ 19 cm

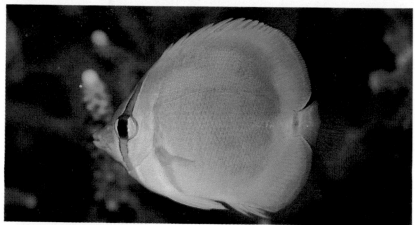

Chaetodon aureofasciatus 338 ♀ ○ ♥ 🖼 🔲 14 cm
Chaetodon ulietensis 338 ∿ ○ ♥ 🖼 🔲 17 cm

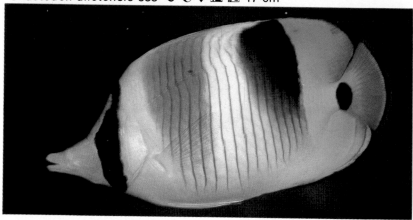

Chaetodon fasciatus 338 ∿ ○ ♥ 🖼 🔲 20 cm

Chaetodon falcula 338 ∿ ◐ ♥ 🖼 ⊟ 29 cm
Chaetodon lunula 338 ∿ ◐ ♥ 🖼 ⊟ 21 cm

Chaetodon kleini 338 ∿ ◐ ♥ 🖼 ⊟ 15 cm

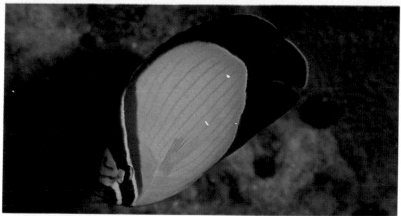

Chaetodon melapterus 338 ～ ☾ ♥ 🖼 ⊟ 14 cm
Chaetodon bennetti 338 ～ ☾ ♥ 🖼 ⊟ 15 cm

Chaetodon citrinellus 338 ～ ☾ ♥ 🖼 ⊟ 13 cm

Chaetodon austriacus 338 ꙅ 🌙 ♥ 🖼 🖂 16 cm
Chaetodon trifasciatus 338 ꙅ 🌙 ♥ 🖼 🖂 17 cm

Chaetodon collare 338 ꙅ 🌙 ♥ 🖼 🖂 18 cm

Chaetodon fremblii 338 ∿ ○ ♥ 🖼 ▣ 18 cm
Chaetodon hemichrysus 338 ∿ ○ ♥ 🖼 ▣ 19 cm

Chaetodon multicinctus 338 ∿ ○ ♥ 🖼 ▣ 15 cm

Chaetodon capistratus 338 ♀ 🌑 ♥ 🖼 🔲 13 cm
Chaetodon striatus 338 ♀ 🌑 ♥ 🖼 🔲 14 cm

Chaetodon miliaris 338 〜 🌑 ♥ 🖼 🔲 17 cm

Chaetodon guttatissimus 338 ∿ ○ ♥ 🖼 ▱ 15 cm
Chaetodon semilarvatus 338 ∿ ○ ♥ 🖼 ▱ 19 cm

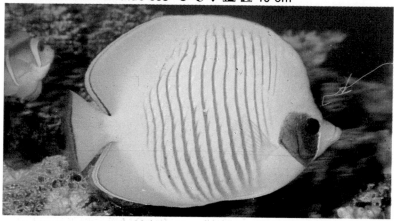

Chaetodon speculum 338 ∿ ○ ♥ 🖼 ▱ 16 cm

Chaetodon semeion 338 ○ ♥ 🖼 ⊟ 22 cm
Chaetodon zanzibariensis 338 ∿ ○ ♥ 🖼 ⊟ 16 cm

Chaetodon vagabundus 338 ∿ ○ ♥ 🖼 ⊟ 17 cm

Chaetodon unimaculatus 338 ∿ ◐ ♥ 🎞 ▭ 20 cm
Chaetodon xanthocephalus 338 ∿ ◐ ♥ 🎞 ▭ 21 cm

Chaetodon auriga 338 ∿ ◐ ♥ 🎞 ▭ 19 cm

Chaetodon marleyi 338 ∿ ◐ ♥ 🖼 ☒ 28 cm
Chaetodon ephippium 338 ∿ ◐ ♥ 🖼 ☒ 21 cm

Chaetodon rafflesi 338 ∿ ◐ ♥ 🖼 ☒ 17 cm

Chaetodon aculeatus 338 ∿ ◑ ♥ 🖼 🖵 12 cm
Chaetodon modestus 338 ∿ ○ ♥ 🖼 🖵 16 cm

Chaetodon tinkeri 338 ∿ ◑ ♥ 🖼 🖵 16 cm

Coradion chrysozonus 338 ⌇ ◐ ♥ 🖼 ⊟ 12 cm
Pseudochaetodon nigrirostris 338 ⌇ ◐ ♥ 🖼 ⊟ 18 cm

Chaetodon aya 338 ⌇ ◐ ♥ 🖼 ⊟ 15 cm

Coradion altivelis 338 ∿ ◑ ♥ 🎞 ⊟ 15 cm
Coradion melanopus 338 ∿ ◑ ♥ 🎞 ⊟ 15 cm

Chaetodon falcifer 338 ∿ ◑ ♥ 🎞 ⊟ 19 cm

Heniochus acuminatus 338 ∿ ○ ♥ 🎞 ⊟ 20 cm
Parachaetodon ocellatus 338 ∿ ○ ♥ 🎞 ⊟ 17 cm

Chelmonops truncatus 338 ∿ ◑ ♥ 🖼 ▭ 23 cm
Chelmon rostratus 338 ∿ ◑ ♥ 🖼 ▭ 20 cm

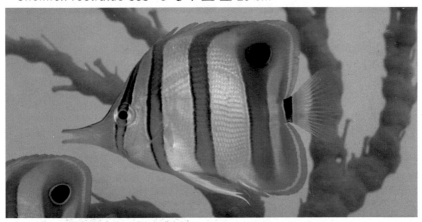

Chelmon rostratus 338 ∿ ◑ ♥ 🖼 ▭ 20 cm

Amphichaetodon howensis 338 ⤳ ◔ ♥ 🖾 ⊟ 20 cm
Chelmonops truncatus 338 ⤳ ◔ ♥ 🖾 ⊟ 23 cm

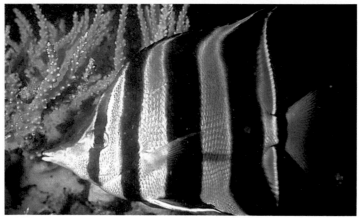

Chelmon muelleri 338 ⤳ ◔ ♥ 🖾 ⊟ 15 cm

Forcipiger longirostris 338 ⤳ ◯ ♥ 🐟 ▤ 27 cm

Forcipiger longirostris 338 ⤳ ◯ ♥ 🐟 ▤ 27 cm

Hemitaurichthys polylepis 338 ⤳ ◑ ♥ 🐟 ▤ 18 cm

Forcipiger longirostris 338 〜 ○ ♥ 🖼 ⊟ 27 cm
Forcipiger flavissimus 338 ⚲ 〜 ○ ♥ 🖼 ⊟ 26 cm

Hemitaurichthys zoster 338 〜 ○ ♥ 🖼 ⊟ 20 cm

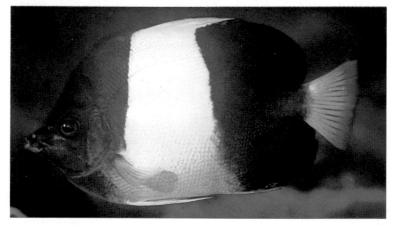

Heniochus chrysostomus 338 ∿ ♥ 🖼 ⊟ 18 cm
Heniochus monoceros 338 ∿ ○ ♥ 🖼 ⊟ 19 cm

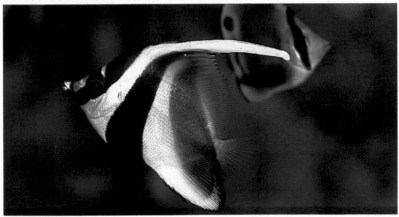

Heniochus varius 338 ∿ ○ ♥ 🖼 ⊟ 17 cm

Heniochus singularius 338 ∿ ◐ ♥ 🐟 ▱ 24 cm
Heniochus varius 338 ∿ ◐ ♥ 🐟 ▱ 17 cm

Heniochus monoceros 338 ∿ ◐ ♥ 🐟 ▱ 19 cm

Apolemichthys trimaculatus 339 ∿ ⚹ ◐ ♥ 🖼 ⊡ 30 cm
Centropyge acanthops 339 ♀ ∿ ⚹ ◐ ♥ 🖼 ⊡ 8 cm

Centropyge bispinosus 339 ♀ ∿ ⚹ ◐ 🖼 ⊡ 12 cm

Centropyge bicolor 339 ♀ ∿ ⚲ ◐ ♥ 🎥 ⊟ 15 cm
Centropyge eibli 339 ♀ ∿ ⚲ ◐ ♥ 🎥 ⊟ 15 cm

Centropyge multispinis 339 ♀ ∿ ⚲ ◐ ♥ 🎥 ⊟ 14 cm

Centropyge loriculus 339 ♀ ∿ ⚡ ◐ ♥ 🖾 ⊟ 12 cm
Centropyge nox 339 ♀ ∿ ⚡ ◐ ♥ 🖾 ⊟ 11 cm

Centropyge heraldi 339 ♀ ∿ ⚡ ◐ ♥ 🖾 ⊟ 12 cm

Centropyge ferrugatus 339 ♀ ∿ ⋆ ◐ ♥ 🎦 ▣ 10 cm
Centropyge potteri 339 ♀ ∿ ⋆ ◐ ♥ 🎦 ▣ 10 cm

Centropyge colini 339 ♀ ∿ ⋆ ◐ ♥ 🎦 ▣ 7 cm

Centropyge resplendens 339 ♀ ∿ ⚚ ◐ ♥ 📷 ▱ 7 cm
Centropyge argi 339 ♀ ∿ ⚚ ◐ ♥ 📷 ▱ 7 cm

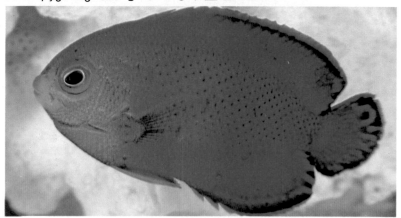

Centropyge hotumatua 339 ♀ ∿ ⚚ ◐ ♥ 📷 ▱ 9 cm

Centropyge fisheri 339 ♀ ⤳ ⚮ ○ ♥ ▣ ▤ 10 cm
Centropyge multicolor 339 ♀ ⤳ ⚮ ○ ♥ ▣ ▤ 8 cm

Centropyge flavicauda 339 ♀ ⤳ ⚮ ○ ♥ ▣ ▤ 8 cm

Paracentropyge multifasciatus 339 ♀ ⌇ ⚲ ○ ♥ 📷 ▭ 15 c₁
Centropyge tibicin 339 ♀ ⌇ ⚲ ○ ♥ 📷 ▭ 19 cm

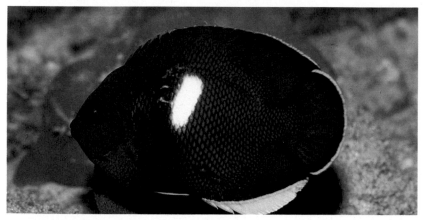

Genicanthus caudovittatus (male) 339 ♀ ⌇ ⚲ ○ ♥ 📷 ▭ 20 cm

Centropyge flavissimus 339 ♀ ⌇ ⚓ ☉ ♥ 🔲 ⊟ 9 cm
Centropyge joculator 339 ♀ ⌇ ⚓ ☉ ♥ 🔲 ⊟ 15 cm

Centropyge vroliki 339 ♀ ⌇ ⚓ ☉ ♥ 🔲 ⊟ 10 cm

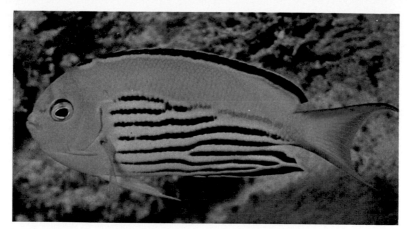

Genicanthus watanabei 339 ♀ ⌇ ⚹ ○ ♥ 📷 ⊟ 10 cm
Genicanthus semifasciatus 339 ♀ ⌇ ⚹ ○ ♥ 📷 ⊟ 10 cm

Genicanthus melanospilus 339 ♀ ⌇ ⚹ ○ ♥ 📷 ⊟ 21 cm

Genicanthus personatus 339 ♀ ∿ ⚹ ◐ ♥ 🔲 ⊡ 21 cm
Genicanthus lamarck 339 ♀ ∿ ⚹ ◐ ♥ 🔲 ⊡ 24 cm

Genicanthus bellus 339 ♀ ∿ ⚹ ◐ ♥ 🔲 ⊡ 13 cm

Chaetodontoplus septentrionalis (juv.) 339 ♀ ∿ ⚓ ○ ♥ 📷 ⊟ 22 cm
Chaetodontoplus duboulayi 339 ♀ ∿ ⚓ ○ ♥ 📷 ⊟ 22 cm

Chaetodontoplus personifer 339 ♀ ∿ ⚓ ○ ♥ 📷 ⊟ 23 cm

Chaetodontoplus melanosoma 339 ♀ ∿ ⋔ ○ ♥ 🎞 ▱ 20 cm
Chaetodontoplus septentrionalis 339 ♀ ∿ ⋔ ○ ♥ 🎞 ▱ 22 cm

Chaetodontoplus chrysocephalus 339 ♀ ∿ ⋔ ○ ♥ 🎞 ▱ 22 cm

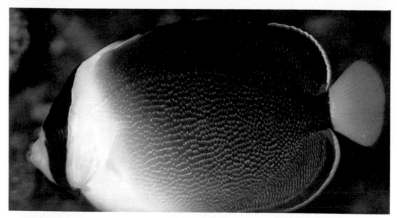

Chaetodontoplus mesoleucus 339 ♀ ∿ ⚲ ◐ ♥ 🔲 ▱ 15 cm
Euxiphipops navarchus 339 (juv.) ♀ ∿ ⚲ ◐ ♥ 🔲 ▱ 25 cm

Euxiphipops xanthometopon 339 ♀ ∿ ⚲ ◐ ♥ 🔲 45 cm

Pomacanthus zonipectus (juv.) 339 ♀ ∿ ⚓ ◐ ♥ 🖾 30 cm
Pomacanthus paru (juv.) 339 ♀ ∿ ⚓ ◐ ♥ 🖾 ⊟ 40 cm

Pomacanthus arcuatus 339 (juv.) ♀ ∿ ⚓ ◐ ♥ 🖾 ⊟ 36 cm

Pomacanthus annularis 339 (juv.) ♀ ∿ ⚲ ◐ ♥ 🔲 ⊡ 25 cm
Pomacanthus imperator 339 (juv.) ♀ ∿ ⚲ ◐ ♥ 🔲 ⊡ 38 cm

Pomacanthus imperator 339 ♀ ∿ ⚲ ◐ ♥ 🔲 ⊡ 38 cm

Pomacanthus chrysurus 339 ♀ ⌇ ⚓ ◐ ♥ 📷 ⊟ 33 cm
Euxiphipops navarchus 339 ♀ ⌇ ⚓ ◐ ♥ 📷 ⊟ 25 cm

Pygoplites diacanthus 339 ♀ ⌇ ⚓ ◐ ♥ 📷 ⊟ 30 cm

Pomacanthus semicirculatus 339 ♀ ∿ ⚓ ◐ ♥ 📷 🖼 45 cm
Euxiphipops sexstriatus 339 ♀ ∿ ⚓ ◐ ♥ 📷 🖼 50 cm

Euxiphipops xanthometapon 339 ♀ ∿ ⚓ ◐ ♥ 📷 🖼 45 cm

Apolemichthys arcuatus 339 ♀ ∿ ⅄ ◐ ♥ 🎴 ⊟ 17.5 cm
Apolemichthys xanthopunctatus 339 ♀ ∿ ◐ ♥ 🎴 ⊟ 25 cm

Holacanthus tricolor (juv.) 339 ♀ ∿ ⅄ ◐ ♥ 🎴 ⊟ 25 cm

Apolemichthys xanthotis 339 ♀ ∿ ⚹ ○ ♥ 📷 ⊟ 20 cm
Sumireyakko venustus 339 ♀ ∿ ⚹ ○ ♥ 📷 ⊟ 11 cm

Holacanthus tricolor 339 ♀ ∿ ⚹ ○ ♥ 📷 ⊟ 25 cm

Holacanthus clarionensis 339 ♀ ∿ ⚲ ◐ ♥ 📷 ⊡ 45 cm
Holacanthus ciliaris 339 (juv.) ♀ ∿ ⚲ ◐ ♥ 📷 ⊡ 25 cm

Holacanthus passer 339 ♀ ∿ ⚲ ◐ ♥ 📷 ⊡ 23 cm

Oplegnathus fasciatus 343 ♀ ⚲ ◑ ✕ 🖾 ☰ 80 cm
Oplegnathus punctatus 343 〜 ◑ ✕ 🖾 ☰ 86 cm

Amphistichus argenteus 345 〜 ◑ ✕ 🖾 ☰ 43 cm

Embiotoca jacksoni 345 ∿ ◑ ✕ 🖼 ⊟ 39 cm
Micrometrus aurora 345 ∿ ◑ ✕ 🖼 ⊟ 18 cm

Damalichthys vacca 345 ∿ ◑ ✕ 🖼 ⊟ 44 cm

Hypsurus caryi 345 ∿ ◐ ✕ 🖼 ▱ 30 cm
Micrometrus minimus 345 ∿ ◐ ✕ 🖼 ▱ 16 cm

Hyperprosopon ellipticum 345 ∿ ◐ ✕ 🖼 ▱ 30 cm

Premnas biaculeatus 346 ♀ ∿ ⚓ ○ ♥ 📷 ⊟ 18 cm
Amphiprion leucokranos 346 ♀ ∿ ⚓ ○ ♥ 📷 ⊡ 12 cm

Amphiprion latezonatus 346 ♀ ∿ ⚓ ○ ♥ 📷 ⊡ 12 cm

Amphiprion percula 346 ♀ ⌇ ⚲ ◑ ♥ 📷 🖼 11 cm
Amphiprion polymnus 346 ♀ ⌇ ⚲ ◑ ♥ 📷 🖼 10 cm

Amphiprion akindynos 346 ♀ ⌇ ⚲ ◑ ♥ 📷 🖼 12 cm

Amphiprion allardi 346 ♀ ⌔ ⚭ ○ ♥ 🔳 ▭ 11 cm
Amphiprion chrysogaster 346 ♀ ⌔ ⚭ ○ ♥ 🔳 ▭ 11 cm

Amphiprion ephippium 346 ♀ ⌔ ⚭ ○ ♥ 🔳 ▭ 7 cm

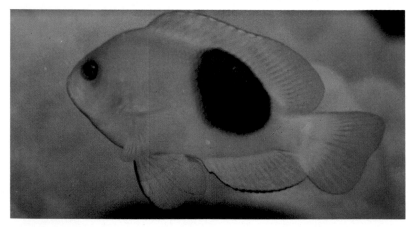

Amphiprion akallopisos 346 ♀ ∿ ⚓ ◑ ♥ 📷 🖵 9 cm
Amphiprion clarkii 346 ♀ ∿ ⚓ ◑ ♥ 📷 🖵 10 cm

Amphiprion chrysogaster 346 ♀ ∿ ⚓ ◑ ♥ 📷 🖵 11 cm

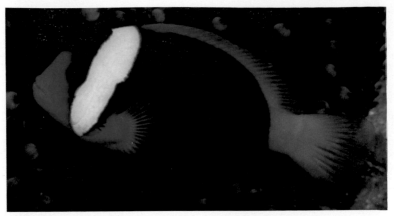

Amphiprion melanopus (var) ♀ ～ ⚲ ◐ ♥ 📷 ▢ 13 cm
Amphiprion ephippium 346 ♀ ～ ⚲ ◐ ♥ 📷 ▢ 10 cm

Amphiprion thiellei 346 ♀ ～ ⚲ ◐ ♥ 📷 ▢ 8 cm

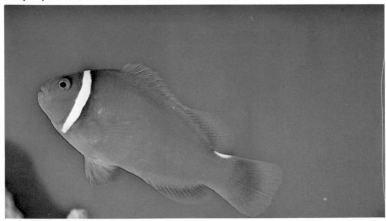

Amphiprion sandaracinos 346 ♀ ～ ⚲ ◐ ♥ ▣ ▭ 14 cm
Amphiprion tricinctus 346 ♀ ～ ⚲ ◐ ♥ ▣ ▭ 14 cm

Amphiprion frenatus 346 ♀ ～ ⚲ ◐ ♥ ▣ ▭ 10 cm

Amphiprion perideraion 346 ♀ ∿ ⚲ ☉ ♥ 📷 🖵 11 cm
Amphiprion chrysopterus 346 ♀ ∿ ⚲ ☉ ♥ 📷 🖵 12.5 cm

Amphiprion rubrocinctus 346 ♀ ∿ ⚲ ☉ ♥ 📷 🖵 14 cm

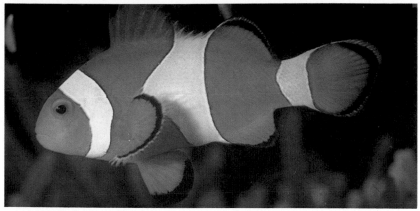

Amphiprion ocellaris 346 ♀ ∽ ⚓ ◐ ♥ 🖼 ⬚ 15 cm
Dascyllus carneus 346 ♀ ∽ ⚓ ◐ ♥ 🖼 ⊟ 10 cm

Dascyllus trimaculatus 346 ♀ ∽ ⚓ ◐ ♥ 🖼 ⊟ 14 cm

Amphiprion sebae 346 ♀ ∿ ⚒ ◐ ♥ 📷 ▭ 14 cm
Dascyllus aruanus 346 ♀ ∿ ⚒ ◐ ♥ 📷 ⊞ 10 cm

Parma bicolor 346 ♀ ∿ ⚒ ◐ ♥ 📷 ⊞ 20 cm

Dascyllus albisella 346 ♀ ∿ ⚡ ◑ ♥ 🔳 ▭ 13 cm
Dascyllus melanurus 346 ♀ ∿ ⚡ ◑ ♥ 🔳 ▭ 10 cm

Dascyllus reticulatus 346 ♀ ∿ ⚡ ◑ ♥ 🔳 ▭ 10 cm

Dascyllus trimaculatus 346 ♀ ∽ ⚲ ○ ♥ 🎥 ⊟ 14 cm
Dascyllus strasburgi 346 ♀ ∽ ⚲ ○ ♥ 🎥 ⊟ 14 cm

Dascyllus marginatus 346 ♀ ∽ ⚲ ○ ♥ 🎥 ⊟ 12 cm

Parma oligolepis 346 ♀ ∿ ⚲ ◐ ♥ 📷 ⊟ 20 cm
Parma microlepis 346 ♀ ∿ ⚲ ◑ ♥ 📷 ⊟ 30 cm

Parma alboscapularis 346 ♀ ∿ ⚲ ◐ ♥ 📷 ⊟ 20 cm

Parma mccullochi 346 ♀ ∿ ⚓ ○ ♥ 🎞 ⊡ 12 cm
Parma microlepis 346 (juv.) ♀ ∿ ⚓ ○ ♥ 🎞 ⊡ 30 cm

Parma victoriae 346 ♀ ∿ ⚓ ○ ♥ 🎞 ⊡ 20 cm

Abudefduf troscheli 346 ♀ ∿ ⊁ ◐ ♥ ▦ ⊟ 23 cm
Abudefduf concolor 346 ♀ ∿ ⊁ ◐ ♥ ▦ ⊟ 25 cm

Abudefduf taurus 346 ♀ ∿ ⊁ ◐ ♥ ▦ ⊟ 25 cm

Abudefduf saxatilis 346 ♀ ∽ ⋇ ◑ ♥ ▧ ⊟ 15 cm
Abudefduf troscheli 346 ♀ ∽ ⋇ ◑ ♥ ▧ ⊟ 23 cm

Abudefduf sordidus 346 ♀ ∽ ⋇ ◑ ♥ ▧ ⊟ 20 cm

Abudefduf margariteus 346 ♀ ∿ ⚲ ◐ ♥ 🎥 ⊟ 16 cm
Abudefduf sexfasciatus 346 ♀ ∿ ⚲ ◐ ♥ 🎥 ⊟ 15 cm

Amblyglyphidodon flavilatus 346 ♀ ∿ ⚲ ◐ ♥ 🎥 ⊟ 25 cm

Abudefduf lorenzi 346 ♀ ∿ ⚹ ☾ ♥ 📷 ⊟ 12 cm
Amblyglyphidodon leucogaster 346 ♀ ∿ ⚹ ☾ ♥ 📷 ⊟ 13 cm

Amblyglyphidodon ternatensis 346 ♀ ∿ ⚹ ☾ ♥ 📷 ⊟ 14 cm

Abudefduf abdominalis 346 ♀ ⌇ ⚲ O ♥ 📷 ⊟ 19 cm
Amblyglyphidodon curacao 346 ♀ ⌇ ⚲ O ♥ 📷 ⊟ 14 cm

Amblyglyphidodon aureus 346 ♀ ⌇ ⚲ O ♥ 📷 ⊟ 13 cm

Chromis atripectoralis 346 ♀ ∿ ⚘ ◐ ♥ 🔍 🔲 13 cm
Chromis viridis 346 ♀ ∿ ⚘ ◐ ♥ 🔍 🔲 10 cm

Chromis elerae 346 ♀ ∿ ⚘ ◐ ♥ 🔍 🔲 10 cm

Chromis dimidiata 346 ♀ ∿ ⚬ ◑ ♥ 📷 ⊟ 9 cm
Chromis klunzingeri 346 ♀ ∿ ⚬ ◑ ♥ 📷 ⊟ 10 cm

Chromis opercularis 346 ♀ ∿ ⚬ ◑ ♥ 📷 ⊟ 16 cm

Chromis enchrysura 346 10 cm

Chromis ovalis 346 19 cm

Chromis westaustralis 346 7 cm

Chromis lineata 346 ♀ ∿ ⚲ ◐ ♥ 📷 ⊡ 9 cm
Chromis acares 346 ♀ ∿ ⚲ ◐ ♥ 📷 ⊡ 9 cm

Chromis analis 346 ♀ ∿ ⚲ ◐ ♥ 📷 ⊡ 13 cm

Chromis vanderbilti 346 ♀ ∿ ⚓ ◑ ♥ 🔳 ⊟ 10 cm
Chromis agilis 346 ♀ ∿ ⚓ ◑ ♥ 🔳 ⊟ 12 cm

Chromis chrysura 346 ♀ ∿ ⚓ ◑ ♥ 🔳 ⊟ 15 cm

Chromis atripes 346 ♀ ⌇ ⚲ ❍ ♥ 🔲 ▣ 5 cm
Chromis lepidolepis 346 ♀ ⌇ ⚲ ❍ ♥ 🔲 ▣ 12 cm

Chromis hypselepis 346 ♀ ⌇ ⚲ ❍ ♥ 🔲 ▣ 13 cm

Chromis nitida 346 ♀ ∿ ⚘ ○ ♥ 📷 ⊟ 12 cm
Chromis margaritifer 346 ♀ ∿ ⚘ ○ ♥ 📷 ⊟ 11 cm

Chromis hanui 346 ♀ ∿ ⚘ ○ ♥ 📷 ⊟ 11 cm

Chromis retrofasciata 346 ♀ ∽ ⚓ ◐ ♥ 🎞 🖼 9 cm
Chromis xanthura 346 ♀ ∽ ⚓ ◐ ♥ 🎞 🖼 16 cm

Chromis iomelas 346 ♀ ∽ ⚓ ◐ ♥ 🎞 🖼 10 cm

Chromis altus 346 ♀ ∿ ⚹ ◐ ♥ 📷 🖼 9 cm
Chromis multilineatus 346 ♀ ∿ ⚹ ◐ ♥ 📷 🖼 16 cm

Chromis viridis 346 ♀ ∿ ⚹ ◐ ♥ 📷 🖼 10 cm

Chromis limbaughi 346 ♀ ～ ↝ ○ ♥ 🔳 ⊟ 10 cm
Chromis altus 346 ♀ ～ ↝ ○ ♥ 🔳 ⊟ 9 cm

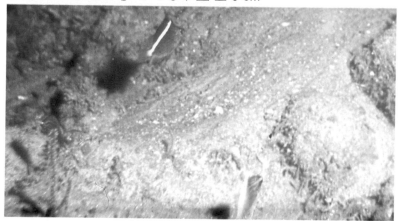

Chromis punctipinnis 346 ♀ ～ ↝ ○ ♥ 🔳 ⊟ 30 cm

Chrysiptera caeruleolineata 346 ♀ ∿ ⚲ ☽ ♥ 🔲 🖼 9 cm
Chrysiptera traceyi 346 ♀ ∿ ⚲ ☽ ♥ 🔲 🖼 10 cm

Chrysiptera annulata 346 ♀ ∿ ⚲ ☽ ♥ 🔲 🖼 7 cm

Chrysiptera leucopoma 346 (juv.) ♀ ∿ ⚲ ○ ♥ 🔒 ⊟ 12 cm
Neoglyphidodon crossi (juv.) 346 ♀ ∿ ⚲ ○ ♥ 🔒 ⊟ 10 cm

Chrysiptera talboti 346 ♀ ∿ ⚲ ○ ♥ 🔒 ⊟ 10 cm

Chrysiptera flavipinnis 346 ♀ ∿ ⚡ ◑ ♥ 📷 ▱ 11 cm
Chrysiptera cyanea 346 ♀ ∿ ⚡ ◑ ♥ 📷 ▱ 8 cm

Chrysiptera taupou 346 ♀ ∿ ⚡ ◑ ♥ 📷 ▱ 75 cm

Chrysiptera starcki 346 ♀ ⌇ ⚹ ○ ♥ 🔒 ⊟ 10 cm
Chrysiptera cyanea 346 ♀ ⌇ ⚹ ○ ♥ 🔒 ⊟ 8 cm

Chrysiptera parasema 346 ♀ ⌇ ⚹ ○ ♥ 🔒 ⊟ 8 cm

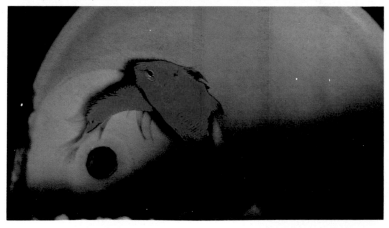

Chrysiptera biocellata 346 ♀ ⤳ ⚹ ◐ ♥ 🎦 ⊟ 12 cm
Chrysiptera unimaculata 346 ♀ ⤳ ⚹ ◐ ♥ 🎦 ⊟ 11 cm

Chrysiptera rollandi 346 ♀ ⤳ ⚹ ◐ ♥ 🎦 ⊟ 9 cm

Chrysiptera glauca 346 ♀ ∿ ⚲ ◐ ♥ 📷 ⊡ 12 cm
Chrysiptera unimaculata 346 ♀ ∿ ⚲ ◐ ♥ 📷 ⊡ 11 cm

Neopomacentrus miryae 346 ♀ ∿ ⚲ ◐ ♥ 📷 ⊡ 11 cm

Plectroglyphidodon johnstonianus 346 ♀ ∿ ⚲ ○ ♥ 🖾 ⊟ 12 cm
Plectroglyphidodon sindonis 346 ♀ ∿ ⚲ ○ ♥ 🖾 ⊟ 15 cm

Pomachromis guamensis 346 ♀ ∿ ⚲ ○ ♥ 🖾 ⊟ 12 cm

Neoglyphidodon melas 346 ♀ ∿ ⚲ ◐ ♥ 🎞 ⊟ 17 cm
Neoglyphidodon polyacanthus 346 ♀ ∿ ⚲ ◐ ♥ 🎞 ⊟ 17 cm

Plectroglyphidodon lacrymatus 346 ♀ ∿ ⚲ ◐ ♥ 🎞 ⊟ 13 cm

Neoglyphidodon nigroris 346 (juv.) ♀ ∿ ⚡ ◐ ♥ 📷 ⊡ 15 cm
Neoglyphidodon melas 346 (juv.) ♀ ∿ ⚡ ◐ ♥ 📷 ⊡ 17 cm

Neoglyphidodon polyacanthus 346 ♀ ∿ ⚡ ◐ ♥ 📷 ⊡ 17 cm

Hypsypops rubicunda 346 (juv.) ♀ ∿ ⚲ ○ ♥ 📷 ▭ 36 cm

Acanthochromis polyacanthus (juv.) 346 ♀ ∿ ⚲ ○ ♥ 📷 ▭ 10 cm

Hemiglyphidodon plagiometopon 346 ♀ ∿ ⚲ ○ ♥ 📷 ▭ *19 cm*

Neopomacentrus taeniurus 346 ♀ ⤳ ⚥ ○ ♥ 📷 ⊟ 11 cm
Neopomacentrus azysron 346 ♀ ⤳ ⚥ ○ ♥ 📷 ⊟ 6 cm

Neopomacentrus nemurus 346 ♀ ⤳ ⚥ ○ ♥ 📷 ⊟ 11 cm

Pomacentrus philippinus 346 ♀ ∽ ⚵ ◐ ♥ 🎥 ⊟ 13 cm
Pomacentrus caeruleus 346 ♀ ∽ ⚵ ◐ ♥ 🎥 ⊟ 10 cm

Pomacentrus alleni 346 ♀ ∽ ⚵ ◐ ♥ 🎥 ⊟ 8 cm

Pomacentrus coelestis 346 ♀ ∿ ⚕ ○ ♥ 🎦 ▦ 12 cm
Pomacentrus reidi 346 ♀ ∿ ⚕ ○ ♥ 🎦 ▦ 14 cm

Dischistodus pseudochrysopoecilus 346 ♀ ∿ ⚕ ○ ♥ 19 cm

Pomacentrus nigromanus 346 ♀ ～ ⋋ ◐ ♥ 🔳 ⊟ 12 cm
Pomacentrus philippinus 346 ♀ ～ ⋋ ◐ ♥ 🔳 ⊟ 13 cm

Pomacentrus stigma 346 ♀ ～ ⋋ ◐ ♥ 🔳 ⊟ 12 cm

Pomacentrus vaiuli 346 ♀ ∿ ⚲ ○ ♥ 🔳 ⊟ 8 cm
Pomacentrus bankanensis 346 ♀ ∿ ⚲ ○ ♥ 🔳 ⊟ 11 cm

Pomacentrus moluccensis 346 ♀ ∿ ⚲ ○ ♥ 🔳 ⊟ 11 cm

Pomacentrus pikei 346 ♀ ∿ ⚸ ◐ ♥ ▣ ⊟ 12 cm
Pomacentrus baenschi 346 ♀ ∿ ⚸ ◐ ♥ ▣ ⊟ 10 cm

Stegastes insularis 346 ♀ ∿ ⚸ ◐ ♥ ▣ ⊟ 9 cm

Pomacentrus trilineatus 346 ♀ ∿ ⚲ ☽ ♥ 📷 ⊟ 7 cm
Stegastes fasciolatus 346 ♀ ∿ ⚲ ☽ ♥ 📷 ⊟ 13 cm

Stegastes pelicieri 346 ♀ ∿ ⚲ ☽ ♥ 📷 ⊟ 8 cm

Stegastes albifasciatus 346 ♀ ⌁ ⚓ ◐ ♥ 🔲 ⊟ 13 cm
Stegastes lividus 346 ♀ ⌁ ⚓ ◐ ♥ 🔲 ⊟ 15 cm

Stegastes sp. 346 ♀ ⌁ ⚓ ◐ ♥ 🔲 ⊟ 12 cm

Stegastes rectifraenum 346 ♀ ∽ ⅄ ◑ ♥ 🎞 ▨ 10 cm
Stegastes leucorus 346 ♀ ∽ ⅄ ◑ ♥ 🎞 ▨ 50 cm

Stegastes emeryi 346 ♀ ∽ ⅄ ◑ ♥ 🎞 ▨ 11 cm

Stegastes partitus 346 ♀ ∿ ⸙ ◑ ♥ 📷 ⊡ 10 cm
Stegastes planifrons 346 ♀ ∿ ⸙ ◑ ♥ 📷 ⊡ 13 cm

Stegastes dorsopunicans 346 ♀ ∿ ⸙ ◑ ♥ 📷 ⊡ 15 cm

Stegastes mellis 346 ⚥ 〜 ⚶ ◑ ♥ 🖼 ⊟ 12.5 cm
Stegastes variabilis 346 ⚥ 〜 ⚶ ◑ ♥ 🖼 ⊟ 13 cm

Stegastes leucostictus 346 ⚥ 〜 ⚶ ◑ ♥ 🖼 ⊟ 10 cm

Microspathodon dorsalis 346 ♀ ∿ ⚓ ○ ♥ 🎞 ⊡ 31 cm
Microspathodon chrysurus 346 ♀ ∿ ⚓ ○ ♥ 🎞 ⊡ 20 cm

Microspathodon bairdi 346 ♀ ∿ ⚓ ○ ♥ 🎞 ⊡ 20 cm

Dischistodus prosopotaenia 346 ♀ ∿ ⚲ ○ ♥ 🖼 ⊡ 15 cm
Dischistodus fasciatus 346 ♀ ∿ ⚲ ○ ♥ 🖼 ⊡ 11.5 cm

Dischistodus melanotus 346 ♀ ∿ ⚲ ○ ♥ 🖼 ⊡ 10 cm

Cirrhitichthys aprinus 348 ♀ ∿ ◐ ♥ 📷 ▭ 7 cm
Cirrhitichthys oxycephalus 348 ♀ ∿ ◐ ♥ 📷 ▭ 8 cm

Cirrhitichthys guichenoti 348 ♀ ∿ ◐ ♥ 📷 ▭ 11 cm

Cirrhitus rivulatus 348 ♀ ∿ ◑ ♥ 🖼 ⬜ 52 cm
Cirrhitus splendens 348 ♀ ∿ ◑ ♥ 🖼 ⬜ 20 cm

Neocirrhitus armatus 348 ♀ ∿ ◑ ♥ 🖼 ⬜ 7.5 cm

Cirrhitops fasciatus 348 ♀ ∿ ◑ ♥ 🖼 ▭ 13 cm
Paracirrhites arcatus 348 ♀ ∿ ◑ ♥ 🖼 ▭ 14 cm

Paracirrhites hemistictus 348 ♀ ∿ ◑ ♥ 🖼 ▭ 18 cm

Paracirrhites forsteri 348 ♀ ⤳ ◐ ♥ 🖻 ▭ 25 cm
Paracirrhites hemistictus 348 ♀ ⤳ ◐ ♥ 🖻 ▭ 18 cm

Oxycirrhitus typus 348 ♀ ⤳ ◐ ♥ 🖻 ▭ 13 cm

Amblycirrhitus pinos 348 ♀ ∿ ⤙ ◑ ✂ 📷 ▭ 68 cm
Amblycirrhitus earnshawi 348 ♀ ∿ ⤙ ◑ ✂ 📷 ▭ 18 cm

Paracirrhites forsteri 348 ♀ ∿ ⤙ ◑ ✂ 📷 ▭ 25 cm

Amblycirrhites bimacula 348 ♀ ∿ ◑ ♥ 🖼 ⬜ 7 cm
Paracirrhites xanthus 348 ♀ ∿ ↘ ◑ ✂ 🖼 ⬜ 11 cm

Paracirrhites forsteri 348 ♀ ∿ ↘ ◑ ✂ 🖼 ⬜ 25 cm

Aplodactylus meandritus 350 ♀ ∿ ➘ ◑ ✄ 🎦 ⬚ 38 cm
Cheilodactylus zonatus 351 ♀ ∿ ◑ ♥ 🎦 ⬚ 45 cm

Nemadactylus macropterus 351 ♀ ∿ ➘ ◑ ✄ 🎦 ⬚ 58 cm

Nemadactylus valenciennesi 351 ♀ ∿ ◑ ♥ 🔲 ⊡ 70 cm
Cheilodactylus fuscus 351 ♀ ∿ ◑ ♥ 🔲 ▭ 46 cm

Cheilodactylus spectabilis 351 ♀ ∿ ◑ ♥ 🔲 ▭ 20 cm

Cheilodactylus gibbosus 351 ♀ ∿ ◑ ♥ 📷 ⬚ 30 cm
Cheilodactylus ephippium 351 ♀ ∿ ◑ ♥ 📷 ⬚ 20 cm

Cheilodactylus vittatus 351 ♀ ∿ ◑ ♥ 📷 ⬚ 25 cm

Cheilodactylus rubrolabiatus 351 ♀ ∿ ◐ ♥ 📷 ☐ 50 cm
Dactylophora nigricans 351 ♀ ∿ ◐ ♥ 📷 ☐ 120 cm

Chirodactylus brachydactylus 351 ♀ ∿ ◐ ♥ 📷 ☐ 40 cm

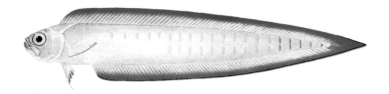

Acanthocepola indica 354 ⤴ ↘ ◑ ✂ 🎴 ⬚ 35 cm
Polydactylus virginicus 357 ⤴ ↘ ◑ ♥ 🎴 ⬚ 30 cm

Polydactylus sexfilis 357 ⤴ ↘ ◑ ♥ 🎴 ⬚ 100 cm

Liza vaigiensis 355 ♀ ∿ ☾ ♥ 🖼 ☐ 60 cm
Mugil cephalus 355 ♀ ∿ ☾ ♥ 🖼 ☐ 50 cm

Liza vaigiensis 355 ♀ ∿ ☾ ♥ 🖼 ☐ 60 cm

Sphyraena jello 356 ↘ ◑ ✕ 🔾 ⊟ 125 cm
Sphyraena barracuda 356 ↘ ◑ ✕ 🔾 ⊟ 200 cm

Sphyraena qenie 356 ↘ ◑ ✕ 🔾 ⊟ 115 cm

Anampses meleagrides 358 ∿ ○ ♥ 🖼 ▦ 22 cm
Anampses cuvier 358 ∿ ○ ♥ 🖼 ▦ 38 cm

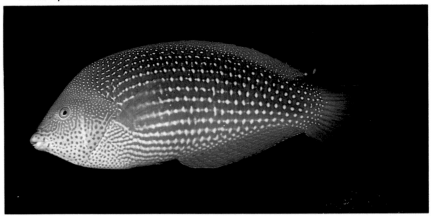

Anampses chrysocephalus 358 ∿ ○ ♥ 🖼 ▦ 16 cm

Anampses lineatus 358 ↝ ◑ ♥ ▥ ▤ 12 cm
Bodianus opercularis 358 ⚲ ◑ ♥ ▥ ▤ 12 cm

Bodianus axillaris 358 (juv.) ⚲ ◑ ♥ ▥ ▤ 20 cm

Anampses lennardi 358 ∿ ○ ♥ 🖼 ⊟ 25 cm
Anampses twisti 358 ∿ ○ ♥ 🖼 ⊟ 18 cm

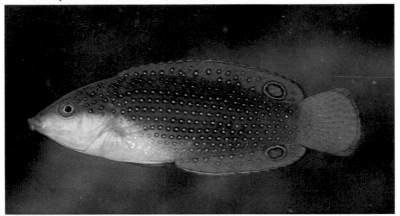

Bodianus bimaculatus 358 ♀ ○ ♥ 🖼 ⊟ 9 cm

Anampses lennardi 358 ∿ ◑ ♥ 🔲 ⊟ 25 cm
Pseudocoris bleekeri 358 ∿ ◑ ♥ 🔲 ⊟ 15 cm

Bodianus sp. 358 ♀ ◑ ♥ 🔲 ⊟ 21 cm

Anampses elegans 358 ⌒ ◑ ♥ 🔍 ⊟ 28 cm
Bodianus oxycephalus 358 ♀ ◑ ♥ 🔍 ⊟ 27 cm

Bodianus masudai 358 ♀ ◑ ♥ 🔍 ⊟ 12 cm

Bodianus bilunulatus 358 ♀ ◐ ♥ 🔳 ⊟ 55 cm
Bodianus mesothorax 358 (juv.) ♀ ◐ ♥ 🔳 ⊟ 30 cm

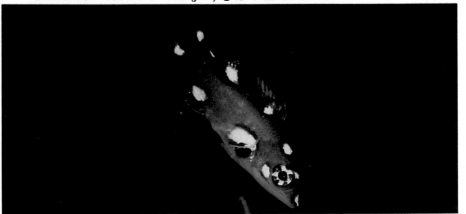

Bodianus diana 358 (juv.) ♀ ◐ ♥ 🔳 ⊟ 25 cm

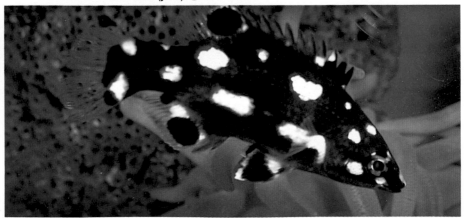

Bodianus bilunulatus 358 (juv.) ♀ ○ ♥ 🔒 ⊡ 55 cm
Bodianus anthioides 358 ∿ ○ ♥ 🔒 ⊡ 21 cm

Bodianus frenchii 358 ♀ ○ ♥ 🔒 ⊡ 50 cm

Bodianus bilunulatus 358 ♀ ○ ♥ 🖼 🖼 55 cm
Bodianus diana 358 ♀ ○ ♥ 🖼 🖼 25 cm

Bodianus macrourus 358 ♀ ○ ♥ 🖼 🖼 40 cm

Bodianus rufus 358 ⌇ ◑ ♥ ▣ ▤ 50 cm
Bodianus pulchellus 358 ⌇ ◑ ♥ ▣ ▤ 23 cm

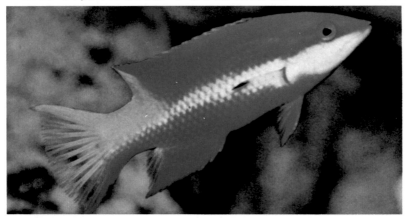

Bodianus eclancheri 358 ⚲ ◑ ♥ ▣ ▤ 25 cm

Choerodon jordani 358 ↘ ◐ ♥ ▣ ⊟ 14 cm
Choerodon schoenleinii 358 ↘ ◐ ♥ ▣ ⊟ 100 cm

Cheilinus unifasciatus 358 ↘ ◐ ♥ ▣ ⊟ 30 cm

Cheilinus bimaculatus 358 〰 ○ ♥ 🖼 ▨ 15 cm
Cheilinus digrammus 358 ↘ ○ ♥ 🖼 ▨ 35 cm

Cheilinus fasciatus 358 ↘ ○ ♥ 🖼 ▨ 35 cm

Cheilinus undulatus 358 ᴠ ⬦ ◐ ✕ 📷 ▭ 230 cm
Cheilinus chlorourus 358 ⬦ ◐ ♥ 📷 ▭ 36 cm

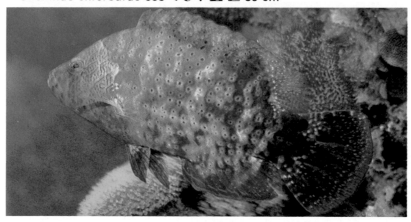

Cirrhilabrus punctatus 358 ᴠ ◐ ♥ 📷 ▭ 15 cm

Cirrhilabrus sp. 358 ∿ ◐ ♥ ▣ ▣ 10 cm
Cirrhilabrus cyanopleura 358 ∿ ◐ ♥ ▣ ▣ 15 cm

Cirrhilabrus exquisitus 358 ∿ ◐ ♥ ▣ ▣ 11 cm

Cirrhilabrus rubriventralis 358 〜 〇 ♥ 🎞 ▣ 7.5 cm
Cirrhilabrus punctatus 358 〜 〇 ♥ 🎞 ▣ 15 cm

Cirrhilabrus laboutei 358 〜 〇 ♥ 🎞 ▣ 10 cm

Cirrhilabrus melanomarginatus 358 ↝ ◑ ♥ 🔳 ⊡ 13 cm
Cirrhilabrus temmincki 358 ↝ ◑ ♥ 🔳 ⊡ 10 cm

Cirrhilabrus sp. "D" 358 ↝ ◑ ♥ 🔳 ⊡ 10 cm

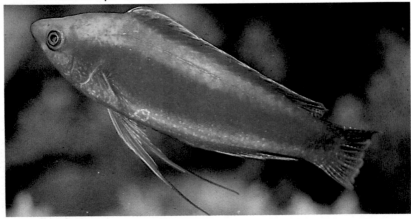

Coris auricularis 358 ♀ ◐ ♥ 🖼 ⊟ 32 cm
Coris aygula 358 (juv.) ♀ ◑ ♥ 🖼 ⊟ 120 cm

Coris schroederi 358 ♀ ◐ ♥ 🖼 ⊟ 20 cm

Coris bulbifrons 358 ♀ ◐ ♥ ▭ ▭ 30 cm
Coris picta 358 ♀ ◐ ♥ ▭ ▭ 25 cm

Coris dorsomacula 358 ♀ ◐ ♥ ▭ ▭ 15 cm

Coris auricularis 358 ♀ ☾ ♥ ▭ ▭ 32 cm
Coris flavovittata 358 ♀ ☾ ♥ ▭ ▭ 45 cm

Coris caudimacula 358 ♀ ☾ ♥ ▭ ▭ 20 cm

Coris gaimard africana 358 (juv.) ♀ ○ ♥ ▱ ▱ 35 cm
Coris formosa 358 (juv.) ♀ ○ ♥ ▱ ▱ 60 cm

Epibulus insidiator 358 ♀ ○ ♥ ▱ ▱ 35 cm

Halichoeres biocellatus 358 ⚲ ◑ ♥ 🔲 ⊟ 10 cm
Halichoeres trispilus 358 ⚲ ◑ ♥ 🔲 ⊟ 10 cm

Halichoeres marginatus 358 ⚲ ◑ ♥ 🔲 ⊟ 18 cm

Halichoeres hartzfeldii 358 ♀ ○ ♥ ▦ ⊟ 18 cm
Halichoeres iridis 358 (juv.) ♀ ○ ♥ ▦ ⊟ 11 cm

Halichoeres brownfieldi 358 ♀ ○ ♥ ▦ ⊟ 15 cm

Halichoeres pelicieri 358 ♀ ○ ♥ ▨ ▤ 15 cm
Halichoeres melasmopomas 358 ♀ ○ ♥ ▨ ▤ 12 cm

Halichoeres cosmetus 358 ♀ ○ ♥ ▨ ▤ 11 cm

Halichoeres timorensis 358 (juv.) ♀ ○ ♥ 🖼 ▤ 20 cm
Halichoeres melanurus 358 ♀ ○ ♥ 🖼 ▤ 12 cm

Halichoeres iridis 358 (juv.) ♀ ○ ♥ 🖼 ▤ 11 cm

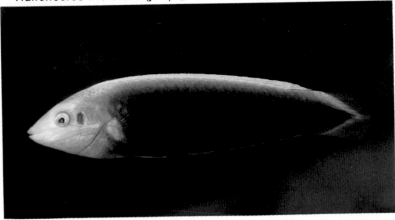

Halichoeres argus 358 (juv.) ♀ ○ ♥ ▭ ▣ 12 cm
Halichoeres biocellatus 358 (juv.) ♀ ○ ♥ ▭ ▣ 10 cm

Halichoeres hoeveni 358 (juv.) ♀ ○ ♥ ▭ ▣ 12 cm

Halichoeres biocellatus 358 ♀ ○ ♥ 🖼 ⊟ 10 cm
Halichoeres melanurus 358 ♀ ○ ♥ 🖼 ⊟ 12 cm

Halichoeres chloropterus 358 ♀ ○ ♥ 🖼 ⊟ 15 cm

Halichoeres poecilopterus 358 ♀ ○ ♥ 🖼 ⊟ 34 cm
Halichoeres margaritaceus 358 ♀ ○ ♥ 🖼 ⊟ 12 cm

Halichoeres trimaculatus 358 ♀ ○ ♥ 🖼 ⊟ 18 cm

Halichoeres hortulanus 358 ♀ ○ ♥ 🖼 ⊟ 22 cm
Halichoeres podostigma 358 ♀ ○ ♥ 🖼 ⊟ 16 cm

Halichoeres ornatissimus 358 ♀ ○ ♥ 🖼 ⊟ 18 cm

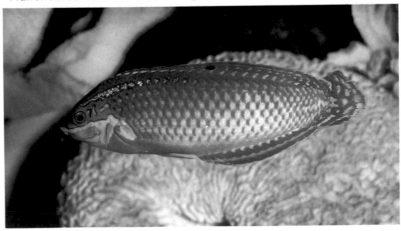

Halichoeres cyanocephalus 358 (juv.) ♀ ○ ♥ 🖼 🔲 30 cm
Halichoeres maculipinna 358 ♀ ○ ♥ 🖼 🔲 18 cm

Halichoeres poeyi 358 (juv.) ♀ ○ ♥ 🖼 🔲 20 cm

Halichoeres garnoti 358 (juv.) ♀ ◐ ♥ 🖼 ⊡ 19 cm
Halichoeres garnoti 358 ♀ ◐ ♥ 🖼 ⊡ 19 cm

Halichoeres bivittatus 358 ♀ ◐ ♥ 🖼 ⊡ 22 cm

Halichoeres chierchiae 358 ♀ ○ ♥ 🔲 🔁 20 cm
Halichoeres nicholsi 358 ♀ ○ ♥ 🔲 🔁 38 cm

Halichoeres sp. 358 ♀ ○ ♥ 🔲 🔁 12 cm

Gomphosus varius 358 ♀ ◐ ♥ ▭ ▭ 30 cm
Gomphosus varius 358 (juv.) ♀ ◐ ♥ ▭ ▭ 30 cm

Epibulus insidiator 358 ♀ ◐ ♥ ▭ ▭ 35 cm

Gomphosus caeruleus 358 ♀ ○ ♥ 🖼 🖾 28 cm
Hologymnosus annulatus 358 (juv.) ♀ ○ ♥ 🖼 🖾 40 cm

Epibulus insidiator 358 (juv.) ♀ ○ ♥ 🖼 🖾 35 cm

Hemigymnus fasciatus 358 (juv.) ♀ ○ ♥ ▣ ▱ 40 cm
Hemigymnus melapterus 358 ♀ ○ ♥ ▣ ▱ 90 cm

Hologymnosus doliatus 358 (juv.) ↝ ↘ ○ ♥ ▣ ▱ 45 cm

Labrichthys unilineatus 358 ♀ ○ ♥ 🔳 ⊡ 18 cm
Larabicus quadrilineatus 358 ♀ ○ ♥ 🔳 ⊡ 12 cm

Macropharyngodon cyanoguttatus 358 ♀ ○ ♥ 🔳 ⊡ 12 cm

Choerodon fasciata (adult) 358 ↘ ◑ ♥ 🔚 🖃 30 cm
Macropharyngodon choati (adult) 358 ↘ ◑ ♥ 🔚 🖃 10 cm

Macropharyngodon geoffroyi 358 ↘ ◑ ♥ 🔚 🖃 15 cm

Labropsis alleni 358 ∿ ○ ♥ ▭ ▱ 8 cm
Labropsis xanthonota 358 ∿ ○ ♥ ▣ ▱ 10 cm

Labropsis manabei 358 ∿ ○ ♥ ▣ ▱ 12 cm

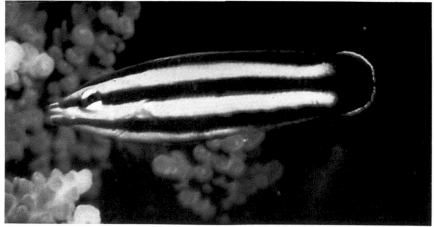

Stethojulis balteata 358 ∿ ○ ♥ 🏞 🖼 14 cm
Stethojulis bandanensis 358 ∿ ○ ♥ 🏞 🖼 15 cm

Stethojulis trilineata 358 ∿ ○ ♥ 🏞 🖼 15 cm

Xyrichthys dea (juv.) 358 ᔓ 〇 ♥ ▱ ▱ 30 cm
Xyrichthys pentadactylus 358 ᔓ 〇 ♥ ▱ ▱ 25 cm

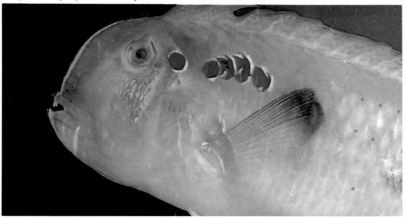

Xyrichthys aneitensis 358 ᔓ 〇 ♥ ▱ ▱ 25 cm

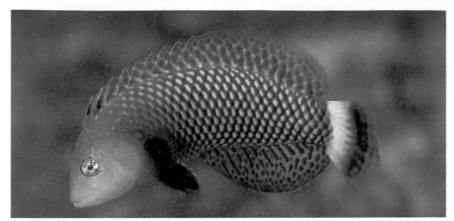

Novaculichthys taeniourus 358 ↘ ◑ ✶ ▱ ▭ 27 cm
Novaculichthys macrolepidotus 358 ↝ ◑ ♥ ▱ ▭ 13 cm

Pseudocheilinus hexataenia 358 ↝ ◑ ♥ ▱ ▭ 10 cm

Novaculichthys taeniourus 358 (juv.) ↘ ○ ✕ 🖼 🔳 27 cm
Pseudocheilinus evanidus 358 ↝ ○ ♥ 🖼 🔳 8 cm

Pseudocheilinus octotaenia 358 ↝ ○ ♥ 🖼 🔳 12 cm

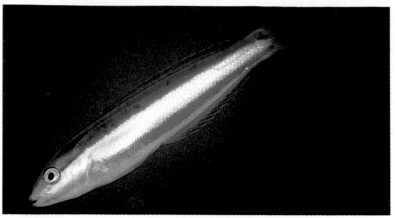

Ophthalmolepis lineolatus 358 ♀ ○ ♥ 🖼 🖼 40 cm
Pseudojuloides erythrops 358 〜 ○ ♥ 🖼 🖼 9 cm

Pseudojuloides cerasinus 358 〜 ○ ♥ 🖼 🖼 12 cm

Pseudojuloides sp. 358 ⤳ ◑ ♥ 🖼 ⊟ 10 cm
Pseudojulis melanotus 358 ⤳ ◑ ♥ 🖼 ⊟ 20 cm

Pseudojulis notospilus 358 ⤳ ◑ ♥ 🖼 ⊟ 25 cm

Pseudojuloides elongatus 358 〜 ◐ ♥ 🖼 ⊟ 14 cm
Pseudojulis notospilus 358 〜 ◐ ♥ 🖼 ⊟ 25 cm

Pseudocheilinus tetrataenia 358 〜 ◐ ♥ 🖼 ⊟ 12 cm

Leptojulis chrysotaenia 358 ∿ ○ ♥ 🔲 🖼 6 cm
Pictilabrus laticlavius 358 ∿ ○ ♥ 🔲 🖼 20 cm

Pictilabrus laticlavius 358 ∿ ○ ♥ 🔲 🖼 20 cm

Semicossyphus reticulatus 358 (juv.) ⌇ ◐ ✕ 🖼 ⊟ 100 cm
Pseudolabrus milesi 358 ⌇ ◐ ♥ 🖼 ⊟ 38 cm

Dotalabrus aurantiacus 358 ⌇ ◐ ♥ 🖼 ⊟ 20 cm

Semicossyphus pulcher 358 91 cm

Semicossyphus reticulatus 358 100 cm

Eupetrichthys angustipes 358 15 cm

Suezichthys gracilis 358 〜 ◑ ♥ 📷 ⊟ 15 cm
Pseudolabrus luculentus 358 ♀ ◑ ♥ 📷 ⊟ 24 cm

Pseudolabrus celidotus 358 ♀ ◑ ♥ 📷 ⊟ 24 cm

Pseudolabrus biserialis 358 ♀ ○ ♥ 🖼 🖼 20 cm
Stethojulis bandanensis 358 ∿ ○ ♥ 🖼 🖼 15 cm

Thalassoma amblycephalum 358 ♀ ○ ♥ 🖼 🖼 16 cm

Stethojulis strigiventer 358 ♀ ○ ♥ 🖼 ⊟ 15 cm
Thalassoma amblycephalum 358 ♀ ○ ♥ 🖼 ⊟ 16 cm

Thalassoma hardwicke 358 ♀ ○ ♥ 🖼 ⊟ 18 cm

Thalassoma lucasanum 358 ♀ ○ ♥ ▭ ▭ 15 cm
Thalassoma bifasciatum 358 ♀ ○ ♥ ▭ ▭ 15 cm

Thalassoma bifasciatum 358 ♀ ○ ♥ ▭ ▭ 15 cm

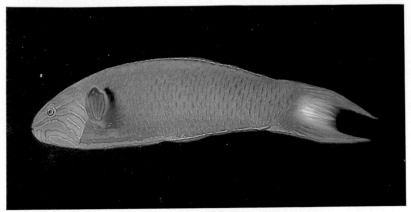

Thalassoma lunare 358 ♀ ◐ ♥ 🖼 🖾 25 cm
Thalassoma quinquevittatum 358 ♀ ◐ ♥ 🖼 🖾 17 cm

Thalassoma genivittatum 358 ♀ ◐ ♥ 🖼 🖾 20 cm

Thalassoma hebraicum 358 ♀ ○ ♥ ▭ ▣ 18 cm
Thalassoma lutescens 358 ♀ ○ ♥ ▭ ▣ 15 cm

Thalassoma klunzingeri 358 ♀ ○ ♥ ▭ ▣ 20 cm

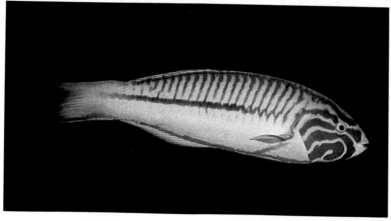

Thalassoma lutescens 358 ♀ ◐ ♥ 🖼 ▭ 15 cm
Thalassoma septemfasciatum 358 ♀ ◐ ♥ 🖼 ▭ 23 cm

Thalassoma trilobatum 358 ♀ ◐ ♥ 🖼 ▭ 30 cm

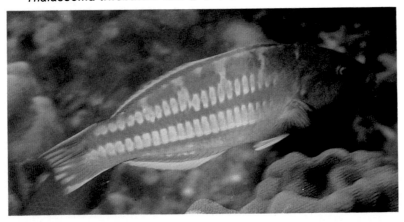

Labroides dimidiatus 358 〜 ○ ♥ ▱ ▱ 12 cm
Labroides rubrolabiatus 358 〜 ○ ♥ ▱ ▱ 6 cm

Labroides bicolor 358 〜 ○ ♥ ▱ ▱ 14 cm

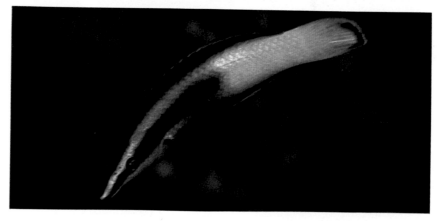

Labroides phthirophagus 358 ꙅ ○ ♥ ▭ ▱ 10 cm
Labroides pectoralis 358 ꙅ ○ ♥ ▭ ▱ 12 cm

Xenojulis margaritaceous 358 ♀ ○ ♥ ▭ ▱ 12 cm

Lachnolaimus maximus 358 ⤏ ○ ✂ 🖼 🔲 91 cm
Clepticus parrae 358 ⤳ ○ ♥ 🖼 🔲 30 cm

Decodon melasma 358 ⤳ ◐ ♥ 🖼 🔲 20 cm

Symphodus ocellatus 358 ♀ ○ ♥ 🐟 ☐ 12 cm
Symphodus tinca 358 ♀ ○ ♥ 🐟 ☐ 35 cm

Symphodus mediterraneus 358 ♀ ○ ♥ 🐟 ☐ 15 cm

Odax acroptilus 359 〜 ○ ♥ 🖼 ☒ 25 cm
Odax cyanomelas 359 〜 ○ ♥ 🖼 ☒ 45 cm

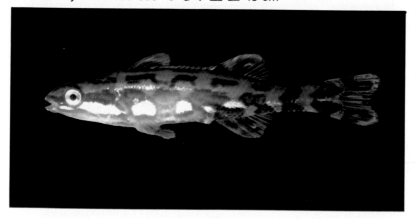

Siphonognathus radiatus 359 〜 ○ ♥ 🖼 ☒ 12 cm

Cetoscarus bicolor 360 (juv.) ᵕ ⚲ ○ ♥ 📷 ▭ 80 cm
Cetoscarus bicolor 360 ᵕ ⚲ ○ ♥ 📷 ▭ 80 cm

Scarus gibbus 360 ᵕ ⚲ ◐ ♥ 📷 ▭ 70 cm

Scarus prasiognathos 360 ∿ ⚓ ◐ ♥ 🖼 ⊟ 70 cm
Scarus niger 360 ∿ ⚓ ◐ ♥ 🖼 ⊟ 40 cm

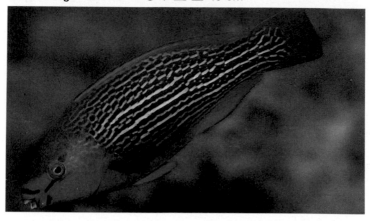

Scarus cyanescens 360 ∿ ⚓ ◐ ♥ 🖼 ⊟ 50 cm

Scarus niger 360 ∽ ⋌ ○ ♥ 🖼 ▱ 40 cm
Scarus psittacus 360 ∽ ⋌ ○ ♥ 🖼 ▱ 27 cm

Scarus rubroviolaceus 360 ∽ ⋌ ○ ♥ 🖼 ▱ 66 cm

Scarus sordidus 360 〜 ⚐ ◐ ♥ 🖼 ⊟ 50 cm
Scarus gibbus 360 〜 ⚐ ◐ ♥ 📷 ⊟ 70 cm

Scarus ghobban 360 〜 ⚐ ◐ ♥ 🖼 ⊟ 80 cm

Scarus brevifilis 360 〜 ⚲ ◐ ♥ 🖼 ▣ 45 cm
Scarus frontalis 360 〜 ⚲ ◐ ♥ 🖼 ▣ 50 cm

Scarus rubroviolaceus 360 〜 ⚲ ◐ ♥ 🖼 ▣ 60 cm

Scarus bleekeri 360 ∿ ⚓ ○ ♥ 📺 ⊟ 25 cm
Scarus perspicillatus 360 ∿ ⚓ ○ ♥ 📺 ⊟ 62 cm

Scarus sp. 360 ∿ ⚓ ○ ♥ 📺 ⊟ 50 cm

Scarus gibbus 360 ∿ ⚲ ○ ♥ 🖼 ▱ 70 cm
Scarus flavipectoralis 360 ∿ ⚲ ○ ♥ 🖼 ▱ 15 cm

Scarus sp. 360 ∿ ⚲ ○ ♥ 🖼 ▱ 25 cm

Scarus ghobban 360 ∿ ↝ ○ ♥ 🔲 ⊡ 80 cm
Scarus prasiognathos 360 ∿ ↝ ○ ♥ 🔲 ⊡ 70 cm

Scarus frenatus 360 ∿ ↝ ○ ♥ 🔲 ⊡ 40 cm

Scarus croicensis 360 ∿ ⚘ ○ ♥ 🖼 🖿 25 cm
Scarus vetula 360 ∿ ⚘ ○ ♥ 🖼 🖿 71 cm

Scarus coelestinus 360 ∿ ⚘ ○ ♥ 🖼 🖿 106 cm

Scarus croicensis 360 ∿ ⚓ ◑ ♥ 🖼 ☒ 25 cm
Scarus vetula 360 ∿ ⚓ ◑ ♥ 🖼 ☒ 71 cm

Scarus taeniopterus 360 ∿ ⚓ ◑ ♥ 🖼 ☒ 30 cm

Sparisoma aurofrenatum 360 ⌇ ◐ ♥ 🎥 ▱ 28 cm
Sparisoma chrysopterum 360 ⌇ ◐ ♥ 🎥 ▱ 50 cm

Sparisoma viride 360 ⌇ ◐ ♥ 🎥 ▱ 51 cm

Calotomus zonarchia 360 ⌇ ⚹ ☾ ♥ 🖼 ⊟ 35 cm
Calotomus spinidens 360 ⌇ ⚹ ☾ ♥ 🖼 ⊟ 19 cm

Nicholsina denticulata 360 ⌇ ⚹ ☾ ♥ 🖼 ⊟ 7 cm

Macrozoarces americanus 362 〰 ⤵ ◑ ✕ 📷 🖵 110 cm
Zoarchias veneficus 362 〰 ⤵ ◑ ✕ 📷 🖵 25 cm

Ernogrammus hexagrammus 363 〰 ⤵ ◑ ✕ 📷 🖵 15 cm

Chirolophis japonicus 363 ♀ ⌇ ⌐ ◐ ✕ ▨ ⬜ 55 cm
Chirolophis nugator 363 ♀ ⌇ ⌐ ◐ ✕ ▨ ⬜ 42 cm

Anarrhichthys ocellatus 366 ♀ ⌇ ⌐ ◐ ✕ ▣ ⬜ 200 cm

Chirolophis decoratus 363 ♀ ∿ ⤙ ◐ ✕ 🎥 🖵 42 cm
Cebidichthys violaceus 363 ♀ ∿ ⤙ ◐ ✕ 🎥 🖵 50 cm

Anarhichas lupus 366 ♀ ∿ ⤙ ◐ ✕ 🎥 🖵 200 cm

Lonchopisthus micrognathus 375 ⌇ ◑ ♥ 🖻 ⊡ 10 cm
Opistognathus lonchurus 375 ⌇ ◑ ♥ 🖻 ⊡ 12 cm

Opistognathus sp. 375 ∿ ◑ ♥ 🔳 ▭ 10 cm
Opistognathus sp. 375 ∿ ◑ ♥ 🔳 ▭ 11 cm

Opistognathus gilberti 375 ∿ ◑ ♥ 🔳 ▭ 5 cm

Opistognathus aurifrons 375 〜 ↖ ◑ ♥ 🔳 ⬜ 10 cm
Opistognathus scops 375 〜 ↖ ◑ ✕ 🔳 ⬜ 15 cm

Opistognathus sp. 375 〜 ◑ ♥ 🔳 ⬜ 11 cm

Blennodesmus scapularis 376 ∿ ❶ ♥ 🎞 ▭ 5 cm
Pholidichthys leucotaenia 380 ♀ ∿ ❶ ♥ 🖼 ▭ 6 cm

Notograptus livingstonei 379 ∿ ❶ ♥ 🖼 ▭ 7 cm

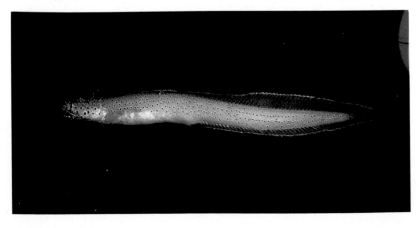

Parapercis sexfasciata 388 ∿ O ♥ 🔊 🖵 12 cm
Parapercis multifasciata 388 ∿ O ♥ 🔊 🖵 15 cm

Parapercis schuinslandi 388 ∿ ↘ O ✕ 🔊 🖵 20 cm

Parapercis bivittata 388 ⌇ ⌐ ◑ ✂ 🎥 ⬜ 20 cm
Parapercis hexophthalma 388 ⌇ ⌐ ◑ ✂ 🎥 ⬜ 25 cm

Parapercis trispilota 388 ⌇ ⌐ ◑ ✂ 🎥 ⬜ 15 cm

Parapercis cephalopunctata 388 ↻ ↘ ◗ ✕ ▣ ▢ 17 cm
Parapercis haackei 388 ↻ ↘ ◗ ✕ ▣ ▢ 15 cm

Parapercis tetracantha 388 ↻ ↘ ◗ ✕ ▣ ▢ 20 cm

Lepidoblennius marmoratus 390 ♀ ⤳ ◑ ♥ 🎞 ▭ 9 cm
Gilloblennius tripennis 390 ♀ ⤳ ◑ ♥ 🎞 ▭ 5 cm

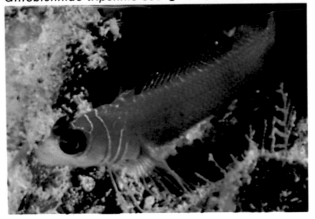

Vauclusella sp. 390 ♀ ⤳ ◑ ♥ 🎞 ▭ 3 cm

Tripterygion tripteronotus 390 ♀ ∿ ◑ ♥ 🖼 ▭ 8 cm
Tripterygion tripteronotus 390 ♀ ∿ ◑ ♥ 🖼 ▭ 8 cm

Tripterygion sp. 390 ♀ ∿ ◑ ♥ 🖼 ▭ 8 cm

Helcogramma striata 390 〰 ◑ ♥ 📷 🖵 3 cm
Helcogramma decurrens 390 〰 ◑ ♥ 📷 🖵 5 cm

Gunnelichthys curiosus 407 〰 ◑ ♥ 📷 🖵 11 cm

Helcogramma decurrens 390 ⤳ ◑ ♥ 📷 ▭ 5 cm
Helcogramma sp. 390 ⤳ ◑ ♥ 📷 ▭ 3 cm

Gunnelichthys monostigma 407 ⤳ ◑ ♥ ▭ ▭ 7.5 cm

Dactyloscopus pectoralis 391 〜 ◑ ♥ 🖼 ⊡ 7.5 cm
Exerpes asper 392 ♀ 〜 ◑ ✕ 🖼 ⊡ 5 cm

Neoclinus bryope 392 ♀ 〜 ◑ ♥ 🖼 ⊡ 6 cm

Labrisomus filamentosus 392 ♀ ∿ ◑ ♥ 🖼 ⬜ 8.0 cm
Labrisomus nigricinctus 392 ♀ ∿ ◑ ♥ 🖼 ⬜ 7.5 cm

Labrisomus kalisherae 392 ♀ ∿ ◑ ♥ 🖼 ⬜ 7.5 cm

Paraclinus mexicanus 392 ↴ ◑ ♥ 🎞 ▭ 4 cm
Malacoctenus margaritae 392 ↴ ◑ ♥ 🎞 ▭ 12 cm

Malacoctenus gilli 392 ♀ ∿ ◐ ♥ 🎞 ⊡ 4.7 cm
Malacoctenus boehlkei 392 ♀ ∿ ◐ ♥ 🎞 ⊡ 6 cm

Dialommus fuscus 392 ♀ ∿ ◐ ♥ 🎞 ⊡ 7.0 cm

Gibbonsia metzi 393 〜 ◐ ♥ 🎞 ☐ 24 cm
Heterostichus rostratus 393 〜 ＼ ◐ ☆ 🔲 ☐ 61 cm

Ophiclinus gracilis 392 〜 ◐ ☆ 🔲 ☐ 6 cm

Heteroclinus sp. cf. *roseus* 393 ∿ ◑ ♥ 🖼 ▭ 8.5 cm
Heteroclinus roseus 393 ∿ ◑ ♥ 🖼 ▭ 7 cm

Heteroclinus adelaidae 393 ∿ ◑ ♥ 🖼 ▭ 7 cm

Acanthemblemaria spinosa 394 ⌇ ◑ ♥ 🖼 ⬜ 5 cm
Acanthemblemaria crockeri 394 ⌇ ◑ ♥ 🖼 ⬜ 5 cm

Lucayablennius zingaro 394 ⌇ ◑ ♥ 🖼 ⬜ 4 cm

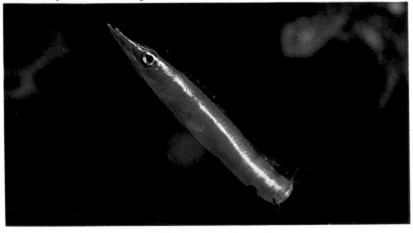

Chaenopsis alepidota 394 ↘ ◐ ♥ ▓ ☐ 15 cm
Ekemblemaria sp. (myersi?) 394 ↘ ◐ ♥ ▓ ☐ 3.5 cm

Chaenopsis ocellata 394 ↘ ◐ ♥ ▓ ☐ 12.5 cm

Emblemaria piratula 394 ∿ ◑ ♥ 🖼 ▢ 5 cm
Emblemaria hypacanthus 394 ∿ ◑ ♥ 🖼 ▢ 5 cm

Emblemaria bottomei 394 ∿ ◑ ♥ 🖼 ▢ 5 cm

Plagiotremus laudandus flavus 395 ⌇ ◑ ♥ 🌿 ▭ 6 cm
Plagiotremus goslinei 395 ⌇ ◑ ♥ 🌿 ▭ 7.5 cm

Plagiotremus rhinorhynchos 395 ⌇ ◑ ♥ 🌿 ▭ 12 cm

Meiacanthus grammistes 395 ⤻ ◑ ♥ 🎞 ⬜ 8 cm
Petroscirtes fallax 395 ⤻ ◑ ♥ 🎞 ⬜ 9 cm

Petroscirtes breviceps 395 ⤻ ◑ ♥ 🎞 ⬜ 10 cm

Meiacanthus oualanensis 395 ᔑ ◑ ♥ 🎞 ⊡ 7.5 cm
Meiacanthus atrodorsalis 395 ᔑ ◑ ♥ 🎞 ⊡ 7.5 cm

Meiacanthus smithii 395 ᔑ ◑ ♥ 🎞 ⊡ 8.5 cm

Aidablennius sphynx 395 ♀ ∿ ⚲ ◐ ♥ 🎞 ⬚ 7 cm
Parablennius tentacularis 395 ♀ ∿ ⚲ ◐ ♥ 🎞 ⬚ 16 cm

Lipophrys adriaticus 395 ♀ ∿ ⚲ ◐ ♥ 🎞 ⬚ 10 cm

Lipophrys nigriceps 395 ♀ ∿ ◑ ♥ 🎦 ⬜ 5.5 cm
Lipophrys canevae 395 ♀ ∿ ◑ ♥ 🎦 ⬜ 8 cm

Parablennius zvonimiri 395 ♀ ∿ ◑ ♥ 🎦 ⬜ 7 cm

Laiphognathus multimaculatus 395 ♀ ∿ ⚘ ◑ ♥ 🖼 ⌷ 4 cm
Salarias sp. 395 ♀ ∿ ⚘ ◑ ♥ 🖼 ⌷ 10 cm

Salarias fasciatus 395 ♀ ∿ ⚘ ◑ ♥ 🖼 ⌷ 15 cm

Salarias guttatus 395 ♀ ∿ ⋏ ◑ ♥ 🎞 ▭ 7 cm
Salarias irroratus 395 ♀ ∿ ⋏ ◑ ♥ 🎞 ▭ 7 cm

Atrosalarias fuscus (juv.) 395 ♀ ∿ ⋏ ◑ ♥ 🎞 ▭ 15 cm

Ophioblennius atlanticus 395 ♀ ∿ ⚲ ◑ ♥ 🖼 ▭ 13 cm
Cirripectes stigmaticus 395 ♀ ∿ ⚲ ◑ ♥ 🖼 ▭ 10 cm

Cirripectes filamentosus 395 ♀ ∿ ⚲ ◑ ♥ 🖼 ▭ 8 cm

Entomacrodus decussatus 395 ♀ ∿ ⅄ ◐ ♥ 🖼 ▭ 17 cm
Entomacrodus nigricans 395 ♀ ∿ ⅄ ◐ ♥ 🖼 ▭ 8 cm

Parablennius tasmanianus 395 ♀ ∿ ⅄ ◐ ♥ 🖼 ▭ 13 cm

Salarias ceramensis 395 ♀ ∿ ⋇ ◑ ♥ 🎞 ⊡ 6 cm
Parablennius gattorugine 395 ♀ ∿ ⋇ ◑ ♥ 🎞 ⊡ 30 cm

Stanulus seychellensis 395 ♀ ∿ ⋇ ◑ ♥ 🎞 ⊡ 3.5 cm

Istiblennius gibbifrons 395 ♀ ⌇ ⅄ ◑ ♥ 🖼 ▭ 13 cm
Istiblennius chrysospilos 395 ♀ ⌇ ⅄ ◑ ♥ 🖼 ▭ 10 cm

Istiblennius lineatus 395 ♀ ⌇ ⅄ ◑ ♥ 🖼 ▭ 8 cm

Aspidontus dussumieri 395 ⌄ ◑ ♥ 🖼 ⊟ 13 cm
Cirripectes variolosus 395 ♀ ⌄ ⚓ ◑ ♥ 🖼 ⊟ 8 cm

Plagiotremus tapeinosoma 395 ⌄ ◑ ♥ 🖼 ⊟ 16 cm

Ecsenius midas 395 ♀ ∿ ❍ ♥ 🖼 ⊟ 13 cm
Ecsenius oculus 395 ♀ ∿ ⚵ ◐ ♥ 🖼 ⊡ 7 cm

Ecsenius lineatus 395 ♀ ∿ ⚵ ◐ ♥ 🖼 ⊡ 8 cm

Istiblennius dussumieri 395 ♀ ᴖ ⚥ ◑ ♥ 🖼 ⬛ 14 cm
Omobranchus elongatus 395 ♀ ᴖ ⚥ ◑ ♥ 🖼 ⬛ 8 cm

Exallias brevis 395 ᴖ ⚥ ◑ ♥ 🖼 ⬛ 15 cm

Istiblennius dussumieri 395 ♀ ᘜ ⚭ ◐ ♥ 🎞 🖵 14 cm
Petroscirtes mitratus 395 ♀ ᘜ ⚭ ◐ ♥ 🎞 🖵 9 cm

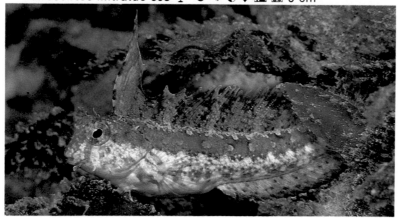

Congrogadus subducens 376 ♀ ⚭ ◐ ✕ 🎞 🖵 50 cm

Neosynchiropus sp. 399 ♀ ∿ ◑ ♥ 🖻 🖵 10 cm
Diplogrammus xenicus 399 ♀ ∿ ◑ ♥ 🖻 🖵 7 cm

Neosynchiropus ocellatus 399 ♀ ∿ ◑ ♥ 🖻 🖵 9 cm

Synchiropus marmoratus 399 〰 ◐ ♥ 📷 🖵 13 cm
Synchiropus sp. 399 〰 ◐ ♥ 📷 🖵 8 cm

Diplogrammus pauciradiatus 399 〰 ◐ ♥ 📷 🖵 5 cm

Trichonotus setigerus 384 ⤳ ◑ ♥ 📷 ⊟ 25 cm
Dactyloptena orientalis 260 ⤳ ◑ ♥ 📷 ⊟ 35 cm

Synchiropus sp. 399 ⤳ ◑ ♥ 📷 ⊟ 8 cm

Pterosynchiropus splendidus 399 6 cm

Synchiropus picturatus 399 7 cm

Synchiropus sp. 399 8 cm

Oxymetopon cyanoctenosum 402 ∿ ◑ ♥ 📷 ▭ 20 cm
Bostrychus sinensis 402 ♀ ∿ ↘ ◑ ✖ 🔲 ▭ 10 cm

Butis amboinensis 402 ∿ ↘ ○ ✖ 🔲 ▭ 8 cm

Amblyeleotris aurora 403 ♀ ⌁ ◑ ♥ 🖼 🖼 9 cm
Amblyeleotris fontanesii 403 ♀ ⌁ ◑ ♥ 🖼 🖼 15 cm

Amblyeleotris steinitzi 403 ⌁ ◑ ♥ 🖼 🖼 🖼 8 cm

Amblyeleotris fasciatus 403 ♀ ∿ ◑ ♥ 🖼 🖵 6 cm
Amblyeleotris periophthalmus 403 ♀ ∿ ◑ ♥ 🖼 🖵 5 cm

Amblyeleotris sungami 403 ♀ ∿ ◑ ♥ 🖼 🖵 10 cm

Amblyeleotris randalli 403 ↜ ◑ ♥ ▱ ▱ 3 cm
Amblyeleotris wheeleri 403 ↜ ◑ ♥ ▱ ▱ 9 cm

Amblyeleotris sp. 403 ↜ ◑ ♥ ▱ ▱ 4 cm

Amblyeleotris sp. aff. *periophthalmus* 403 ∿ ◑ ♥ 🖼 ☐ 5 cm
Amblyeleotris wheeleri 403 ∿ ◑ ♥ 🖼 ☐ 9 cm

Amblyeleotris guttata 403 ∿ ◑ ♥ 🖼 ☐ 7 cm

Amblyeleotris steinitzi 403 ⤳ ◐ ♥ 📷 ▭ 8 cm
Amblyeleotris steinitzi 403 ⤳ ◐ ♥ 📷 ▭ 8 cm

Ctenogobiops aurocingulus 403 ⤳ ◐ ♥ 📷 ▭ 4 cm

Cryptocentrus cinctus 403 ∿ ◑ ♥ 🖾 ⊡ 8 cm
Cryptocentrus aurora 403 ∿ ◑ ♥ 🖾 ⊡ 9 cm

Ctenogobiops aurocingulus 403 ∿ ◑ ♥ 🖾 ⊡ 4 cm

Ctenogobiops tangaroai 403 ∿ ◑ ♥ ▱ ▱ 4 cm
Cryptocentrus ambanora 403 ∿ ◑ ♥ ▱ ▱ 5 cm

Amblyeleotris sp. cf *japonica* 403 ∿ ◑ ♥ ▱ ▱ 6 cm

Cryptocentrus cinctus 403 ∿ ◑ ♥ ▱ ▱ 8 cm
Cryptocentrus fasciatus 403 ∿ ◑ ♥ ▱ ▱ 8 cm

Cryptocentrus strigilliceps 403 ∿ ◑ ♥ ▱ ▱ 5 cm

Fusigobius sp. *(neophytus?)* 403 〜 ◑ ♥ 🖼 ⬚ 3(?) cm
?Eviota sp. 403 〜 ◑ ♥ 🖼 ⬚ 3 cm

Yongeichthys criniger? 403 〜 ◑ ♥ 🖼 ⬚ 13 cm

Signigobius biocellatus 403 ∿ ◑ ♥ 🖼 ▭ 5 cm
Eviota sp. 403 ∿ ◑ ♥ 🖼 ▭ 3.5 cm

Amblygobius rainfordi 403 ∿ ◑ ♥ 🖼 ▭ 10 cm

Yongeichthys criniger 403 ~ ◐ ♥ ▣ ☐ 15 cm

Oxyurichthys papuensis 403 ~ ◐ ♥ ▣ ☐ 17 cm

Stiphodon sp. 403 ~ ◐ ♥ ▣ ☐ 4.5 cm

Tridentiger trigonocephalus 403 ⌇ ◑ ♥ 📷 🖵 4 cm
Asterropteryx sp. 403 ⌇ ◑ ♥ 📷 🖵 7 cm

Lotilia graciliosa 403 ⌇ ◑ ♥ 📷 🖵 4 cm

Stonogobiops xanthorhinica 403 ∿ ◑ ♥ 🖼 ▱ 10 cm
Mahidolia mystacina 403 ∿ ◑ ♥ 🖼 ▱ 8 cm

Gobulus myersi 403 ∿ ◑ ♥ 🖼 ▱ 3 cm

Amblygobius hectori 403 ∿ ◑ ♥ 📷 🖵 5 cm
Amblygobius phalaena 403 ∿ ◑ ♥ 📷 🖵 10 cm

Amblygobius sphynx 403 ∿ ◑ ♥ 📷 🖵 18 cm

Amblygobius decussatus 403 ∿ ◐ ♥ 📷 🖵 7.5 cm
Amblygobius nocturnus 403 ∿ ◐ ♥ 📷 🖵 10 cm

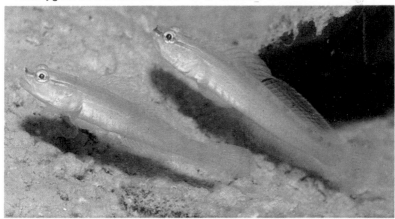

Amblygobius semicinctus 403 ∿ ◐ ♥ 📷 🖵 10 cm

Cryptocentrus leptocephalus 403 ∿ ◑ ♥ 📷 ⬚ 10 cm
Cryptocentrus fasciatus 403 ∿ ◑ ♥ 📷 ⬚ 8 cm

Cryptocentrus niveatus 403 ∿ ◑ ♥ 📷 ⬚ 11 cm

Cryptocentrus sp. 403 〜 ◑ ♥ ▭ ▭ 7 cm

Cryptocentrus leptocephalus 403 〜 ◑ ♥ ▭ ▭ 10 cm

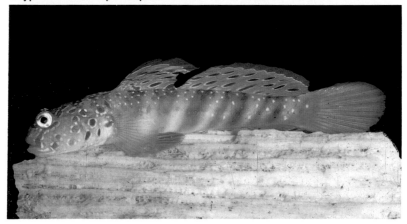

Cryptocentrus cinctus 403 〜 ◑ ♥ ▭ ▭ 8 cm

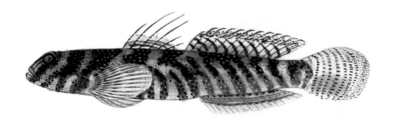

Cryptocentroides insignis 403 ∿ ◑ ♥ 🖼 🖼 9 cm
Cryptocentrus albidorsus 403 ∿ ◑ ♥ 🖼 🖼 6 cm

? Unidentified Goby 403 ∿ ◑ ♥ 🖼 🖼 11 cm

Gobius niger 403 ᔕ ◑ ♥ 🔳 ☐ 15 cm
Gobius vittatus 403 ᔕ ◑ ♥ 🔳 ☐ 4 cm

Caffrogobius caffer 403 ᔕ ◑ ♥ 🔳 ☐ 18 cm

Gobius cruentatus 403 ∿ ◐ ♥ ▣ ▢ 18 cm
Gobius auratus 403 ∿ ◐ ♥ ▣ ▢ 12 cm

Gobius elanthematicus 403 ∿ ◐ ♥ ▣ ▢ 12 cm

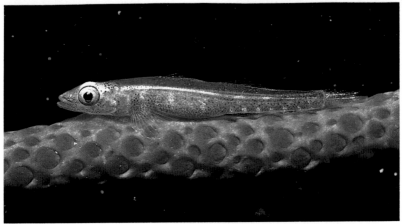

Bryaninops amphis 403 ∿ ◑ ♥ 🎞 ▭ 3 cm
Bryaninops natans 403 ∿ ◑ ♥ 🎞 ▭ 3 cm

Parioglossus formosus 403 ∿ ◑ ♥ 🎞 ▭ 2.8 cm

Bryaninops sp. 403 ⌇ ◑ ♥ 🖼 ▭ 3 cm
Bryaninops tigris 403 ⌇ ◑ ♥ 🖼 ▭ 3 cm

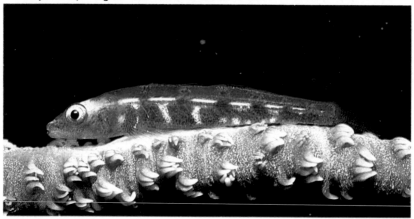

Parioglossus lineatus 403 ⌇ ◑ ♥ 🖼 ▭ 3 cm

Nemateleotris helfrichi 403 ∿ ◐ ♥ 🐟 ▤ 5 cm
Ptereleotris monoptera 403 ∿ ◐ ♥ 🐟 ▤ 11 cm

Ptereleotris zebra 403 ∿ ◐ ♥ 🐟 ▭ 10 cm

Callogobius snelliusi 403 ∿ ◑ ♥ 🖼 ⬛ 5 cm
Callogobius mucosus 403 ∿ ◑ ♥ 🖼 ⬛ 6 cm

Arenigobius brifrenatus 403 ∿ ◑ ♥ 🖼 ⬛ 8 cm

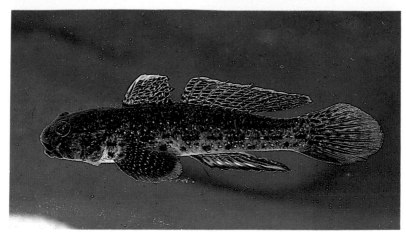

Istigobius ornatus 403 ∿ ◑ ♥ 🔳 ⬜ 8 cm
Istigobius nigroocellatus 403 ∿ ◑ ♥ 🔳 ⬜ 8 cm

Istigobius goldmanni 403 ∿ ◑ ♥ 🔳 ⬜ 8 cm

Exyrias belissimus 403 〜 ◑ ♥ 📷 ⬜ 11 cm
Exyrias puntang 403 〜 ◑ ♥ 📷 ⬜ 13 cm

Barbulifer pantherinus 403 〜 ◑ ♥ 📷 ⬜ 5 cm

Pterogobius zonoleucus 403 ॱᔓ ◑ ♥ 🎥 ▭ 9 cm
Pterogobius elapoides 403 ॱᔓ ◑ ♥ 🎥 ▭ 8.5 cm

Gobionellus stigmaticus 403 ॱᔓ ◑ ♥ 🎥 ▭ 8 cm

Valenciennea puellaris 403 ⌇ ◐ ♥ 📷 ⬜ 12 cm
Valenciennea sexguttata 403 ⌇ ◐ ♥ 📷 ⬜ 13 cm

Valenciennea muralis 403 ⌇ ◐ ♥ 📷 ⬜ 10 cm

574

Valenciennea helsdingenii 403 ∿ ◑ ♥ 📷 ⬚ 16 cm
Valenciennea puellaris 403 ∿ ◑ ♥ 📷 ⬚ 12 cm

Valenciennea strigata 403 ∿ ◑ ♥ 📷 ⬚ 18 cm

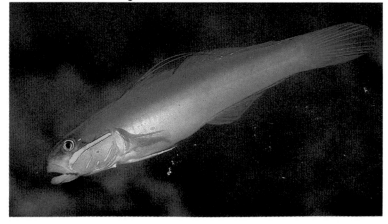

575

Gymneleotris seminudus 403 ⤳ ◑ ♥ 📷 ⬜ 5 cm
Chriolepis zebra 403 ⤳ ◑ ♥ 📷 ⬜ 6 cm

Priolepis cinctus 403 ♀ ⤳ ◑ ♥ 📷 ⬜ 6 cm

Pleurosicya sp. 403 〰 ◑ ♥ 🎞 ⊡ 2.5 cm
Ptereleotris evides 403 〰 ◑ ♥ 🎞 ⊟ 12 cm

Ptereleotris microlepis 403 〰 ◑ ♥ 🎞 ⊟ 11 cm

Ioglossus helenae 403 ⌇ ◐ ♥ 🔲 🔲 24 cm
Ioglossus calliurus 403 ⌇ ◐ ♥ 🔲 🔲 10 cm

Microgobius carri 403 ⌇ ◐ ♥ 🔲 🔲 7.5 cm

Fusigobius sp. 403 ♀ ∿ ◑ ♥ 📷 ▱ 6 cm
Gobiodon citrinus 403 ♀ ∿ ◐ ♥ 📷 ▱ 6 cm

Nemateleotris decora 403 ♀ ∿ ◐ ♥ 📷 ▱ 7.5 cm

Fusigobius neophytus 403 ♀ ~ ◐ ♥ 🔳 🔲 7.5 cm
Gnatholepis sp. 403 ♀ ~ ◐ ♥ 🔳 🔲 5 cm

Nemateleotris magnifica 403 ♀ ~ ◐ ♥ 🔳 🔲 9 cm

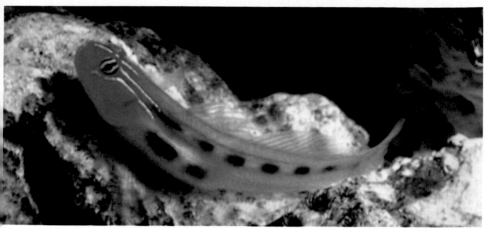

Gobiosoma puncticulatus 403 ♀ ∿ ◑ ♥ 📷 🖵 3 cm
Gobiosoma macrodon 403 ♀ ∿ ◑ ♥ 📷 🖵 5 cm

Gobiosoma aff. *limbaughi* 403 ♀ ∿ ◑ ♥ 📷 🖵 3 cm

Gobiosoma saucrum 403 ♀ ∿ ◐ ♥ 📷 ⬚ 3 cm
Gobiosoma multifasciatum 403 ♀ ∿ ◐ ♥ 📷 ⬚ 4.5 cm

Gobiosoma digueti 403 ♀ ∿ ◐ ♥ 📷 ⬚ 3 cm

Gobiosoma illecebrosum 403 ♀ ∿ ◑ ♥ 📷 ▭ 3.5 cm
Gobiosoma evelynae 403 ♀ ∿ ◑ ♥ 📷 ▭ 6 cm

Gobiosoma xanthiprora 403 ♀ ∿ ◑ ♥ 📷 ▭ 3.7 cm

Gobiosoma oceanops 403 ♀ ∿ ◑ ♥ 📷 🖵 5 cm
Gobiosoma randalli 403 ♀ ∿ ◑ ♥ 📷 🖵 2.5 cm

Gobiosoma evelynae 403 ♀ ∿ ◑ ♥ 📷 🖵 6 cm

Coryphopterus dicrus 403 ♀ ∿ ◑ ♥ 📷 🖵 5 cm
Coryphopterus personatus 403 ♀ ∿ ◑ ♥ 📷 🖵 3 cm

Coryphopterus sp. 403 ♀ ∿ ◑ ♥ 📷 🖵 3 cm

Coryphopterus lipernes 403 ♀ ∿ ◐ ♥ 📷 ⬜ 3 cm
Coryphopterus glaucofrenum 403 ♀ ∿ ◐ ♥ 📷 ⬜ 7.5 cm

Coryphopterus nicholsi 403 ♀ ∿ ◐ ♥ 📷 ⬜ 15 cm

Lythrypnus dalli 403 ♀ ⌇ ◐ ♥ 📷 ▱ 3 cm
Lythrypnus pulchellus 403 ♀ ⌇ ◐ ♥ 📷 ▱ 2.5 cm

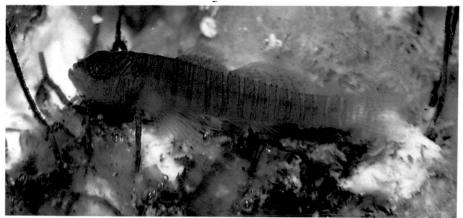

Lythrypnus nesiotes 403 ♀ ⌇ ◐ ♥ 📷 ▱ 1.5 cm

Eviota sp. 403 ♀ ∿ ◑ ♥ 📷 ⬛ 3 cm
Eviota sp. 403 ♀ ∿ ◑ ♥ 📷 ⬛ 3 cm

Eviota fasciola 403 ♀ ∿ ◑ ♥ 📷 ⬛ 3 cm

Trimma striata 403 ♀ ⌇ ◑ ♥ ▦ ▭ 3 cm
Trimma sp. 403 ♀ ⌇ ◑ ♥ ▦ ▭ 3.5 cm

Trimma sp. 403 ♀ ⌇ ◑ ♥ ▦ ▭ 3.5 cm

Gobiodon atrangulatus 403 ♀ ∿ ◑ ♥ 🎥 ☐ 3.5 cm
Gobiodon histrio 403 ♀ ∿ ◑ ♥ 🎥 ☐ 7 cm

Gobiodon multilineatus 403 ♀ ∿ ◑ ♥ 🎥 ☐ 3.5 cm

Gobiodon rivulatus 403 ♀ ∿ ◑ ♥ 📷 🖵 5 cm

Paragobiodon xanthosomus 403 ♀ ∿ ◑ ♥ 📷 🖵 3.5 cm

Gobiodon quinquestrigatus 403 ♀ ∿ ◑ ♥ 📷 🖵 4 cm

Periophthalmus sp. 403 ♀ ∿ ○ ♥ ▭ ▭ 12 cm
Mudskipper 403 ♀ ∿ ◑ ♥ ▭ ▭ 20 cm

Periophthalmus regius 403 ♀ ∿ ○ ♥ ▭ ▭ 10 cm

Brachyamblyopus coecus 404 ◑ ♥ 🖼 ⬜ 7 cm
Gunnelichthys monostigma 407 ∿ ◐ ♥ 🖼 ⬜ 9 cm

Gunnelichthys curiosus 407 ∿ ◐ ♥ 🖼 ⬜ 13 cm

Acanthurus coeruleus 409 ♀ ∿ ⋌ ◑ ♥ 🖼 ▭ 23 cm
Acanthurus blochii 409 ♀ ∿ ⋌ ○ ♥ 🖼 ▭ 42 cm

Acanthurus achilles 409 ♀ ∿ ⋌ ○ ♥ 🖼 ▭ 28 cm

Acanthurus bahianus 409 ⚲ ∿ ⋔ ◑ ♥ 🖼 ▭ 30 cm
Acanthurus coeruleus 409 ⚲ ∿ ⋔ ◑ ♥ 🖼 ▭ 23 cm

Acanthurus japonicus 409 ⚲ ∿ ⋔ ◑ ♥ ▭ 20 cm

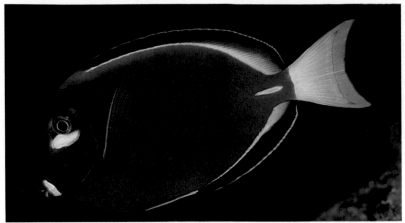

Acanthurus nigricans 409 ♀ ∿ ⚲ (♥ 🐟 ▭ 17 cm
Acanthurus guttatus 409 ♀ ∿ ⚲ (♥ 🐟 ▭ 23 cm

Acanthurus leucosternon 409 ♀ ∿ ⚲ (♥ 🐟 ▭ 23 cm

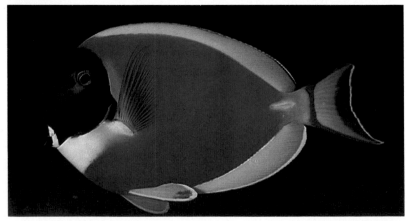

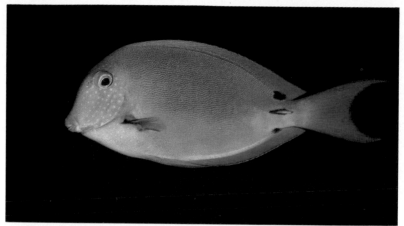

Acanthurus nigrofuscus 409 ♀ ∿ ⚓ ☉ ♥ ▨ ▤ 21 cm
Acanthurus pyroferus 409 ♀ ∿ ⚓ ☉ ♥ ▨ ▤ 19 cm

Acanthurus sohal 358 ♀ ∿ ⚓ ☉ ♥ ▨ ▤ 40 cm

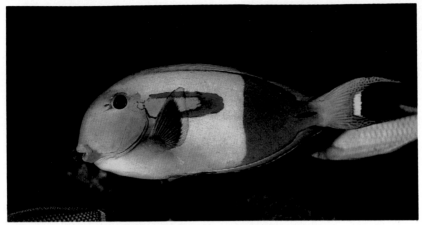

Acanthurus olivaceus 409 ♀ ∿ ⚭ ◐ ♥ 🖼 ▭ 25 cm
Acanthurus pyroferus 409 ♀ ∿ ⚭ ◐ ♥ 🖼 ▭ 19 cm

Acanthurus tennenti 409 ♀ ∿ ⚭ ◐ ♥ 🖼 ▭ 31 cm

Acanthurus lineatus 409 ♀ ∿ ⚲ ◐ ♥ 🖼 ⊡ 38 cm

Acanthurus pyroferus 409 ♀ ∿ ⚛ ◐ ♥ ▨ ▱ 19 cm
Acanthurus dussumieri 409 ♀ ∿ ⚛ ◐ ♥ ▨ ▱ 54 cm

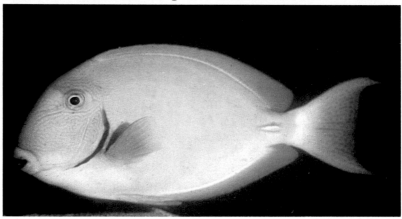

Acanthurus olivaceous 409 ♀ ∿ ⚛ ◐ ♥ ▨ ▱ 25 cm

Prionurus laticlavius 409 ♀ ∿ ⚓ ○ ♥ 🎴 ▱ 50 cm
Acanthurus leucopareius 409 ♀ ∿ ⚓ ○ ♥ 🎴 ▱ 20 cm

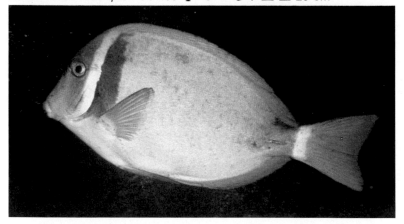

Acanthurus nigroris (juv.) 409 ♀ ∿ ⚓ ○ ⚓ ♥ ▱ 20 cm

Prionurus punctatus 409 ♀ ∿ ⚓ ☽ ♥ 🖼 ▭ 60 cm
Prionurus scalprus 409 ♀ ∿ ⚓ ☽ ♥ 🖼 ▭ 40 cm

Acanthurus triostegus 409 ♀ ∿ ⚓ ☽ ♥ 🖼 ▭ 27 cm

Paracanthurus hepatus 409 ♀ ∿ ⚘ ○ ♥ 🖼 ⊟ 26 cm
Ctenochaetus marginatus 409 ∿ ⚘ ○ ♥ 🖼 ⊟ 20 cm

Ctenochaetus tominiensis (juv.) 409 ∿ ⚘ ○ ♥ 🖼 ⊟ 15 cm

Ctenochaetus strigosus 409 ∿ ⚓ ◐ ♥ 🖼 ⊟ 18 cm
Paracanthurus hepatus 409 ♀ ∿ ⚓ ◐ ♥ 🖼 ⊟ 26 cm

Naso brevirostris 409 ♀ ∿ ⚓ ◐ ♥ 🖼 ⊟ 60 cm

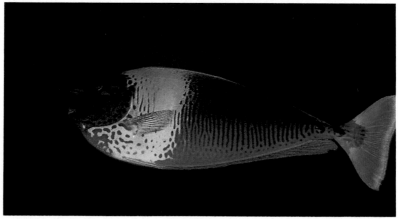

Naso tuberosus 409 ♀ ∿ ⚲ ◐ ♥ 🖼 ▤ 60 cm
Naso lituratus 409 ♀ ∿ ⚲ ◐ ♥ 🖼 ▤ 45 cm

Naso vlamingi 409 ♀ ∿ ⚲ ◐ ♥ 🖼 ▤ 60 cm

Naso hexacanthus 409 ♀ ∿ ⚓ ◐ ♥ 🖼 ⊟ 75 cm
Naso lituratus 409 ♀ ∿ ⚓ ◐ ♥ 🖼 ⊟ 45 cm

Naso unicornis 409 ♀ ∿ ⚓ ◐ ♥ 🖼 ⊟ 70 cm

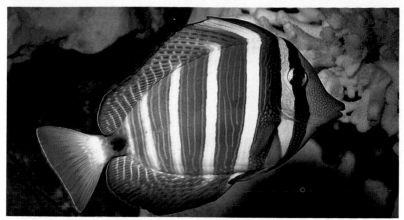

Zebrasoma veliferum 409 ♀ ∿ ⚡ ⦾ ♥ 🖼 ⊡ 40 cm
Zebrasoma gemmatum 409 ♀ ∿ ⚡ ◯ ♥ 🖼 ⊡ 22 cm

Zebrasoma flavescens 409 ♀ ∿ ⚡ ◯ ♥ 🖼 ⊡ 15 cm

Zebrasoma rostratum 409 ♀ ∿ ⚘ ○ ♥ 🖼 ⊟ 20 cm
Zebrasoma veliferum 409 ♀ ∿ ⚘ ○ ♥ 🖼 ⊟ 40 cm

Zebrasoma flavescens 409 ♀ ∿ ⚘ ○ ♥ 🖼 ⊟ 15 cm

Zebrasoma desjardinii 409 ♀ ∿ ⚘ ☉ ♥ 🎞 ▣ 40 cm
Zebrasoma xanthurus 409 ♀ ∿ ⚘ ☉ ♥ 🎞 ▣ 22 cm

Zanclus canescens 409 ♀ ∿ ⚘ ☽ ♥ 🎞 ▣ 22 cm

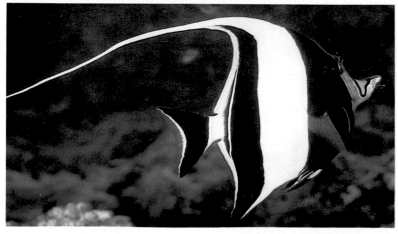

Lo vulpinus 410 ♀ ∿ ⚓ ◐ ♥ 🖼 ⊟ 19 cm
Lo uspi 410 ♀ ∿ ⚓ ◐ ♥ 🖼 ⊟ 18 cm

Siganus canaliculatus 410 ♀ ∿ ⚓ ◐ ♥ 🖼 ⊟ 30 cm

Lo magnifica 410 ♀ ∽ ⚹ ☽ ♥ ▨ ▣ 18 cm
Siganus guttatus 410 ♀ ∽ ⚹ ☽ ♥ ▨ ▣ 35 cm

Siganus corallinus 410 ♀ ∽ ⚹ ☽ ♥ ▨ ▣ 25 cm

Siganus lineatus 410 ♀ ⌇ ⚘ ◐ ♥ 🐟 ▭ 40 cm
Siganus javus 410 ♀ ⌇ ⚘ ◐ ♥ 🐟 ▭ 45 cm

Siganus punctatus 410 ♀ ⌇ ⚘ ◐ ♥ 🐟 ▭ 40 cm

Siganus doliatus 410 ♀ ∿ ⋔ ◐ ♥ 🖼 ▭ 30 cm
Siganus stellatus 410 ♀ ∿ ⋔ ◐ ♥ 🖼 ▭ 40 cm

Siganus trispilos 410 ♀ ∿ ⋔ ◐ ♥ 🖼 ▭ 25 cm

Siganus virgatus 410 ♀ ∿ ⚵ ○ ♥ 🖼 ▭ 30 cm
Siganus puellus 410 ♀ ∿ ⚵ ○ ♥ 🖼 ▭ 30 cm

Siganus punctatus 410 ♀ ∿ ⚵ ○ ♥ 🖼 ▭ 40 cm

Nomeus gronovii 420 〜 ◐ ✕ ▱ ▭ 39 cm
Ariomma indica 420 〜 ↘ ◐ ✕ ▱ ▭ 25 cm

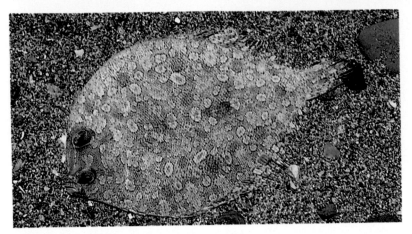

Bothus leopardus 434 〰 ◑ ♥ ▱ ▱ 20 cm
Bothus pantherinus 434 〰 ◑ ♥ ▱ ▱ 30 cm

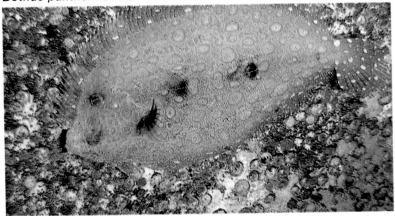

Bothus lunatus 434 〰 ◑ ♥ ▱ ▱ 45 cm

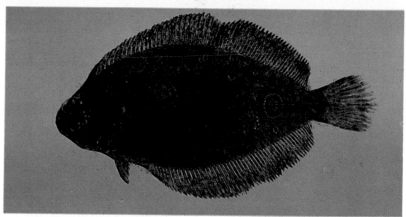

Xystreurys liolepis 434 〜 〜 ◑ ✖ 🖼 🖼 53 cm
Pseudorhombus jenynsii 434 〜 ◑ ♥ 🖼 🖼 27 cm

Zeugopterus punctatus 434 〜 ◑ ♥ 🖼 🖼 25 cm

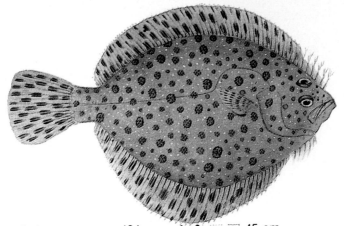

Scopthalmus aquosus 434 〜 ➘ ● ☠ ▦ ▭ 45 cm
Ancylopsetta dilecta 434 〜 ● ♥ ▦ ▭ 25 cm

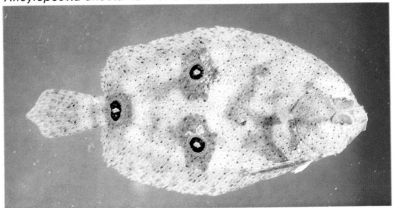

Pleuronectes americanus 435 〜 ➘ ● ☠ ▦ ▭ 64 cm

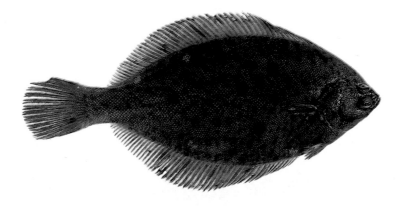

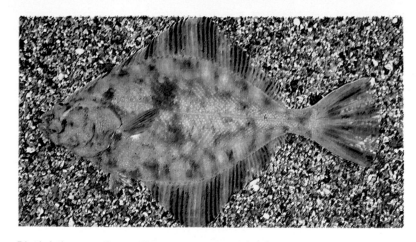

Platichthys stellatus 435 〰 ➘ ◐ ✕ ▭ ▭ 91 cm
Platichthys flesus 435 〰 ➘ ◐ ✕ ▭ ▭ 51 cm

Pleuronichthys cornutus 435 〰 ◐ ♥ ▭ ▭ 30 cm

Pleuronectes platessa 435 ∿ ◑ ♥ ▱ ▱ 71 cm
Psettichthys melanostictus 435 ∿ ◑ ♥ ▱ ▱ 63 cm

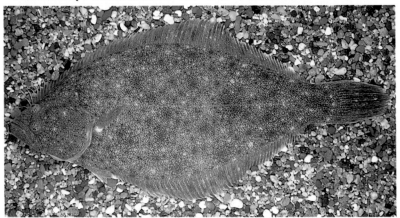

Pleuronichthys coenosus 435 ∿ ◑ ♥ ▱ ▱ 36 cm

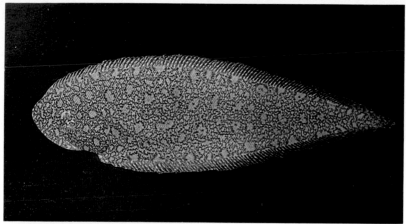

Paraplagusia bilineata 436 ᵔ ◑ ♥ 🖼 🖼 30 cm
Symphurus fasciolaris 436 ᵔ ◑ ♥ 🖼 🖼 25 cm

Symphurus arawak 436 ᵔ ◑ ♥ 🖼 🖼 51 cm

Zebrias zebra 437 ∿ ◑ ♥ 🖼 🖼 19 cm
Pardachirus pavoninus 437 ∿ ◑ ♥ 🖼 🖼 25 cm

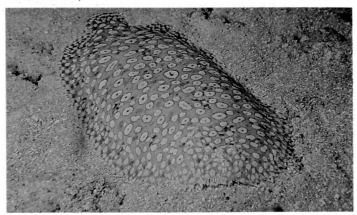

Trinectes maculatus 437 ∿ ◑ ♥ 🖼 🖼 20 cm

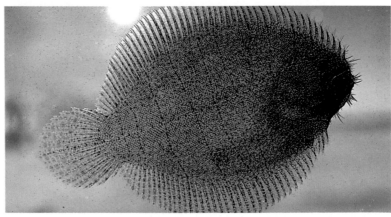

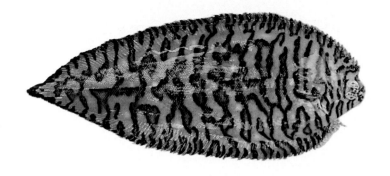

Soleichthys heterorhinos 437 〜 ◑ ♥ 🔲 🔲 11 cm
Achirus lineatus 437 〜 ◑ ♥ 🔲 🔲 10 cm

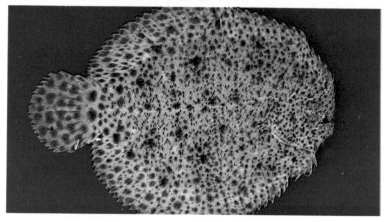

Trinectes maculatus 437 〜 ◑ ♥ 🔲 🔲 20 cm

Johnsonina eriomma 438 ∿ ◑ ♥ 🎞 ⊡ 13 cm
Triacanthodes anomalus 438 ∿ ◑ ♥ 🎞 ⊡ 10 cm

Tripodichthys blochii 439 ∿ ◑ ♥ 🎞 ⊡ 15 cm

Balistoides conspicillum 440 ♀ ∿ ↘ ◑ ✕ ▣ ⊟ 50 cm
Melichthys niger (var.) 440 ♀ ∿ ↘ ◑ ✕ ▣ ⊟ 35 cm

Melichthys indicus 440 ♀ ∿ ↘ ◑ ✕ ▣ ⊟ 25 cm

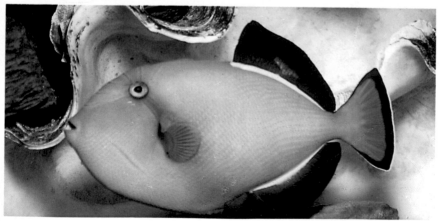

Balistapus undulatus 440 ♀ ∿ ↘ ◐ ✕ ▣ ⊟ 30 cm
Melichthys vidua 440 ♀ ∿ ↘ ◐ ✕ ▣ ⊟ 40 cm

Melichthys indicus 440 ♀ ∿ ↘ ◐ ✕ ▣ ⊟ 25 cm

Xanthichthys ringens 440 ♀ ∿ ↘ ◐ ✕ 📷 ▭ 25 cm
Sufflamen sp. (juv.) 440 ♀ ∿ ↘ ◐ ✕ 📷 ▭ 25 cm

Sufflamen albicaudatus 440 ♀ ∿ ↘ ◐ ✕ 📷 ▭ 30 cm

Xanthichthys mento 440 ♀ ⌇ ↘ ◑ ✳ 🔲 ⊟ 30 cm
Sufflamen chrysopterus (juv.) 440 ♀ ⌇ ↘ ◑ ✳ 🔲 ⊟ 30 cm

Sufflamen fraenatus 440 ♀ ⌇ ↘ ◑ ✳ 🔲 ⊟ 40 cm

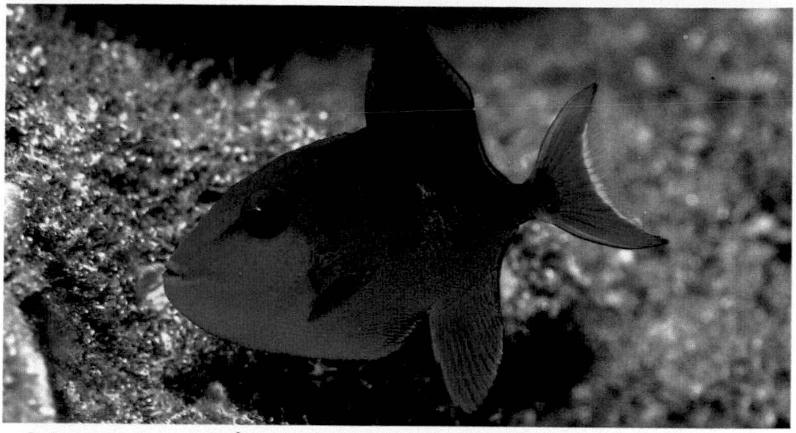

Odonus niger 440 50 cm

Balistes polylepis 440 76 cm

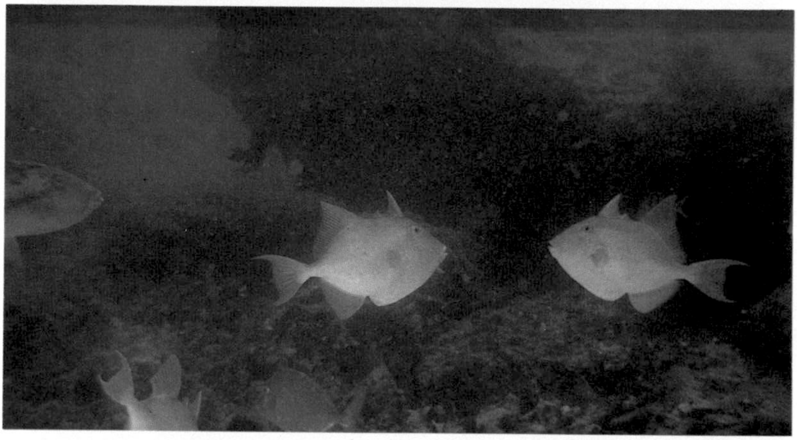

Balistes forcipitus 440 44 cm

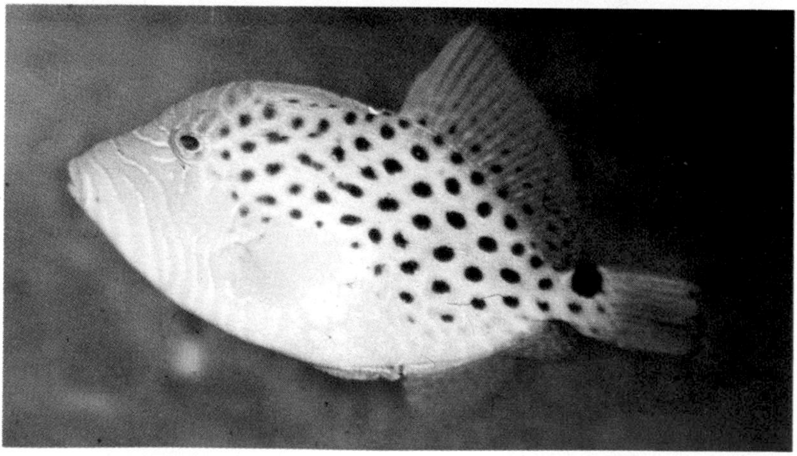

Canthidermis sufflamen 440 ♀ 〜 丶 ◑ ✖ 🖻 ⊟ 25 cm
Balistes capriscus 440 ♀ 〜 丶 ◑ ✖ 🖻 ⊟ 30 cm

Balistes vetula 440 ♀ 〜 丶 ◑ ✖ 🖻 ⊟ 60 cm

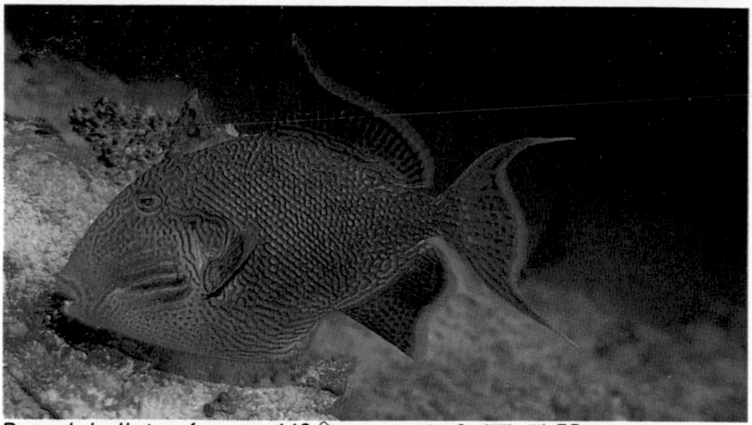

Pseudobalistes fuscus 440 55 cm

Rhinecanthus aculeatus 440 30 cm

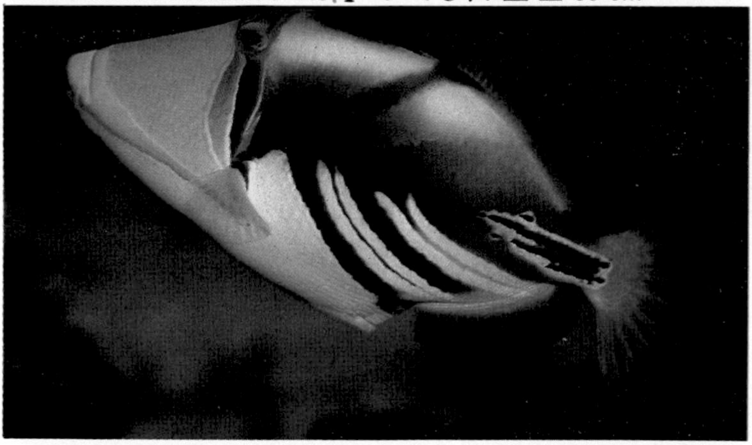

Rhinecanthus rectangulus 440 30 cm

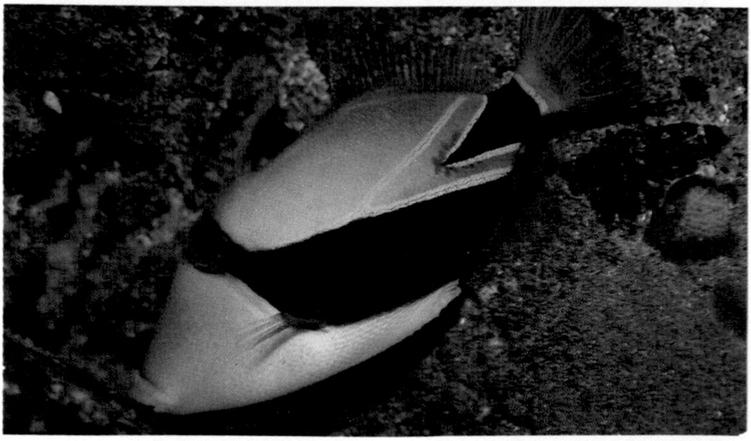

Sufflamen bursa 440 ♀ ∿ ↖ ◑ ✕ 🎞 ⊟ 25 cm
Sufflamen fraenatus 440 ♀ ∿ ↖ ◑ ✕ 🎞 ⊟ 40 cm

Amanses scopas 440 ♀ ∿ ↖ ◑ ✕ 🎞 ⊟ 20 cm

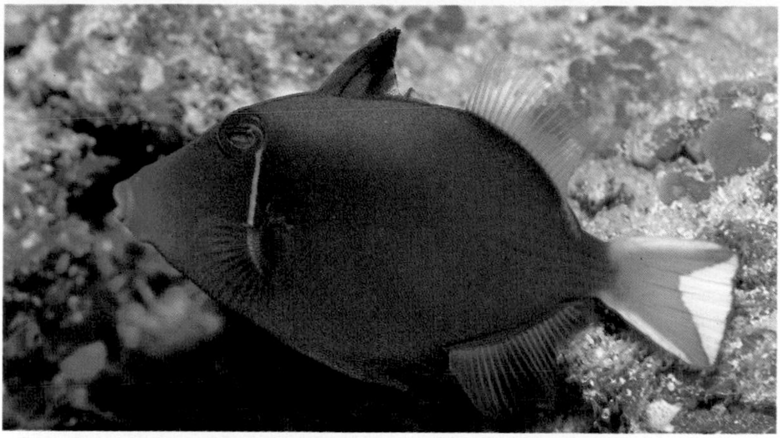

Sufflamen chrysopterus 440 30 cm

Odonus niger 440 50 cm

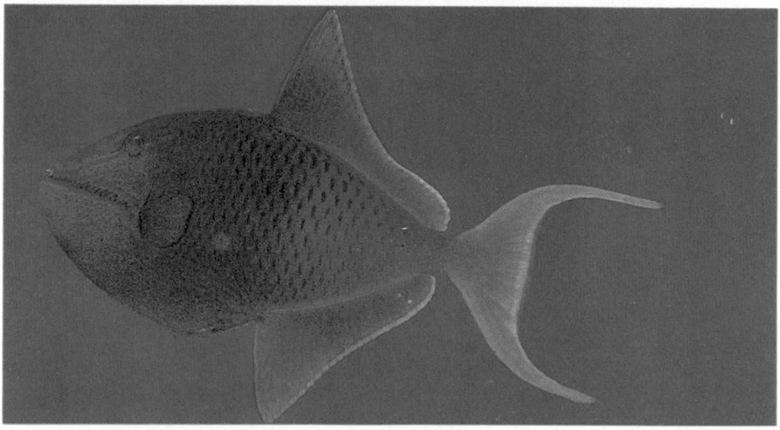

Cantherhines pardalis 440 27 cm

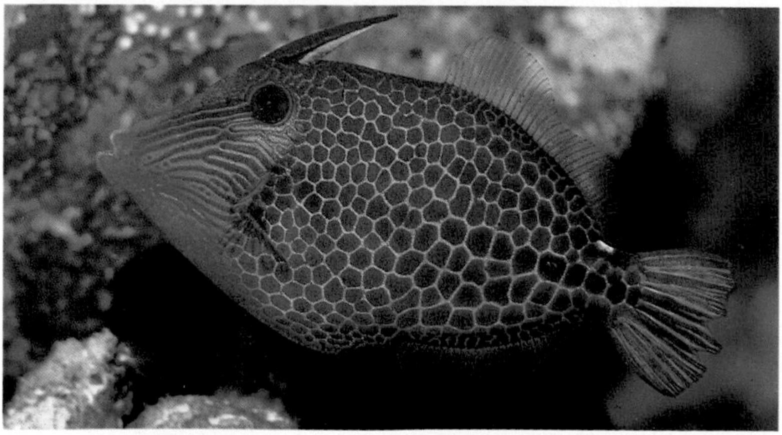

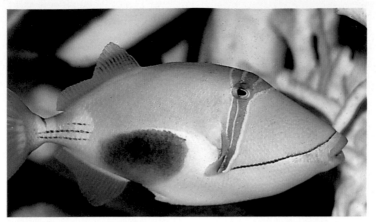

Rhinecanthus verrucosus 440 ♀ ∿ ↘ ◐ ✕ 🎞 ⊟ 30 cm
Monacanthus tuckeri 440 ♀ ∿ ↘ ◐ ✕ 🎞 ⊟ 9 cm

Monacanthus ciliatus 440 ♀ ∿ ↘ ◐ ✕ 🎞 ⊟ 20 cm

Monacanthus filicauda 440 ♀ ∿ ⚮ ◑ ♥ 🔳 ▭ 25 cm
Cantherhines sandwichensis 440 ♀ ∿ ⚮ ◑ ♥ 🔳 ▭ 13 cm

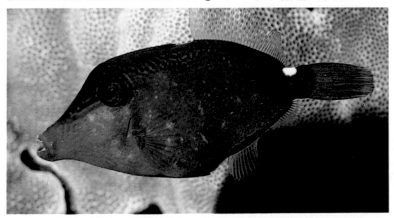

Cantherhines macroceros 440 ♀ ∿ ⚮ ◑ ♥ 🔳 ▭ 46 cm

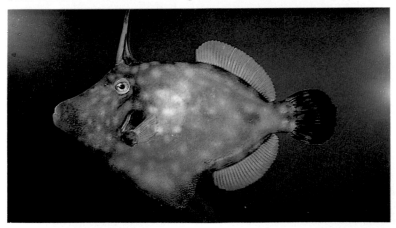

Monacanthus chinensis 440 ♀ ⌇ ⸙ ◐ ♥ 🖼 ⊟ 25 cm
Cantherhines pullus 440 ♀ ⌇ ⸙ ◐ ♥ 🖼 ⊟ 20 cm

Monacanthus chinensis 440 ♀ ⌇ ⸙ ◐ ♥ 🖼 ⊟ 25 cm

Meuschenia flavolineata 440 ♀ ∿ ⚲ ◑ ♥ 🖼 ⊡ 20 cm
Meuschenia hippocrepis 440 ♀ ∿ ⚲ ◑ ♥ 🖼 ⊡ 50 cm

Pervagor melanocephalus 440 ♀ ∿ ⚲ ◑ ♥ 🖼 ⊡ 16 cm

Meuschenia galii 440 ♀ ⌇ ⚓ ◑ ♥ 🔳 ⊟ 23 cm
Oxymonacanthus longirostris 440 ⌇ ◑ ♥ 🖼 ⊟ 7 cm

Scobinichthys granulatus 440 ♀ ⌇ ⚓ ◑ ♥ 🔳 ⊟ 25 cm

Cantherhines verrucundus 440 ♀ ∿ ⚓ ◑ ♥ 🖻 ⊟ 12.5 cm
? *Paramonacanthus japonicus* 440 ♀ ∿ ⚓ ◑ ♥ 🖻 ⊟ 16 cm

Meuschenia hippocrepis 440 ♀ ∿ ⚓ ◑ ♥ 🖻 ⊟ 50 cm

Cantheschenia grandisquamis 440 20 cm

Cantherhines fronticinctus 440 14 cm

Meuschenia flavolineata 440 20 cm

Pervagor janthinosoma 440 ♀ ∿ ⚡ ◐ ♥ 🖼 ⊟ 16 cm
Pervagor aspricaudus 440 ♀ ∿ ◐ ♥ 🖼 ⊟ 15 cm

Rudarius ercodes 440 ♀ ∿ ◐ ♥ 🖼 ⊟ 10 cm

Pervagor janthinosoma 440 ♀ ∿ ◑ ♥ 🖼 ⊟ 16 cm
Pervagor spilosoma 440 ♀ ∿ ◑ ♥ 🖾 ⊟ 13 cm

Acreichthys tomentosus 440 ♀ ∿ ◑ ♥ 🖾 ⊟ 9 cm

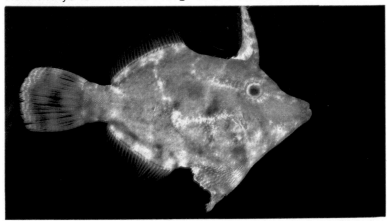

Paramonacanthus barnardi 440 ♀ ∿ ⚲ ◐ ♥ 🖼 ▣ 9 cm
Chaetodermis pencilligerus 440 ♀ ∿ ⚲ ◐ ♥ 🖼 ▣ 25 cm

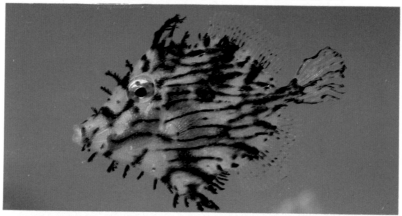

Eubalichthys mosaicus 440 ♀ ∿ ⚲ ◐ ♥ 🖼 ▣ 30 cm

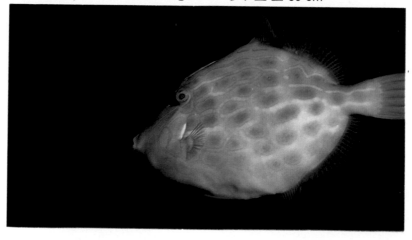

Aluterus schoepfi 440 ♀ ∿ ⅄ ◑ ♥ 🎞 ▱ 60 cm
Aluterus monoceros 440 ∿ ⅄ ◑ ♥ 🎞 ▱ 75 cm

?Aluterus sp. 440 ♀ ∿ ⅄ ◑ ♥ 🎞 ▱ 40 cm

Thamnaconus modestus 440 ♀ ᴧ⟋ ⋇ ◑ ♥ 🐟 ⊟ 30 cm
Aluterus monoceros 440 ᴧ⟋ ⋇ ◑ ♥ 🐟 ⊟ 75 cm

Pseudalutarius nasicornis 440 ♀ ᴧ⟋ ⋇ ◑ ♥ 🐟 ⊟ 18 cm

Thamnaconus modestus 440 ♀ ∿ ⚲ ◐ ♥ 🖻 ⊟ 30 cm
Nelusetta ayraudi 440 ♀ ∿ ⚲ ◐ ♥ 🖻 ⊟ 50 cm

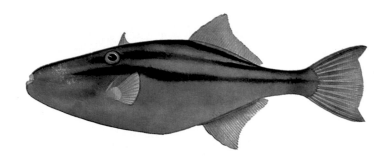

Aluterus scriptus 440 ♀ ∿ ⚲ ◐ ♥ 🖻 ⊟ 100 cm

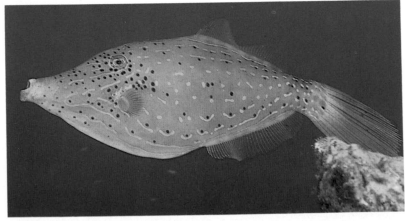

Thamnaconus australis 440 ♀ ∿ ⚘ ◑ ♥ 🖼 🔲 27.5 cm

Parika scaber 440 ♀ ∿ ⚘ ◑ ♥ 🖼 🔲 20 cm

Paraluterus prionurus (juv.) 440 ♀ ∿ ⚘ ◑ ♥ 🖼 🔲 9.5 cm

Ostracion cubicus 441 ♀ ∿ ◑ ♥ 🖼 ⊟ 45 cm
Ostracion meleagris meleagris 441 ♀ ∿ ◑ ♥ 🖼 ⊟ 15 cm

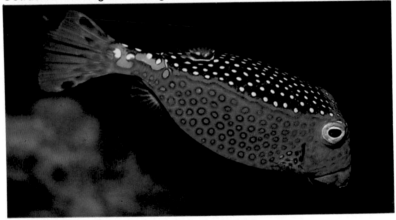

Ostracion trachys 441 ♀ ∿ ◑ ♥ 🖼 ⊟ 15 cm

Ostracion meleagris meleagris 441 ♀ ∿ ◐ ♥ 🔀 ⊟ 15 cm
Tetrosomus gibbosus 441 ♀ ∿ ◐ ♥ 🔀 ⊟ 30 cm

Lactoria diaphanus 441 ♀ ∿ ◐ ♥ 🔀 ⊟ 25 cm

Strophiurichthys robustus 441 ♀ ∿ ◑ ♥ 🐟 ⊟ 25 cm
Lactoria cornuta 441 ♀ ∿ ◑ ♥ 🐟 ⊟ 50 cm

Ostracion cubicus 441 ♀ ∿ ◑ ♥ 🐟 ⊟ 45 cm

Ostracion solorensis 441 ♀ ∿ ◐ ♥ 🖼 ▣ 15 cm
Ostracion whitleyi 441 ♀ ∿ ◐ ♥ 🖼 ▣ 13 cm

Ostracion meleagris camurum 441 ♀ ∿ ◐ ♥ 🖼 ▣ 14 cm

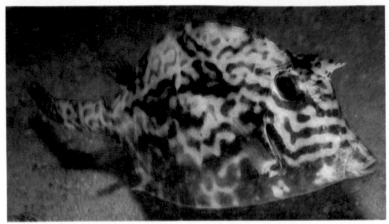

Lactophrys quadricornis 441 ⚲ ∿ ◑ ♥ 🖼 ⊟ 43 cm
Lactophrys polygonia 441 ⚲ ∿ ◑ ♥ 🖼 ⊟ 48 cm

Lactophrys bicaudalis (juv.) 441 ⚲ ∿ ◑ ♥ 🖼 ⊟ 43 cm

Lactophrys polygonia 441 ♀ ∿ ◐ ♥ 🖼 ▤ 48 cm
Lactophrys bicaudalis 441 ♀ ∿ ◐ ♥ 🖼 ▤ 43 cm

Lactophrys triqueter 441 ♀ ∿ ◐ ♥ 🖼 ▤ 28 cm

Amblyrhynchotes sp. 443 ∿ ↘ ◑ ✕ 📷 ⊟ 25 cm

Amblyrhynchotes hypselogenion 443 ∿ ↘ ◑ ✕ 📷 ⊟ 18 cm

Omegophora armilla 443 ∿ ↘ ◑ ✕ 📷 ⊟ 14 cm

Arothron meleagris 443 15 cm

Arothron meleagris 443 15 cm

Arothron nigropunctatus (gray) 443 40 cm

Arothron meleagris 443 ～ ↘ ◑ ✄ 📷 ⊟ 15 cm
Arothron nigropunctatus 443 ～ ↘ ◑ ✄ 📷 ⊟ 40 cm

Arothron hispidus (var.) 443 ～ ↘ ◑ ✄ 📷 ▭ 50 cm

Arothron nigropunctatus 443 ∿ ↘ ◑ ✕ 🔳 ⊟ 40 cm
Arothron stellatus 443 ∿ ↘ ◑ ✕ 🔳 ⊟ 100 cm

Omegophora cyanopunctata 443 ∿ ↘ ◑ ✕ 🔳 ⊟ 12 cm

Arothron nigropunctatus 443 〜 ↘ ◑ ☆ 📷 ⊟ 40 cm
Arothron inconditus 443 〜 ↘ ◑ ☆ 📷 ⊟ 40 cm

Amblyrhinchotes hypselogenion 443 〜 ↘ ◑ ☆ 📷 ⊟ 18 cm

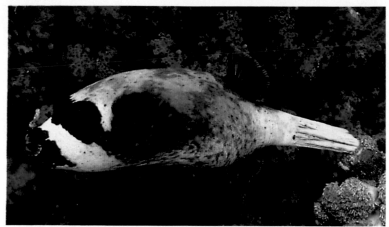

Arothron diadematus 443 ～ ↘ ◐ ✕ 🖾 ⊟ 30 cm
Arothron stellatus 443 ～ ↘ ◐ ✕ 🖾 ⊟ 90 cm

Arothron stellatus 443 ～ ↘ ◐ ✕ 🖾 ⊟ 90 cm

Arothron mappa 443 〜 ⟍ ◑ ✕ 🐟 ⊟ 70 cm
Arothron stellatus 443 〜 ⟍ ◑ ✕ 🔲 ⊟ 90 cm

Arothron reticularis 443 〜 ⟍ ◑ ✕ 🔲 ⊟ 35 cm

Arothron manilensis 443 ♀ ∿ ◑ ✕ 🔳 ☒ 50 cm
Arothron immaculatus 443 ♀ ∿ ◑ ✕ 🔳 ☒ 30 cm

Takifugu pardalis 443 ♀ ∿ ◑ ✕ 🔳 ☒ 35 cm

Sphoeroides annulatus 443 〜 ↘ ◑ ✕ 🔲 ▱ 10 cm
Sphoeroides erythrotaenia 443 〜 ↘ ◑ ✕ 🔲 ▱ 9 cm

Chelonodon patoca 443 〜 ↘ ◑ ✕ 🔲 ▱ 20 cm

Canthigaster amboinensis 443 〜 ⬅ ◐ ✕ 📷 ▣ 15 cm
Canthigaster bennetti 443 〜 ⬅ ◐ ✕ 📷 ▣ 10 cm

Canthigaster epilamprus 443 〜 ⬅ ◐ ✕ 📷 ▣ 7 cm

Canthigaster coronata 443 ᵔᵁ ᕁ ◐ ✕ 🔳 ⊟ 13 cm
Canthigaster janthinoptera 443 ᵔᵁ ᕁ ◐ ✕ 🔳 ⊟ 9 cm

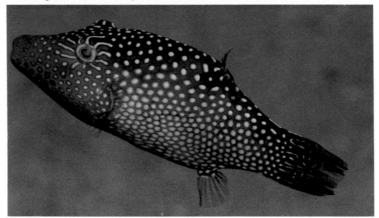

Canthigaster smithae 443 ᵔᵁ ᕁ ◐ ✕ 🔳 ⊟ 13 cm

Canthigaster punctatissimus 443 ᭡ ↘ ◑ ✕ ▣ ⊟ 7.5 cm
Canthigaster compressus 443 ᭡ ↘ ◑ ✕ ▣ ⊟ 10 cm

Canthigaster bennetti 443 ᭡ ↘ ◑ ✕ ▣ ⊟ 10 cm

665

Canthigaster callisterna 443 〜 ↘ ◐ ✕ 🔲 ⊡ 23 cm
Canthigaster rostrata 443 〜 ↘ ◐ ✕ 🔲 ⊡ 11 cm

Canthigaster valentini 443 〜 ↘ ◐ ✕ 🔲 ⊡ 20 cm

Canthigaster solandri 443 11 cm

Chilomycterus affinis 444 51 cm

Diodon hystrix 444 60 cm

Canthigaster natalensis 443 ⌇ ↘ ◖ ✕ 🎦 ▣ 15 cm
Canthigaster tyleri 443 ⌇ ↘ ◖ ✕ 🎦 ▣ 14 cm

Cyclichthys orbicularis 444 ⌇ ↘ ◖ ✕ 🎦 ▣ 14 cm

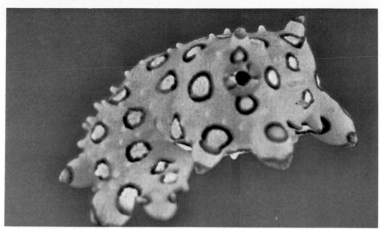

Chilomycterus antillarum 444 〜 ↘ ◑ ✗ 📷 🔲 25 cm
Chilomycterus schoepfi 444 〜 ↘ ◑ ✗ 📷 🔲 25 cm

Chilomycterus antennatus 444 〜 ↘ ◑ ✗ 📷 🔲 23 cm

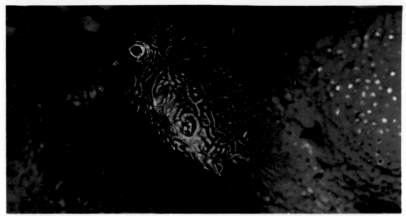

Chilomycterus schoepfi 444 〜 ↘ ◑ ✕ 🔳 ⊟ 25 cm
Chilomycterus antennatus 444 〜 ↘ ◑ ✕ 🔳 ⊟ 23 cm

Chilomycterus spilostylus 444 〜 ↘ ◑ ✕ 🔳 ⊟ 34 cm

Diodon holacanthus 444 〜 ↘ ◑ ✕ 📷 ⊟ 50 cm
Lophodiodon calori 444 〜 ↘ ◑ ✕ 📷 ⊟ 20 cm

Diodon hystrix 444 ♀ 〜 ↘ ◑ ✕ 📷 ⊟ 60 cm

Diodon holacanthus 444 ⤳ ↘ ◑ ☠ 📷 ⊟ 50 cm
Diodon liturosus 444 ⤳ ↘ ◑ ☠ 📷 ⊟ 50 cm

Diodon myersi 444 ⤳ ↘ ◑ ☠ 📷 ⊟ 41 cm

MARINE AQUARIUM
Set-Up and Maintenance Section
Contents

The Marine Aquarium

Aquaria for keeping marine fishes and invertebrates must be fully resistant to salt water. Thus old style stainless steel framed tanks are out. There may be stainless steel exterior decorations, but the strength of the tank must lie in the bonding of the glass or plastic by salt-resistant sealants, such as silicones strong enough to act on their own. Unless you are accustomed to making aquaria, it rarely pays you to attempt the job. Sealants are hard to manage and you can probably buy an aquarium for less than it would cost to make.

You may need to order a special tank to suit your particular requirements. Even then you will get a better article to fit if it is made professionally. I have kept around a hundred tanks in a long life of aquarium keeping and have never made one of them.

Except for special purposes, marine aquaria should be fairly large. You cannot keep nearly as many fishes, particularly tropicals, as in a similar-sized freshwater tank. Small aquaria are also liable to fluctuations in temperature and water quality that can be harmful to the inhabitants, causing distress or disease. If a heater fails, for example, a small tank falls rapidly in temperature while a larger one takes many hours to change and thus gives a margin of safety. Small means less than about 25 U.S. gallons. A good size for most purposes is between 50 and 100 U.S. gallons, the smaller rather than the larger for beginners. Over 100 gallons, various factors begin to operate such as cost, difficulties of servicing if the tank is too deep, and the cost of whatever goes into it.

The shape of a tank is a compromise between biological and aesthetic considerations. Biology demands a shallow vessel with a large water surface for gas exchange. Because such a tank looks clumsy, 'low' aquaria are not too popular. 'High' ones are favored because they display their contents better even though they cannot support the same number of fishes or other creatures, gallon for gallon. The usual solution is to have the height a few inches greater than the width, say 36 x 16 x 20 inches (length x width x height). This tank would

Although the marine aquarium can be placed anywhere, be sure items adjacent to it can not have adverse effects.

contain just under 50 gallons of water. Larger tanks do not usually have a much greater depth for the reasons already mentioned, so that one holding approximately 100 gallons would be, say, 60 x 18 x 22 inches. Calculate the volume of a tank by multiplying the three dimensions together and dividing by 231 for U.S. gallons.

Furnishing a marine aquarium often costs much more than the cost of the tank itself. Even if it holds only fishes, decorations such as cured coral and gravel are needed. So are the usual heating, filtering, and aerating equipment, since most marine aquaria are tropical. In many areas, a successful temperate or coldwater tank needs cooling equipment instead. Add to all this the cost of the inhabitants and you have a tidy sum to invest. This is why I advise an aquarium holding around 50 gallons. This is its theoretical capacity; by the time the contents have been added the actual water volume may be nearer 40 gallons.

Tank Construction

A well-made tank is a very safe proposition and rarely leaks. However, it must have certain attributes for this to be true. It must be made of new, unscratched glass or plastic thick enough to withstand the pressure of the water it will hold. Glass thickness depends on depth, and, to a lesser extent, on the length of the tank. Up to 12 inches deep and 30 inches long it can be made of ³⁄₁₆ inch glass with preferably a ¼ inch plate base. From 12 to 15 inches deep ¼ inch plate is needed with preferably a ³⁄₈ inch plate base. If longer than 30 inches, cross supports are needed at the top to prevent bowing of the long sides. From 15 to 18 inches deep, ³⁄₈ inch plate should be used throughout. Above 18 inches it is best to use ½ inch plate, but most tanks up to 24 inches in depth are made of ³⁄₈ inch plate for economy and seem to be safe enough. Thinner glass

Aquaria in a bathroom would be subject to temperature fluctuations and would fog up like mirrors.

Aquaria make excellent room dividers. They can be set up for one- or two-sided viewing.

than that just recommended is quite often used. Do not buy such a tank! The chance of breakage is perhaps still low, but not low enough. The thickness of plastics needed depends on the type used.

Large aquaria need one or more supports across the top: one in a 36-inch tank, across the center, 3 inches or more wide; two at one third of the way along the tank for a 60-inch tank or more, each at least 4 inches to 6 inches wide. They should be countersunk about ½ inch below the top of the tank to prevent overflow of splash or drips. They can be used as rests for top covers if these are employed. In all but reef aquaria such covers are usual and should be made of strong glass. Things will rest on them from time to time and may even be dropped on them. They should have small cut-outs for air lines and heater leads and handles for convenience. Really

677

Two ideas for setting up a marine display tank. The round bar-base tank provides more area of unhampered viewing.

elegant aquaria will have the rests for such covers countersunk so that they are flush with the cross-supports. They will also have seams lined by thin glass rods to strengthen them, forming a seal with the silicone or other sealant.

Top covers stop loss by evaporation from being excessive, fishes from jumping out, and unwanted things from falling in. However, they also prevent free circulation of air. For that reason they are omitted in reef aquaria where absolute maximum aeration is desired. The

Another room divider setup but with three viewing sides. The decorations must be placed accordingly.

hood over the lights is usually a secondary cover preventing too much risk of losing fishes. If overhead lighting is used, a completely open top is practical, except for the cross-supports. When covers are used, countersinking makes it easy to keep the top clean and free of salt evaporation. Fresh water used for making up evaporation loss can be poured over them and will drip into the tank without overflowing. Best to dust them first, however! Very little salt-creep occurs in an open tank, one of its advantages.

It does not matter whether the bottom glass fits inside the upright ones or they rest upon it. What does matter is that everything should be absolutely square, that is at right angles. The sealant should have no bubbles or channels in it that could lead to a leak. The edges of all glass panels, where exposed, should have been honed so as not to remain sharp. A refinement I have not yet seen would be to have the front panel of non-reflecting glass to make for better vision and possibly easier photography.

The Stand

Small aquaria up to about 50 gallons can safely be placed on almost any piece of furniture. Larger ones must be specially supported. Water weighs 8¼ lbs. per U.S. gallon. The addition of sand, coral, and rocks increases the weight even more. With the weight of the tank itself plus contents we are looking at something in excess of 500 lbs in the case of a 36-inch aquarium. This is getting toward the limit of strength of domestic furniture and floors except near walls or over joists. Concrete floors are, of course, much stronger. So calculate the strength of your floors before installing a really large aquarium. Such an aquarium is by far best placed on a strong cabinet that can hold equipment and will make an attractive addition to the home.

Whatever the aquarium sits on, it must be absolutely level. Make checks in both directions, lengthwise and from front to back. To be sure of this, use a spirit level. A very slight tilt looks dreadful when the tank is filled as the water level will not correspond to the top of the glass. You will not be happy until you have put

The cover of this reef tank stand has been removed to show the wet/dry filter with a double spiral system.

things right. A serious tilt will also stress the tank and could lead to cracking or leaks. The surface upon which the tank rests must be smooth. To ensure safety it should be covered by ¼ to ½ inch of styrofoam or some other water resistant and compressible but firm material. This will take up any small departures from absolute flatness of either tank or stand. It is possible to place styrofoam over closely placed cross-bars in a regular type of stand, but not preferable.

Again, for elegance, the top of the cabinet or stand should be countersunk so that the aquarium base is hidden; about 1 inch is sufficient. Leave the back open so that the aquarium can be slid into position. If the surface is not countersunk, it is strongly advisable to tape the styrofoam temporarily in position so that it stays put as you place the tank in position upon it. This is best done by at least two people with a tank of any but very small size. The styrofoam, unless countersunk, should be cut a little larger than the tank base and can be trimmed afterward. Most tanks have a definite front and back determined by slots and handles on the top. If not, make sure that you place the best panel to the front if the glass has any minor flaws.

This filter compartment contains plastic balls that provide increased surface area for biological filtration.

The same setup as on page 681 but with the cover replaced, making the aquarium base appear more like a piece of furniture.

Lighting

The type and intensity of lighting in a marine aquarium depends on the nature of its contents. Fishes need enough light to feel comfortable, feed, and find their way around. Some algae need more light than that and will only flourish with the correct spectrum. Some invertebrates that house unicellular algae in their tissues need quite intense light. Just how intense depends on the spectrum of the light once more. In the ocean, in clear conditions, the red end of the spectrum is rapidly absorbed while the blue end penetrates much deeper. Of the total radiation from the sun in the visible spectrum, about 75% remains at 3-4 feet, 50% at 6-8 feet, and 20% at 15-20 feet. All of this is toward

A reef tank combines fishes with invertebrates and algae for a very natural look.

the blue end since at 15 feet nearly all of the red is gone. The home aquarium is never deep enough to absorb much of the light unless the water is very yellow or turbid, so a different effect is seen. Sunlight comes in parallel rays, but aquarium lights usually produce divergent beams. The intensity of light near the source is much greater than it is a foot or two into the tank, even with reflectors.

The absorption of light by chlorophylls, the green pigments that enable plants to build up organic material from simple inorganic compounds like carbon dioxide and water,

Your pet shop can supply complete setups including tank, base, and cover. Filters, pumps, heaters, etc., may or may not be included. Decorations and fishes can be added once the tank is ready for them.

is in the red and blue end of the spectrum. Plants look green because that is the color reflected from their surface. Even when they do not look green they still possess chlorophylls. The chlorophylls are masked by other pigments that may absorb other colors and pass the energy gained on to the chlorophylls. So, for the health of the algae it is necessary either to supply fairly intense overall light, containing enough of the reds and blues to suffice, or to supply plenty of red and/or blue even if the light is deficient elsewhere. This is particularly true of the algae, *Zooxanthellae* or *Zoochlorellae*, embedded in the cells of corals and other invertebrates, where light is further absorbed by the tissues of the host.

Two systems of lighting have been found that fulfill the requirements above. **Metal halide** lamps give out a very intense wide-spectrum light that resembles daylight. They are placed high above the water surface because they are hot. They also produce too much ultraviolet light that must be screened off by glass either in the filament or elsewhere. Most **actinic fluorescent** lamps give out an intense blue light. They do not seem very bright to us, so a mixture of actinic and white or daylight fluorescents is used to produce a pleasant overall effect. Other lighting systems are available, but they do not seem to be as satisfactory as those just described.

If the aquarium is to house only fishes and perhaps a few tough invertebrates, such as crabs, other crustaceans, and some types of coelenterates, practically any light sufficient for them to see and be seen is adequate. This can be provided by a single fluorescent tube running the length of the tank, usually rating 10 watts per foot. Some algae, mostly encrusting or filamentous types, will flourish. Even some of the fronded macroalgae (seaweeds) will thrive depending on which tubes are used. Algal growth can be stimulated by plant growth fluorescents that emit mainly in the red and blue end of the spectrum. They shed a purplish light that actually enhances the colors of fishes as well. Various white or daylight fluorescents provide enough blue light to

Some fishes are gregarious and do poorly if purchased singly for your tank. Seen here is an aggregation of *Anthias evansi* on the outer reef slope of Christmas Island.

encourage plant growth, although very little red. So if your interest is primarily in fishes, you can choose from a wide variety of tubes, keep some invertebrates, and grow some algae to decorate the tank and feed some of its inhabitants. Two tubes, one plant-growth and one white or daylight, can give ample light and encourage quite a luscious growth of algae.

Aquaria should be lit for around 12 hours per day. Up to 14 hours is acceptable, but as most of the marine aquarium fishes are tropicals, they are not accustomed to longer days. It is best to avoid plunging the fishes suddenly into daylight or darkness. Marine fishes, in particular, like to settle down in the evening in a familiar place. However, I must confess to using timers in some tanks. The inhabitants get used to them. If possible, however, see that room lights are on or it is daylight before the aquarium lights go on and after they go off. Keep to a regular schedule to which fishes will become adapted. They will settle down at the appropriate time as 'lights off' approaches. You will wish to enjoy your aquarium in the evening, so a routine such as 8 a.m. to 10 p.m. would be in order. Avoid too much direct sunlight. It can overheat even quite a large tank. Aquaria placed near windows are also subject to dangerous drafts. Never place marine tanks near uncontrolled sources of both light and heat, if at all possible.

Heating

Most marine aquaria house tropical fishes. They must be provided with a source of heat. Make sure that any heater you purchase is suitable for a marine tank, preferably a heater that is totally submersible. This means that you do not have to worry about water levels or splashes even if the head is above water. Most modern heaters are of glass and include a thermostat controlled from the top of the heater. Countries vary in safety requirements, but the equipment is generally very robust and reliable. It is best placed in the aquarium near the bottom. Here the submergible heater operates most efficiently, causing a rising stream of heated water that helps to circulate the contents. Such a position also guards against

Surge zone fishes may require higher oxygen levels than those in more quiet surroundings. The animals seen on the bottom in this wave battered area are filter-feeding soft corals.

forgetting to switch off the heater when making water changes and shattering a hot glass instrument when filling up again. The heater should be kept from contacting sand or the tank sides, whether by suction cups or supports of some kind. Suction cups need frequent careful checking.

We do not want a heater that can boil the fishes if anything goes wrong and it sticks in the **ON** position. Nor do we want one that is so powerful that it switches on and off all the time. Both requirements dictate a careful choice of heater wattage. There is usually a pilot light that tells you when the heater is on or off so you can easily check the performance of the one you have. A heater capable of raising the tank temperature 20°F above room temperature is ample in all normal circumstances except for unheated rooms in really cold climates. An overnight fall in room temperature usually does not matter. A reasonably sized tank does not lose heat all that quickly. Of course, if you have permanent temperature controlled air-conditioning, settle for a rise of only around 10°F. Or,

do without tank heating at all, depending on your choice of room temperature. A typical tropical marine aquarium should be kept at 74°-78°F.

Claims of some heating systems to control the water temperature to within a fraction of a degree, even if true, are quite unnecessary. Practically all of our marine aquarium fishes are accustomed to some temperature changes. Sensitive fishes, like open ocean schooling species, are not kept in home aquaria. Reef fishes predominate and these experience quite large swings at times. Control within ±1.5°F is adequate and avoids frequent thermostat operation that tends to wear out the instrument.

Small tanks lose heat more rapidly than large ones because the surface area per gallon is greater. This means the surfaces of both the water and the glass. To maintain a given temperature small tanks need more watts per gallon than large tanks. **Table 1** gives the wattage requirements for different sizes of aquaria on the assumption that a 20°F differential is to be the

Table 1
The Theoretical Wattage of Heaters for Different Sizes of Tanks

Tank length in inches	Gallon Capacity	Watts per Gallon	Total Watts
12	1.9	8.0	15
18	6.3	5.3	33
24	15.0	4.0	60
30	30	3.2	96
36	50	2.6	136
42	80	2.3	183
48	120	2.0	240
Above 48*	—	2.0	—

* A tank over 48 inches long will not usually be more than 24 inches deep and so the watts per gallon stays at 2.0 approximately.

maximum heating potential, and that the heater is placed within a 'normal' type of tank. Aquaria that are of an unusual shape may need additional heat, but they have to be quite unusual for the difference to matter. If the heater is not in the aquarium, but in an outside filter or other piece of equipment, additional wattage is needed. No exact figure can be given since the extra heat loss depends on circumstances, but a fairly safe figure would be one third extra wattage.

Naturally, you will have to select a heater as near as possible to the wattage required. Take one above the actual requirement rather than one below it unless the one below is much nearer to the theoretical figure. In **Table 1**, the 12- and 18-inch tanks are included to show how the necessary wattage increases per gallon with small tanks. You are not expected to employ them unless for special purposes. Years ago, heaters were available in series like 20, 30, 40 watts, etc. Nowadays, there are practically only 50, 100, 150, 200 watts, etc. available. Thus small tanks have to utilize excess heating.

A thermometer should always be placed in any aquarium, heated or not. Two types are recommended: an alcohol thermometer to go inside the tank or a stick-on liquid crystal type to be attached to the outside glass. Stick-on thermometers show the approximate temperature and are slightly influenced by room temperature. So for accuracy an inside thermometer is preferred. Mercury thermometers are best avoided; in the event of a breakage a toxic amount of mercury can be released into the water. This is particularly dangerous in sea water where it will be rapidly taken up. Clock-type thermometers tend to be unreliable, and since they are usually metallic should not be used in a marine aquarium. Any new thermometer should be checked against a reliable one. Your pharmacist can help if you do not have one of your own. Even reputable brands can be a few degrees off.

Aeration

In a marine tank, aeration may be partly or wholly integrated with filtration. This is especially so in reef tanks where airstones are not used in the aquarium itself. The main source of gas exchange within the aquarium is the water surface, not the bubbles. Their main function is to stir the water, bringing it up from the bottom to the surface so that it may lose unwanted gases and absorb oxygen. On the way up, the bubbles exchange some of their oxygen. However, unless they are much more numerous than is usual, this exchange is small compared to the surface exchange. To rival it, the mass of bubbles would have to be unpleasant to the eye and would create a cauldron in the water.

Bubbles of between $\frac{1}{30}$th to $\frac{1}{50}$th of an inch in diameter give the best stirring action. Larger ones 'glug' up to the surface without carrying much water with them. Smaller ones form a fine mist that also fails to do the job. There is no need for an over-vigorous stream of bubbles. Quite a modest output will give adequate water movement without creating undue splash. It is better to have two or three such streams than one very strong one. When we discuss filtration you will see that there are better methods of aeration in the marine tank than airstones within it. It is a good idea, though, to have airstones ready to be placed in the aquarium in case of failure of other equipment.

If you are accustomed to keeping freshwater fishes, you know that aeration can double the holding capacity of a tank. This is not so in marine aquaria. Marine fishes need practically all the oxygen they can get. Therefore, any marine set-up must provide for as near saturation as possible. Additional airstones will not do much to increase it. If no other aeration is provided, as in a natural system tank, the airstones are essential to the set-up. They are not something to be provided to increase its carrying capacity. The stirring of the water in a natural system aquarium is also a necessity for the health of the fishes. They are accustomed to it in nature.

The best airstones for a marine tank are of fused, small-grain ceramic or similar safe material. They

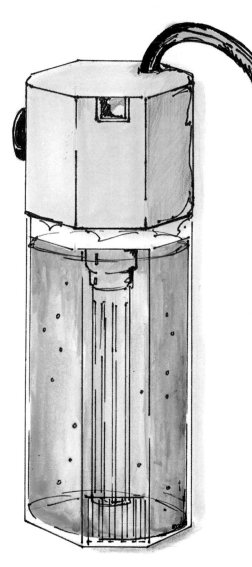

Inside corner filters are popular but do take up tank space and usually must be hidden by decorations.

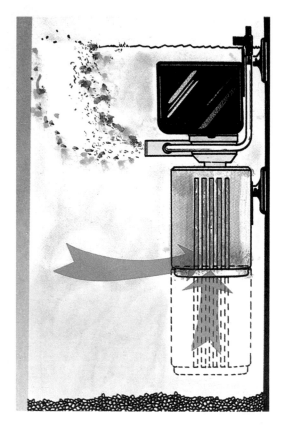

This inside power filter draws water to be filtered from the bottom of the tank. It also provides a slight current as well as aeration by its discharge in the upper layers.

give smaller bubbles in salt water than in fresh water, and of just around the right size. Wooden ones give very fine bubbles but tend to clog up rapidly. They are used in equipment, such as undergravel filters or protein skimmers (discussed later), where the bubbles are confined within tubes and cannot spread out ineffectively.

There are many types of air pumps. Modern diaphragm pumps are powerful, reasonably quiet,

reliable, and not too expensive. It is best to purchase one with an output control. This way the volume of air it delivers can be nicely adjusted to the needs of the equipment. The pump is not subjected to undue back pressure which can cause the diaphragm to malfunction or rupture. A non-return valve can be fitted to the main airline. In this way the pump can be placed below the aquarium without danger of water

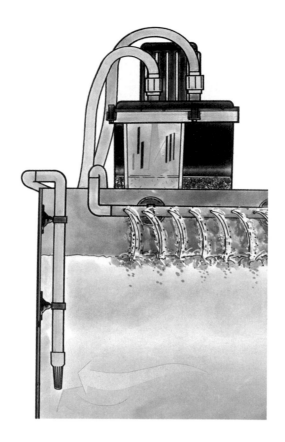

Outside canister filters also draw the water to be filtered from the bottom but return it from above the surface, providing surface agitation and aeration.

siphoning back if the pump is turned off or fails. Then should come a set of gang-valves to distribute air to various pieces of equipment. Do not forget a spare line for aerating new arrivals or a container of newly made-up water.

Be sure to use soft, pliable tubing that gives no resistance to being placed where you want it. Check that airstones and other equipment allow free passage of air so that the pump is not working against undue resistance and connections will not blow. If you can blow easily through the arrangement, so can the pump. If you cannot, something is wrong. The maximum desirable pressure is 4 lbs per square inch, about ¼ atmosphere. Pressure gauges are available, but not usually necessary.

Streamlined modern equipment for larger aquaria is available. It is always good aquarium practice to be systematic as well as neat and tidy.

Filtration

Nearly all marine aquaria have filters of one kind or another. Often there is more than one kind. Only in the *natural* system in its original form is a filter omitted. Many present-day aquarists use an underground filter for better results. There are so many types of filters that it seems best to classify them into broad groups.

To begin with, let us consider what filters are supposed to do. First, they remove particles of uneaten food, feces, and other gross particulate debris. Two, if they are fine enough in structure and power drive, they remove algae and bacteria. Three, they remove coloring matter and harmful chemicals, produced by the metabolism of the inhabitants. Four, they help to maintain a suitable pH. Five, they act as a substrate for helpful bacteria that convert waste products into harmless chemicals.

The first two functions are mechanical actions. Functions three and four are part mechanical and part chemical. The fifth function is biological. Initially filters were thought of as performing functions one and two. In large public aquaria, beds of sand or other fine materials were used. In home aquaria, filter floss or pads replaced the sand and are still used. The filter floss or pads are changed or cleaned frequently to remove the materials collected during the filtration process. Such pads often form the mechanical part of a complex filter. They are situated so they can be replaced easily. When function two predominates, the set-up is normally in the form of a power filter. The water is forced through a capsule with such fine pores that little but pure sea water gets through. Such a filter is driven by an electric water pump. Before entering the capsule, the water may be filtered for large particles by a simple mechanical filter.

Functions three and four are performed by chemical filters or the chemical part of a complex filter. The most common such filter is activated carbon, or charcoal. This is backed up

For those interested in hi-tech marine aquarium keeping, there are many gadgets that are designed for the hobby or that can be adapted for use in the hobby.

by crushed coral or other limey material designed to keep up the pH. The extent to which these pH substances really work is questionable.

Function five is that of a biological filter. This type has undergone great development. Modern marine aquaria depend on the functioning and upkeep of biological filters. They were invented in the U.S. and promoted for marine use by the late R. P. Straughan in the form of an undergravel filter. Straughan thought of the underground filter as a mechanical one and was not convinced of its potential biological actions.

Activated Carbon

Properly manufactured activated carbon is heated to 1650°F. This process causes the development of fine pores and channels within the charcoal grains. An enormous surface area is thus created for the uptake of chemical substances. Useful carbon comes in fine, dull grains about the size of a pinhead. The chunky, shiny filter 'coals' are of little use. There are no reliable tests by which the efficiency of activated carbon in the aquarium can be judged. You must depend on the supplier to guarantee the product as suitable for aquarium use. The rather expensive gas grade filter

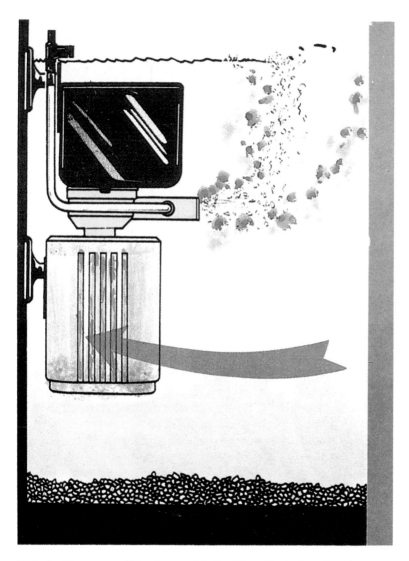

This inside power filter is provided with activated carbon in the upper section, but the biological filtration occurs in the lower section.

carbons found in chemical supply houses usually do a good job. Statements that activated carbon can be reactivated after being heated in an oven are incorrect. Neither can it be reactivated by steam. After three months of use, throw the carbon away.

High-grade carbon can absorb over half its own weight of waste materials dissolved in the water. Coloring matter, heavy metals, antibiotics, noxious gases, sugars, amino acids, and most medications are taken up quite rapidly and rendered harmless. Products of the nitrogen cycle (ammonia, nitrites, nitrates) are not absorbed in significant amounts, but vitamins and some trace elements are. The pH is likely to fall when using carbon, so activated carbon does have its pitfalls. This is particularly so in marine tanks where the pH must be kept around 8.2 and there are many natural elements for the carbon to take up. The carbon is so useful, though, that the drawbacks are ignored or precautions are taken against them. In particular, fishes are more dependent on vitamins, etc., in the food than in the water. More likely it is the invertebrates that will be affected.

Wash the carbon in some salt water before using it to remove debris and toxic materials. Gas grade carbon is safe with a brief rinse or two. Commercial grades must be washed more thoroughly by circulating a few gallons of sea water through them for 24 hours. There is no need for more than 4 oz. of carbon in a 20-30 gallon tank. If the tank is understocked, the carbon may not even have to be changed every three months. Test the water by adding just enough methylene blue to tint the water. It should disappear quickly. Eventually, the carbon gets coated with bacteria, absorbs less and less, and becomes an ineffective biological filter.

Even new, highly active carbon does not remove everything. For example, copper is reduced in a day or so to less than 0.1 ppm (parts per million), but only slowly beyond that level. This is adequate for the safety of the fishes, but not for all the invertebrates. Do not forget that gas grade carbon can take up materials from the air and become tainted with sprays and fumes. This is one reason for the washing.

Keep the carbon tightly sealed until use.

A simple test for whether a filter carbon is exhausted, or whether or not it is effective, is to place some of the filter carbon in a glass and add a drop of methylene blue or any marine medication containing a dye. The water should become clear of the dye in a short time, say an hour, but this of course depends upon the strength of the dye, the amount of filter carbon that you use, and many other factors.

Ion Exchange Resins and the Like.

These resins, in fresh water, exchange sodium for any heavier metal. Thus they can be used to soften the water by exchanging sodium for calcium, or to purify it of copper, zinc, iron, etc. This could be important because many flake foods are excessively laden with iron and zinc. In the hydrogen phase, they may even be used to exchange hydrogen for sodium, thereby acidifying the water. For example,

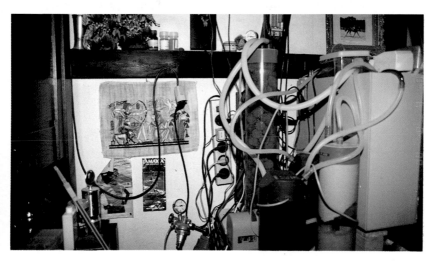

Viewed from the front the marine aquarium presents a beautiful scene. But few people realize how much equipment there may be working behind the scenes to keep it that way.

$NaCl$ becomes HCl (hydrochloric acid). What they do in sea water is another matter, and greatly debated. A sodium-charged resin will exchange its sodium for calcium or magnesium, and any of

Carbon dioxide diffusion is now being used by advanced hobbyists.

these for heavier metals like copper. So the resins can still be used, like carbon, to take up heavy metals. They will not do much more, though, unless different types of resin or resin-like materials are used.

Products are on the market that contain mixtures of resins and activated carbon. Resin-like materials are available in sheetform. It is claimed that these materials absorb various chemicals, including ammonia, nitrites, and nitrates above certain levels. Reports on such products in the aquarium literature are favorable. It is doubtful, though, that they make it possible to forego water changes as some claim. Seawater is such a complex mixture and undergoes such changes in an aquarium that it is just not credible that it can remain healthy indefinitely without a gentle turnover.

Water polishers cause fine suspended particles in the water to clump together. The particles can no longer pass through an ordinary filter, including an undergravel filter. Polishers are harmless in the recommended dosage. They do away with the need

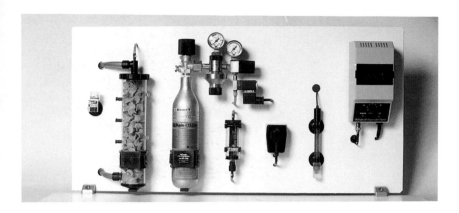

A setup for carbon dioxide diffusion. This is used to aid the growth of corals and other animals and plants that need greater amounts of this gas.

for a power filter. It might not be a good idea to use them continuously if the water keeps clouding up.

Simple Filters

The simplest filter is just a pad or wad of something that removes coarse particles. It is used as a pre-filter. The water passes through it before entering a more complicated filter set-up. This prevents the main filter from clogging up with the stuff the pre-filter removes. A simple filter used alone, or in conjunction only with an undergravel filter, is usually provided with a layer of activated carbon or other material to act as a chemical filter. This is sandwiched between two

pads of synthetic filter material so that the granules of carbon are retained.

Such a filter may be inside or outside the aquarium. If inside, it is common to place it in a back corner in the shape of a ¼ segment of a round fruitcake. It is then hidden by tank decorations. It will have a perforated lid through which the water passes through filter pads, etc., and then back to the aquarium via a perforated bottom plate. This leads to an airlift rising through the middle of the arrangement. The airlift may be a narrow tube with bubbles rising in it, or it may be provided with an airstone. Alternatively, the filter can sit at the top of the tank, have water passed into it by an airlift, and drain back into the tank through a perforated base. This

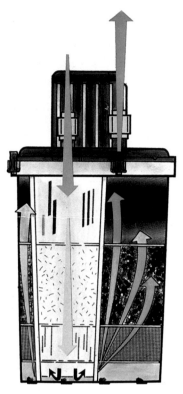

One of the more modern canister filters. Different filtering materials can be placed in the various compartments.

Outside filters are usually hung on the side of the aquarium. They are designed so that water is pumped out of them by an airlift or a power head (water pump). One or more siphons with guards to prevent the passage of small fishes, etc., feed the aquarium water into the filter. The water level in the filter is maintained below that of the aquarium by the outflow pump. The level rises to that of the aquarium if the pump fails or is turned off, but the water does not overflow the filter. There are many designs for such filters, some of which are quite complex. Power filters that stand by themselves, which may also have complicated interiors, simply suck water from the aquarium. They filter the water and return it to the tank via another tube. They are safe unless a leak occurs. This is because the water never leaves the enclosed circuit.

simple type has a disadvantage in that it is difficult to hide from view. Both types of inside filters are popular because they cannot cause water to flow out of the aquarium. In combination with an undergravel biological filter, good results can be achieved at low cost.

BIOLOGICAL FILTERS

Mr. Straughan's introduction of the undergravel filter to marine aquarium keeping was a great step forward. Even though he did not use it correctly, he had good results. He could keep fishes healthy longer than before. This was because he left it in the tank long enough for cultures of bacteria to grow on the gravel or coral sand. These bacteria became a factory for the conversion of ammonia to less harmful products. Ammonia is the most toxic and most common of the breakdown products of protein metabolism and is an unavoidable pollutant in any aquarium. In the freshwater tank, it was not recognized as a source of trouble for some time. This is because ammonia becomes more dangerous as the pH rises. Ammonia causes little trouble at pH 7 or below, which is that of a typical freshwater aquarium. In the marine aquarium, at pH 8 or above, it is ten times as toxic.

Dissolved ammonia takes the following forms in water:

The higher the pH, the more the equation above is pushed toward the right hand side. This results in the freeing of more ammonia gas. Dissolved ammonia gas is lethal in fractions of a ppm (part per million) to most fishes and some invertebrates. An increase in one unit of pH around neutral (pH 7) results in the production of about ten times the amount of free ammonia gas. There are small effects due to temperature and salinity, but, as seen in **Table 2**, these are unimportant in comparison with the effect of pH.

Ammonia is formed in the aquarium from any proteins: those in fish foods which must be rich in protein to be nutritious; those in the tissues of the fish even if it is not fed and becomes thin; and those of plants and other living creatures as they turn over the contents of their bodies, or decay. Eaten and uneaten food contributes equally to protein breakdown unless the consumer is actively growing. Then some of the protein will be temporarily

$$NH_4OH \longleftrightarrow NH_4^+ + OH^- \longleftrightarrow NH_3 + H_2O$$

| AMMONIUM HYDROXIDE | AMMONIUM & HYDROXYL IONS | AMMONIA GAS & WATER |

Table 2

The percentage of free ammonia gas at different temperatures and pH in sea water.

pH	Temp: 60°F	70°F	80°F
6.5	0.09	0.13	0.19
7.0	0.28	0.42	0.60
7.5	0.88	1.32	1.88
8.0	2.75	4.07	5.42
8.5	8.25	11.82	16.03

incorporated. Although some algae may utilize ammonia, its elimination is dependent on chemical removal, water changes, or bacterial action. The first two are inadequate unless there is a continuous flow of new water or impracticably frequent changes are made. Instead, the aquarist must depend on oxidation by bacteria.

The Nitrogen Cycle

The nitrogen cycle (proteins contain nitrogen) describes the natural course of nitrogen metabolism in nature. Nitrogen is converted into ammonia as proteins break down and are acted on by aerobic bacteria (those consuming oxygen) of the genus *Nitrosomonas*. The nitrogen is converted into nitrites, salts of nitrous acid (HNO_2). The nitrites are converted by other bacteria, genus *Nitrobacter*, to nitrates, salts of nitric acid (HNO_3). Nitrites are less toxic to most organisms than ammonia. They are still poisonous if present in any quantity, whereas nitrates are almost harmless unless in concentrations of hundreds of parts per million. Nitrates are easily utilized by algae that build them up again to proteins. Flourishing algae in the aquarium can keep them at a low level. Water changes also help. Other bacteria can convert nitrates to free nitrogen gas. This does not usually happen in the aquarium unless there are

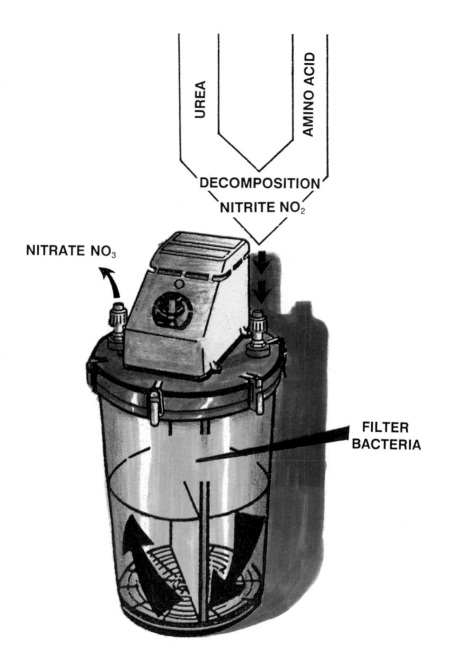

UREA

AMINO ACID

DECOMPOSITION

NITRITE NO$_2$

NITRATE NO$_3$

FILTER BACTERIA

Biological filters convert the more dangerous nitrogenous compounds into the less dangerous nitrates. Nitrates also can be utilized by the algae.

A "Dutch" miniature reef aquarium setup showing the four drip trays and other filters.

dead spots where, being anaerobic (not needing oxygen), the bacteria can flourish.

A lethal concentration of ammonia has been found to be different for different fish species, varying from 0.07 to 1.4 ppm. The safe limit is usually quoted as 0.05-0.1 ppm, measured as NH_3-nitrogen. The atomic weight of nitrogen is 14, that of hydrogen is approximately 1. Hence, 0.1 ppm of NH_3-nitrogen is equivalent to 0.12 ppm of ammonia itself, not much of a difference. Oddly, the reason for the high toxicity of ammonia is not fully understood. It is thought that since fishes excrete it through their gills, even a low concentration in the water causes a backing-up and poisons the fishes with their own excretory products. Certainly, gill disease and actual death of the gill tissues are effects of ammonia poisoning. This leads to respiratory distress, liver degeneration, and susceptibility to various diseases. The majority of marine invertebrates appear to be much less affected by ammonia than are fishes.

Nitrites combine with hemoglobin, the red pigment in the blood that carries oxygen and converts it to methemoglobin. Methemoglobin does not carry oxygen and so an affected fish dies from lack of oxygen. Nevertheless, it is a lot less toxic than is ammonia, particularly to marine fishes. Some authorities doubt whether it is really even a threat in the marine aquarium because the salinity considerably lessens its activity. Thus, a juvenile chinook salmon is killed by 19 ppm in fresh water, but it takes 1070 ppm to kill it in salt water. Chinooks, however, are tough, whereas rainbow trout are killed by 0.2-0.6 ppm. To be safe, various levels from 0.1-0.25 ppm are recommended, measured as NO_2-nitrogen. The molecular weight of NO_2 is 46, that of N is still 14. So 0.1 ppm of NO_2-nitrogen is equivalent to 0.33 ppm of NO_2. This is a big difference. With both ammonia and nitrites, it is really best to regard any measurable amount as dangerous. Take appropriate action.

Nitrates have very low toxicity. Public aquaria have reported up to 300 ppm in old recirculated water without obvious ill effects. However, some fishes and invertebrates, such as corals, are known to be susceptible to high nitrate levels. In the home aquarium, we have come to regard between 20 and 40 ppm as the limit to be allowed. Less is allowed for corals. Apart from hard corals and a few other invertebrates, there is no firm knowledge of the toxicity of either nitrites or nitrates. The effects of nitrites on hemoglobin may not be paralleled in the non-iron blood pigments of most invertebrates that have a blood circulation.

Undergravel Filters

These were first used in fresh water. They were introduced to the hobby in the early 1950's by Norman Hovlid of California. It was found that pulling the water down through a few inches of sand or gravel resulted in sparkling clear water and a very healthy aquarium. Since some plants did not like having their roots aerated, the filter usually occupied only the center front of the tank. The filter was in the form of a V with the point at the back. The airlift was also placed here. This was adequate in freshwater

A three-chambered filter. The arrows indicate the water movement through the filter from the intake (left) to the exit (right). The compartments can contain various filter media as needed.

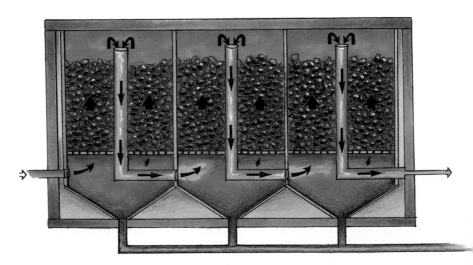

Another design that provides aeration for the incoming water and only two filter media compartments. The stopcocks on the bottom of each compartment are for removing material that settled out by gravity.

tanks. In a marine tank, however, it is necessary to cover the whole base. Undergravel filters were standard in the U.S. and Britain until recently. The introduction of reef aquaria began to displace them in marine aquaria and even in some freshwater setups. They were never popular in continental Europe, where outside filters of various designs were extensively used many years ago. Since undergravel filters are so inexpensive and work so well, they no doubt will continue in use. It is only when the aquarist wishes to keep very delicate organisms and to crowd them to produce the typical reef effect that they become inadequate. However, even in a fish-only tank the installation of a wet/dry filter (see below) can brighten up the fishes perceptibly. I shall describe a typical undergravel filter of the most common and efficient design.

The filter consists of a perforated plate raised by downward projections about ½ inch above the base of the aquarium. You can buy almost any size. Commercial ones are so inexpensive and efficient that it hardly pays to make

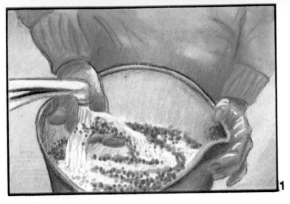

Steps in setting up a simple marine aquarium. 1. Wash the gravel or sand thoroughly before placing it in the tank. A plastic bucket is ideal for this job. 2. Place the undergravel filter in the tank first and cover it with a layer of plastic mesh to keep the sand from clogging up the filter. 3. The bottom material can then be added. Grade this material from front to back and from middle to the sides. 4. The decorations can be placed in the tank after being rinsed thoroughly. Some hobbyists may prefer to do this after the water is added.

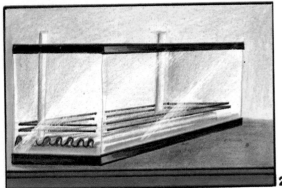

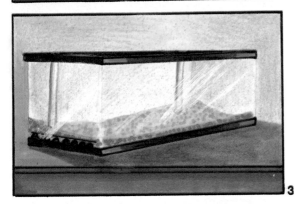

5. The proper amount of sea salt can be added. This must be determined according to the volume of the tank. Be sure to follow the specific instructions on the package. 6. Clean fresh water can be added at this point. It is best to place a dish or plastic bowl in the tank so that the sand and decorations are not disturbed by the incoming water. Some hobbyists prefer to premix the water for their tank in glass carboys, adjusting the parameters before adding it to the tank. 7. Filters, heaters, etc., can now be added and adjusted to the proper levels.

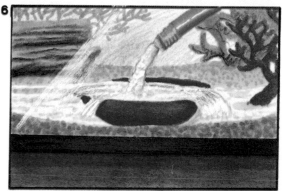

your own. Care should be taken to select the best since some are not designed as well as others. The deficiencies usually lie in the structure of the airlifts that operate the equipment, pulling water down through the gravel and returning it to the top of the tank. Cheaper undergravel filters are also made of soft, flexible plastic that are vacuum formed instead of being injection molded. A good one has a wide tube with an airstone at the base. This is served preferably by an airline that runs down inside the tube. This gives a good uplift and allows easy access to the airstone. The top of the airlift should have an elbow bend to direct the flow across the water's surface to give maximum water movement and gas exchange. Avoid types with an airline leading to the base of the airlift from the outside with no airstone. These are less efficient and clog up easily. They also are difficult to service.

Instead of an airlift, a power head may be used to advantage. However, there is no aeration unless it directs the water across the surface so as to cause maximum turbulence. The water from a biological filter of this type has been depleted of oxygen by the bacteria. It must be given every chance of taking it up again. Luckily, oxygen moves quickly from air to water and so a surface current is enough. A desirable flow rate varies with other factors, but around three tank volumes per hour is normal. Too slow a rate fails to oxygenate the water adequately; too fast a rate gives the bacteria insufficient time to do their stuff and can exhaust the fishes as well. Power heads are rated at so many gallons per hour. They may also be adjustable. You have to make your own estimate with airlifts. To do this, run the upcoming water into a vessel of known volume and time it. Soon you will find the best bubble rate for your purpose.

The 'gravel' of an undergravel marine filter is usually calcareous. Coral sand, shell grit, and dolomite are popular. Dolomite is suspected of being toxic, though. Otherwise it is an attractive substrate because it contains magnesium as well as calcium. Actually, it is not clear to what extent calcareous substrates act to maintain pH or the calcium content of the water.

Inert materials are used in many modern biological filters. Whatever is used, grain size is most important. If the filter slots are too wide, small grains will pass through them and come to rest under the filter and cause trouble. This is easily guarded against by placing a layer or two of plastic fly screen over the whole surface before adding the substrate. We want as small a grain size as possible so as to give a maximum area for the bacteria to settle upon, but not so minute as to clog up and allow too little water to pass. About $\frac{1}{10}$ to $\frac{1}{5}$ of an inch is suitable. Whatever the nature of the substrate, the bacteria will cover it so that little exchange can occur between it and the water once it is fully functional.

In a tank with a power head instead of airlifts, a problem may arise. If the aquarium is fairly well covered, it is possible to get a layer of carbon dioxide sitting over the surface of the water. This is because carbon dioxide is heaver than air. Airlifts will blow the carbon dioxide away, but even air-cooled power heads are unlikely to shift it much. The carbon dioxide is not in itself the real danger in a marine tank. The absence of oxygen is the problem. An airstone may be added for safety, or a small fan can be used to blow over the surface of the water. The fan is preferable so as to avoid the splash of an airstone in salt water. Some power heads have an aeration arrangement built into them. This could be used to avoid the problem, as long as the air is sucked in from outside the aquarium.

An undergravel filter can be left for a year or two undisturbed, but there comes a time when it gets clogged. Gurgling airlifts, swirling mulm at the bottom of the tank, and an obvious slowing of the circulation rate give notice that something needs attention. These symptoms can be put off by disturbing the surface of the substrate sufficiently to loosen it. Debris will be sucked up when making water changes. Eventually more is needed. It is fatal to remove and wash too much substrate at once. The biological action is then impaired. No more than $\frac{1}{4}$ to $\frac{1}{3}$ of it should be renewed at a time, with a week or two between times so as to allow the bacteria time to fill the gap.

Some aquarists have experimented with different layers in the filter. Perhaps a top layer of coral sand is divided from a layer of plastic shapes or balls below by fly screen or such. This makes for complications when renewal is needed. It has the advantage of preventing channels from developing that allow an easy flow of water bypassing the substrate. It also prevents burrowing creatures, like worms and wrasses, from helping to create them. Others have reversed the flow of the water so that it passes upward through the filter. They forget that it is delivering deoxygenated water to the bottom of the aquarium where it cannot take oxygen up again until it is stirred by some means to reach the top. Unless a pre-filter of some kind removes debris from the water as it is pumped under the filter, there will also be problems of inaccessible clogging below the filter bed.

Divided Tank Filtration

If you want a 'window on the sea' effect in your aquarium, it is necessary to

Most aquarists have a desire to create a tank that imitates or reproduces the living reef as closely as possible. New products and techniques are providing the means with which to do this.

get rid of airlifts, other tubing, internal box filters, heaters, etc. This can easily be done by giving the tank a false back. An opaque back hides all the equipment that is placed behind it. A space of about 4 inches wide is all that is needed. The opaque partition is best if it is dark blue or black. It can be of thinner glass than the rest of the aquarium because it will not have to bear much water pressure. The tank is best made 4 inches wider than usual so as to give a normal front space. Any aquarium larger than 24 x 16 x 16 inches can accommodate the necessary equipment and function well.

Instead of airlifts or risers for water pumps, the undergravel filter delivers its flow via a slot at the base of the back partition. About ½ inch is enough. This has the advantage that a very free flow of water is available and the filter bed can be left undisturbed for a long time. I left one aquarium for eight years before touching it! In this simple form, heaters, carbon filters, and an aeration pump can all sit in the back compartment. All that is seen from the front is a small return pipe sending the filtered water across the top of the tank. The false back must come right up under the cover glasses so that nothing can get over the top into the back compartment. Any slots, such as for the return pipe, must fit snugly around their contents. The advantage of this type of aquarium is that there are no external pieces of equipment. Even the return pump can be submerged. There is nothing to leak and nothing to see.

You need not keep things quite so simple. For instance, the rear compartment can be divided into further segments. A very workable design for a 36 x 18 x 18 inch aquarium would be a 24-inch section receiving the water from the undergravel filter plus two 6-inch sections. The one next to the 24-inch section is divided from it by a partition that stops an inch or two from the top. The water flows into it from the larger section. Then it passes down through a carbon filter (or any other you prefer) and into the final 6-inch section via a slot at the base of the partition between them.

This partition reaches right to the top, but not to

Sometimes it is necessary to place a glass partition in the tank to prevent harmful aggression.

the bottom. The final 6-inch section houses the return pump (or airlift in a small tank). The filter material should sit on a segment of undergravel filter base so as to enable easy flow of the water. Heating and any other equipment is placed in the 24-inch section where there is plenty of room for it. An airstone is a good idea to keep the atmosphere above the back compartments free of excess carbon dioxide. Since all the water has to pass up over the partition between the 24-inch and the 6-inch compartment next to it and then over the surface of the aquarium, a good gas exchange is assured. The oxygen-depleted water from the undergravel filter gets plenty of reoxygenation.

A really elaborate arrangement has worked very well for me. This was a 72-inch aquarium. It had the same general design duplicated on each side of a central 40-inch compartment receiving the water from the undergravel filter. Water was returned via a tube at each back

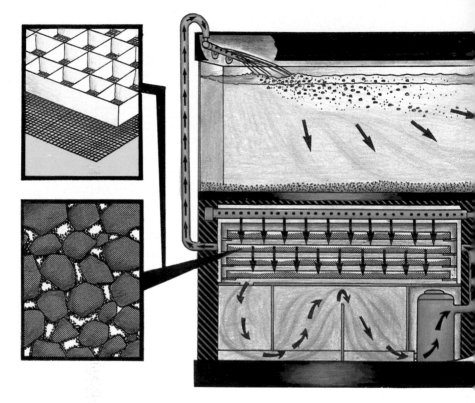

corner. The two return
submersible pumps could
be on together or
separately so as to give a
back-and-forth flow in the
tank. An electrician friend
devised a control that could
switch from one pump to
the other at any interval
between 10 seconds and 5
minutes. This was the tank
that went undisturbed for 8
years. It had flourishing
hard corals in it at the end
when it was dismantled to
be replaced by a reef
aquarium. Another 36-inch
aquarium of the 3-
compartment design is still

going strong after 10 years.
It was taken down about
seven years ago in order to
move it.

This type of biological
filtration probably sees the
best that an undergravel
filter can offer. It is
relatively cheap to set up,
has no external tubing as a
potential for leaks or other
troubles, and comes close
to the efficiency of a reef
aquarium. However, it
cannot rival it because a
fully built reef would cover
too much of the filter. That
is its biggest drawback.
Otherwise, it has beautiful

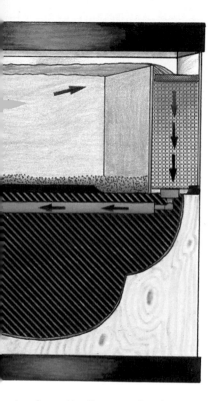

A schematic diagram showing the workings of a trickle filter.

'Dry' Filtration

You can move the biological filter away from the aquarium. Here it is more flexible and easier to service. If it is merely shifted to a separate tank, it can be made larger, deeper, or combined with carbon or other types of filtration. It will not become more efficient per volume or area of surface, and it will still deliver deoxygenated water. It can be made lighter by using matting or plastic shapes instead of sand or gravel. This may or may not be more efficient than an aquarium substrate depending on surface area.

What the originators of the 'dry' filter did was to bring in a new concept of aerating the filter instead of leaving it totally under water. Originated in Europe, this technique has revolutionized biological filtration and enabled the development of the reef aquarium. Basically, although the filter bed must be kept moist, that is all that is needed. The so-called 'dry' filter is not completely dry. It has plenty of air available in addition to the aquarium water that is trickled through it. A better name is **trickle filter**. Its great

built-in checks as the various water levels in the different compartments tell you what is happening. If the level in the pump compartment falls, the carbon filter needs attention. If the level in the main back compartment starts to fall, the biological filter needs attention, and so on.

advantage is that the biological processes that need oxygen do not have to extract it from the water. They draw it directly from the surrounding air, leaving the water well oxygenated. Allegedly the efficiency is increased nearly threefold.

The original Dutch miniature reef has a filter that is quite elaborate. Water from the tank is sprayed over a stack of trays. Each tray has a shallow layer of coral gravel and a perforated base. The water drips from one to the next, getting plenty of oxygen while the bacteria are at work. It is then led into compartments located under the stack of trays. The first two trays contain more coral gravel, which is under water like a conventional gravel filter. The water flows through a carbon filter, and then into a pump compartment. From here it is returned to the aquarium. Hence, the 'wet/dry', or really 'dry/wet' filter. As originally conceived, it has what seems to be unnecessary features. It also tends to be noisy. Better versions, both simpler and quieter, are now available.

Since the trickle filter does practically all of the work, it is not necessary to have the 'wet' part at all, except for perhaps the carbon filter and the pump. There must be a depth of water for the pump to function, but it does not have to contain more gravel. Also, there are plastic substitutes for the gravel in the 'dry' filter that may be more efficient and have the advantage of lightness. As indicated earlier, they come in various designs, all aimed at increasing surface area and efficiency. In fact, most of them would not appear to have a greater surface area, when used in the amounts recommended, than an undergravel filter. Their advantages lie elsewhere: lightness, efficiency, and length of service undisturbed. It seems likely that the efficiency of most of the designs now available is so good that they offer excess capacity in most situations. That means that they could hold a greater bacterial population than is demanded of them by the contents of the aquarium. This is an excellent state of affairs.

An attractive and very efficient design is offered by the rotating sprinkler arm. It's jets are so arranged that the water pressure causes the sprinkler to revolve over the top of the 'dry' filter. The sprinkler can be composed of any of the above materials, but it is frequently in the form of a double-layered spiral (DLS). This is a plastic mat with a layer of dense material and another of more open structure. It is springy, so that when it is coiled into a spiral about 12 inches deep, like a rolled-up carpet stood on end, it resists collapsing. It remains an open, well aerated filter bed. This and any other dry filter can have an airstone beneath it to increase aeration. In practice, this hardly seems necessary.

In order to preserve the efficiency of any external filter, such as those described above, and to prevent clogging of the jets of the sprinkler system, the water entering it must be pre-filtered. There are various ways to do this, depending on the overall design of the set-up. In a miniature reef, the pre-filter is built into a corner of the aquarium. It leads to a hose at its base that runs via a hole to the filter in the cabinet below. In adaptations designed to convert an existing aquarium to a reef type, siphons may run into an external pre-filter.

The preferred flow rate through a dry filter is faster than through an undergravel one because it can cope with it. Five to six tank volumes per hour is acceptable. The longer period for which the water is in contact with the filter bed allows it. Either there is a stack of filter beds, as in the Dutch miniature reef, or a deep filter bed, as in other designs. However, practical observation shows that a slower flow is quite effective, down to even 1-2 tank volumes per hour in large tanks. I suspect that this is because the water/biomass (weight of living creatures) ratio is usually greater in large aquaria.

Denitrifying Filters

We have been discussing aerobic filters so far that work in the presence of oxygen whether they need it (biological filters) or not (mechanical and chemical filters). To create an anaerobic filter, we must exclude oxygen. The processes that go on are the reverse of those in an aerobic filter. Thus, some bacteria then reduce

Denitra filters are becoming more and more popular. They can be obtained in kit form from your local pet shop. The action is similar to that of the biological filter, the difference being in the slow rate of flow and the periodic addition of a source of organic carbon to the denitra filter.

nitrates to nitrites and ammonia. Others will reduce them to nitrous oxide and free nitrogen gas. In sea water, *Pseudomonas* and *Vibrio* species do this. Even the same bacteria that oxidize ammonia to nitrites and nitrates can act as denitrifiers if placed in an anaerobic environment. To do the job, denitrifiers usually need a source of organic carbon. This is supplied in aquarium filters in the form of lactose, citrates, or malates. The organic carbon must be metered into the system so that just the right amount is supplied and little flows back into the aquarium.

A *denitra* filter, a term used for shortness, must act in a water flow of its own. Often it is bled off from the main filter and returned to it as it delivers water back to the aquarium, or before the water passes through the normal biological filter. Since the flow through the denitra filter must be quite slow, the deoxygenated water it produces adds only a small quota to the total flow and gets reoxygenated on the way. The structure of the denitra filter is just the same as a normal biological filter. The slow rate of flow and the drip-wise or periodic addition of the source of organic carbon make the difference between the two.

Denitrifying processes often occur in the aquarium, such as in static parts of living rock or bypassed areas of an undergravel filter. They are fortuitous and uncontrolled. They may produce noxious gases like hydrogen sulfide, as well as contribute significantly to the reduction of nitrates. The nitrate filter box, a recent introduction, imitates these processes. It contains a suitable substrate buried in the gravel or just placed on it. The contents become anaerobic and nitrogen bubbles up from them. Growing algae are excellent denitrifiers. They incorporate the nitrates into their own tissues. A type of denitra filter that takes advantage of this is an algal filter where water from the aquarium is passed over a bed of growing algae, such as *Ulva*, the sea lettuce. The algae is cropped regularly to remove the nitrates. Such a filter needs bright illumination and can be difficult to maintain. It seems to be favored mainly by large commercial aquaria.

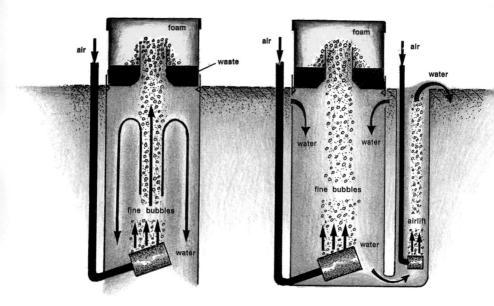

Two types of protein skimmers (also called air-strippers or foam fractionators). Protein skimmers use a rising stream of bubbles to carry contaminants into a collecting chamber via a foam.

AUXILIARY EQUIPMENT

There are various items of equipment favored by different aquarists. This equipment is intended to increase the quality of the water by removing pollutants, and killing bacteria and other organisms. How useful they are in any particular set-up depends upon the efficiency of the existing filters. A biological plus a carbon filter may be doing all that is needed. The addition of further equipment in that particular case will not be seen to convey any obvious benefit. When, as in a reef aquarium, we need absolutely every refinement that will add to water quality, additional equipment may make a big difference. It will rarely be the case that everything I am about to describe will be needed together. The addition of one or two extra pieces of gadgetry is likely to be critical.

Protein Skimmers

A **protein skimmer**, or foam fractionator, removes many substances from the water. It collects the foam that forms when a column of bubbles rise through a polluted sample. It can be of very simple design. It may be just a wide glass tube with an airstone at its base, opening above water level into a collecting cup where the foam collapses and the residue can be removed periodically. If the water is pure, very little foam collects. What there is forms a clear liquid. If it is charged with pollutants, there is a lot of sticky foam that collapses into a brown liquid. It is formed by proteins and their break-down products, phenols and other surfactants. Such substances are attracted to interfaces, in this case the air/liquid interface of bubbles with aquarium water. This gets rid of substances that will eventually break down further and form ammonia. Thus it relieves the biological filter of some of its work. At the pH of sea water, protein skimmers do not remove ammonia, nitrites, or nitrates.

As used in an aquarium or as part of an outside filter, a skimmer is only a foot or so in height. It does not do a very good job, but can be improved by using the countercurrent principle. The water and the air flow in opposite directions, lengthening contact time and bringing fresh, new water in at the top of the column. Here it meets bubbles already loaded with organic substances. This is a more efficient extraction process. Even this can be improved by having a much longer column in a skimmer outside the aquarium, serviced by a power head. The head turns over more air and water as well. However, a countercurrent skimmer in the cabinet below an aquarium is nicely hidden away. It is also adequate in most circumstances. Its efficiency can be increased by using ozone—charged air (discussed below). Many aquarists find this unnecessary.

Ozone

Ozone is tri-atomic oxygen. It is O_3 instead of the usual O_2. Oxygen in this form is highly reactive. It is ready to rid itself of the third atom and revert to the normal form. It will oxidize many substances, kill small organisms, like bacteria and protozoa, and even larger ones if provided in sufficient quantities. It is a danger in the aquarium and even to ourselves if not properly handled. Oddly, it does not do much to ammonia or nitrites, although it might be expected to. In the past, it has been metered into the aquarium directly. It is recommended that administration should only be via a protein skimmer so that any excess is not passed into the aquarium. For complete safety, the water from the skimmer can be passed through a carbon filter before being returned to the aquarium.

Ozonizers are available that either deliver a fixed dose or that have a variable control. They are high tension electric discharge tubes. They should be supplied with dry air or else they quickly deteriorate. This is managed by passing the air over absorbent crystals of anhydrous calcium chloride before it enters the ozonizer. If you can smell ozone, an 'electrical' smell, there is too much of it. Reduce the dose. If it enters the aquarium in more than about 0.2 mg per gallon per hour, it can cause distress to the fishes. It can damage their skin by stripping off the protective mucus and then attacking the skin itself. It can also damage invertebrates and the bacteria of the biological filter.

Ultraviolet Light

Light of a wave-length between 2000 and 3000 angstroms is lethal to bacteria and other small organisms. It is also harmful to algae, fishes, and invertebrates. Therefore, it cannot be directed into the aquarium in sufficient intensity to be bactericidal. As with ozone, treatment must be external to the tank. It is most effective when the water is passed over UV tubes lengthwise so that a thin layer receives adequate treatment. The tubes should be shielded from the observer so that he does not receive a harmful dose. A flow of 10 gallons per minute over 66 inches of UV tubes, with an ⅛ inch layer of water around the tubes, is said to sterilize completely. Less efficient sterilizers are available for aquarium use, though they may do a good enough job.

The drawback to both ozone and UV sterilization is that they can only reduce the count of unwanted organisms since not all of them are obliging enough to leave the aquarium. Those that remain on fishes and surfaces within the tank are not affected. Hence, no pretense of complete eradication is possible. Still, the swarming stages of diseases like *Oodinium* and *Cryptocaryon* may be significantly reduced.

Ultraviolet light has other interesting effects. At shorter wavelengths than are used for killing bacteria, it generates ozone. This is not desirable if the water passes back from the UV equipment straight into the tank. It also causes various organisms to fluoresce, including some corals, which may show different beautiful colors when irradiated with it. The wavelengths that do this are clearly able to penetrate several feet of water. This can be seen in public aquaria that demonstrate the effect. This gives lie to the common belief that all UV is absorbed by a thin layer of water. UV rays are dangerous to your eyes. Never look into the rays without wearing protective UV-shielding glasses.

Combined Systems

There are quite a number of systems combining biological filters with combinations of some of the other types of equipment just described.

You can build your own arrangement or you can buy a complete outfit. There are some very good systems that sit over the aquarium. You can also get a miniature reef system that either sits below in a cabinet or beside the aquarium. My own choice is the reef aquarium with cabinet. This hides all the works and makes a really attractive piece of furniture. When a reef aquarium is arranged by adapting an existing tank without boring a hole in the base, the equipment necessarily becomes more obtrusive. It can still be placed behind the tank in many circumstances, though.

As an example of a popular type of combination, take a typical reef tank. After passing through a pre-filter within the tank, the water flows via the bottom hose into a dry/wet sprinkler arrangement. This is the main biological filter. Part of the flow may be diverted via a slow denitra filter. It can rejoin the main current either before or after the dry part of the main filter. The water then collects in a fairly deep compartment. Here it is heated and further treated by a protein skimmer. Somewhere along the line there will also be a carbon filter. If the protein skimmer is supplied with ozone, the carbon filter should follow it. Alternatively, it could precede it, or even be situated in a small bubble-up filter of its own, sitting in the same compartment that also houses the pump. Therefore, it must be quite large—around 12 inches square and 12 inches deep is common. The pump must be robust and capable of raising the water several feet up into the aquarium and maintaining a suitable turnover rate.

With such a system, the wise aquarist insures his considerable investment by using two heaters and having spares of all parts likely to give trouble if a failure occurs. Both heaters would have to fail to do serious damage. A spare pump is mandatory unless you are quite sure of immediate assistance in the event of a breakdown. Provide similarly with the air pump that may run the protein skimmer, although this is a less serious matter.

MEASURING WATER QUALITY

Various instruments and kits are necessary for keeping checks on the state of the aquarium. This is done more so in the case of a marine aquarium than in a freshwater one. The number considered advisable increases as we learn more about what happens in the aquarium. Refinements in technology make it possible to measure various things without investing in a scientific laboratory.

Specific Gravity

The salts dissolved in sea water make it heavier per unit volume than fresh water. Water is peculiar in that, as with only a few other substances, it is lighter in the frozen state than when liquid. Hence, ice floats. So does iron when it melts. We do not often see that. Water is densest at 4°C (39°F). Strictly speaking, specific gravity is the ratio of equal volumes of any fluid and pure water at 4°C. This direct comparison is rarely made. Instead, we have instruments that give the same answer, or very nearly so.

The commonly used instrument is a **hydrometer**.

It consists of a hollow glass bulb with a slender stem. It is carefully weighted so that markings on the stem give the specific gravity when it floats in the fluid to be measured. In our case, this is marine aquarium water. For oceanographers it has become the practice to measure specific gravity at 15°C (59°F, usually approximated to 60°F) because that is a typical temperate sea temperature. Most hydrometers do that. Some, though, specially made for marine aquarists, measure it at 24°C (75°F). This causes some confusion. I shall talk in terms of 15°C. This is when the specific gravity of sea water is 1.025. At 24°C, the same specimen of sea water would have a specific gravity of 1.023 or more accurately, 1.02275.

The general rule, when using a normal hydrometer at a higher temperature than the 15°C it is set for, is to add 0.00025 for every 1°C (1.8°F) the actual temperature is above 15°C or 60°F. So if your aquarium water is at 78°F and the hydrometer reads 1.0225, just about the limit of its accuracy, you must add 10 x 0.00025 = 0.0025 to the estimate to get 1.025. Then the specific gravity is

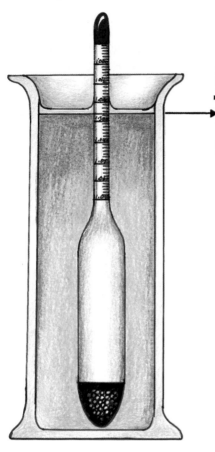

A simple but very useful aquarium accessory is the hydrometer. It is used to measure the specific gravity of the water.

just right for normal ocean water. When using the instrument, read the scale on the stem just under the water's surface, not above it, to get a true reading. Some aquarists take the precaution to siphon some of the water into a jar and take the reading and temperature there. Leave the hydrometer floating in the water for long enough to settle down and also to equilibrate in temperature with the water.

There are different instruments for measuring specific gravity than hydrometers. A recently available one has the water poured into it, or it can be situated in the aquarium and give a pointer reading. There are also other measures in use that you may come across now and again. The density of sea water is its weight in grams per milliliter. It is almost the same as specific gravity. However, other terms, such as salinity, chlorinity, and chlorosity measure quite different aspects of the dissolved salts, such as the total solids per kilogram or the halogen (chlorine, bromine, etc.) content of the water. These are not in normal aquarium usage.

pH

The pH scale measures the acidity or alkalinity of the water. The scale runs from 0 to 14, with 7 being neutral. For sea water, a range of 7.6 to 8.6 is adequate. The aquarium range measures very mildly alkaline waters. However, each unit of pH denotes a 10-fold decrease in the number of hydrogen (acid) ions present. This represents a profound biological change even if not a very profound chemical one. We have seen what a difference one unit makes to the effects of ammonia, for example.

Sea water is buffered, mainly by bicarbonates and carbonates. They interact with carbon dioxide and prevent a fall in pH that it would cause if they did not. Only an excess of carbon dioxide can overcome the buffering effect. It can be corrected for by adding more of the salts mentioned, usually sodium bicarbonate. This is the best salt to use because after adding it the pH slowly rises over the next two days. The pH then settles down for a period. Very approximately, it takes 0.08 grams of sodium bicarbonate to raise the pH of one gallon of sea water 0.1 pH unit. If your tank is at pH 7.9 and you wish to raise it to 8.3, use 0.32 grams per gallon. Avoid sodium carbonate or other harsher substances, as they cause a rapid rise in pH followed by a fall. This tends to stress the inhabitants. Further, always dissolve the bicarbonate in some tank water first. Add it gently to the aquarium. Better still, add it to the water in the pump compartment of an outside filter.

Scientifically, pH is measured electrically by a pH meter. This consists of a pair of electrodes that are immersed in the water. A cable is connected to the meter that reads off the pH on a direct scale. Most aquarists use chemical methods. They use substances that change color with changes in pH. For general purposes, wide range kits are available. They measure perhaps from pH 3 to pH 10 or 11. Kits for aquarium use are confined to a much narrower band, say pH 5-9 for fresh water and pH 7.5-

8.6 for sea water. Phenol red, for instance, is yellow at pH 7.4, pink at 7.7, light red at 8.0, and dark red at 8.3. The best kits are quite accurate. They usually consist of two tubes of aquarium water. One is backed by colored slides and the other receives drops of the indicator chemical. This is to eliminate any color due to the water itself. Less accurate kits compare a tube of water plus indicator with a color chart. Less accurate ones still have slips of paper previously soaked in indicator solution to be dipped into the water. These can be quite misleading and should be avoided.

Carbonate Hardness

Sea water is extremely hard. There is no point in measuring total hardness as with fresh water. Carbonate hardness, often referred to as alkalinity, measures the amount of acid needed to neutralize the carbonates, bicarbonates, and any other alkaline salts present in the water. Different systems express the result in different ways, the two most common being degrees of carbonate hardness (DKH) and milliequivalents per liter

(meq/l). Fifty parts per million of calcium carbonate is 2.8 DKH or 1 meq/l. Sea water has 2.0-2.5 meq/l or 5.6-7.0 DKH.

Carbonate hardness is measured by a simple kit. An acid solution containing an indicator chemical is added drop by drop to a few ml of the water to be tested until a color change occurs. Typically, each drop indicates 1.0 DKH. Aquarium water tends to drop in both pH and carbonate hardness, which are related to one another. The routine addition of suitable amounts of sodium bicarbonate takes care of both. A high alkalinity usually is accompanied by a high pH. Particularly in reef aquaria, care must be taken to keep both within narrow limits. The pH should be between 8.1 and 8.3. The carbonate hardness should be between 7 and 11 DKH, a little above normal limits. Higher levels of carbonate hardness are thought to be detrimental to corals and some other invertebrates. However, high levels of carbon dioxide production or its purposeful addition to improve plant growth may demand a higher alkalinity than usual to mop up the excess. Fishes are not much affected by wider changes in pH or alkalinity. It is the invertebrates we must consider.

Nitrogen Cycle Kits

Simple kits are available for measuring ammonia, nitrites, and nitrates. It is important to purchase kits that clearly state exactly what they are measuring. Textbooks always discuss these compounds in terms of the nitrogen content of each of them. This becomes progressively different from the amount of the compound itself as we go through the series. Thus, ammonia is NH_3, nitrite is NO_2, and nitrate, NO_3. The amount of nitrogen is 82%, 30%, and 24% of the chemicals, respectively. When discussing tolerances, I shall always talk about the nitrogen content of each substance, not the total amount. The nitrogen content is indicated by NH_3-N, NO_2-N, and NO_3-N.

Ammonia is difficult to measure accurately. Fortunately, this hardly matters in practice, as the detection of any of it is a warning sign. There are two occasions when ammonia levels are important: when first setting up a tank and when it is suspected that all is not well at any later stage. The most reliable kits appear to be those using capsules of the dry chemicals involved. The dry chemicals keep better than liquid solutions. This is true of nitrites and nitrates as well.

Nitrites can be measured reasonably accurately by the kits available. As with ammonia, when all is well the reading should be virtually zero. It is best to measure the nitrite level routinely. This gives assurance that nothing is going wrong with the biological filter. The ammonia part of the cycle can be ignored unless there are symptoms of its presence, such as fishes breathing heavily, inflamed gills, etc.

Nitrates are easy to measure. The amounts in question are much higher than with the other components of the cycle. They will accumulate in the aquarium unless they are being removed by denitra equipment or algae. Water changes are rarely sufficient to keep them very low. Most kits are calibrated in steps of 5 or 10 ppm, up to around 100 ppm. Special ones are obtainable that read down to 1 ppm. There is really no point in aiming for such accuracy. Check your tap water to see that it is not an unwanted source of nitrates. It can easily start you off with 50 ppm in some areas.

Oxygen

Until recently, the measurement of oxygen in the water was a tedious, multistage operation or required expensive equipment. There are now one-stage oxygen kits that make it easy. Marine aquaria need a high level of oxygen saturation (80-100%). Without it, both the inhabitants and the biological filter are liable to suffer. It is lucky that oxygen is rapidly taken up by the water. A reasonable exposure of the circulating water to the air as it returns to the tank or is passing over a dry filter ensures a high level of oxygen. Reef aquaria are so designed as to give really maximal oxygen levels that keep their dense populations happy.

Redox Potential

The measurement of redox potential requires careful calibration of the instruments used. Carefully cleaned electrodes are immersed in the tank water after it has been filtered. Preferably they are left in position semi-permanently.

Water conditioners and other chemicals upset the measurement. However, it is becoming popular with really serious aquarists as it gives an indication of the purity of the water. The name stands for oxidation-reduction potential. It refers in effect to the capacity of the water to conduct electricity. The maximum expected is 450 millivolts, 350 millivolts being regarded as the minimum reading desirable.

Phosphates

Kits measuring the phosphate content of aquarium water are available. They are rarely stocked by pet shops, though. When hair algae start to build up, it is a sign that the water is becoming eutrophic, or overloaded with nutriments. An important part of this is phosphates. They are released by both animals and algae. They are, in part, removed by protein skimmers, by precipitation, or by aeration. As the air bubbles burst, phosphates are released into the air. With overfeeding, undetected deaths, or decay these sources of phosphate reduction become insufficient and phosphates build up. Sea water has rarely more than 0.06 ppm, but aquaria can rise to 100 times that figure. Anything above 2 ppm of phosphate-phosphorus (PO_4-P, is regarded as dangerous and is likely to promote red and green microalgal outbursts.

different trace elements and so a good variety is necessary. Recent tests of flake foods indicate concentrations of 124 ppm. This is 1,240 times higher than the concentration recommended for marine water and more than 6,000 times the natural amount found in the sea.

Iron

Algae need iron to grow. Many artificial salt mixes have more of it in their composition than natural sea water. The bacteria in biological filters are also stated to benefit from a higher iron level than normal. Kits for adding and measuring the iron content in the aquarium can be very useful. This is particularly so if you want to cultivate macroalgae (higher fronded algae like *Caulerpa*, etc.). A concentration of 0.1 ppm is recommended, some fifty times the natural amount. Some mixes have 0.6 ppm. This is regarded as excessive if all stays in solution. Iron is usually included in algae-stimulating preparations, together with various other minerals and vitamins known to be required. Different algae require

Copper

Copper kits are most frequently used to monitor the dosage when attempting to cure some diseases of fishes. They should also be used to monitor the domestic water supply. It is often high in copper because of copper piping or storage tanks. It is essential to remove it from the aquarium water as it is a poison. It kills many invertebrates and even fishes if they are left in it too long. Medicinal levels, to be maintained for only a week or so, are around 0.15-0.3 ppm. This level is toxic to crustaceans, etc., but not to fishes. Natural sea water has only a minute amount of copper although it is needed by some creatures. Anything over 0.005-0.01 ppm should be regarded as dangerous to invertebrates and in the long term to fishes.

Copper kits measure simple salts of copper, such as sulfate or citrate with reasonable accuracy. They do not measure complex preparations correctly and should not be relied upon if you use them. Those using solid chemicals, as with nitrogen cycle kits, are the most reliable and have longer shelf lives than the rest. In testing some flake foods, it was found that some contained as much as 6.8 ppm, or almost 700 times as much as should be tolerated in the aquarium water.

Siphons, Nets, etc.

You will need various pieces of equipment to assist in the running of an aquarium. The choice of the right type can make a big difference to efficiency. All of them can be purchased from your local pet shop. In the case of a siphon, I prefer to make my own. This consists of a length of glass or rigid plastic tubing, about ½-¾ inch internal diameter, and some rubber tubing, not plastic. The rigid siphon should be just over the tank depth in length and have a rubber tip. This way it will damage neither itself nor the tank contents. The rubber tubing should be a little longer than the tank-top-to-the-floor distance to simplify siphoning into a bucket. Rubber is used because of ease of handling. A good quality tubing will last for years if kept out of the light when stored away. Some pet shops have equal quality siphons but they are usually made of plastic.

A pair of tongs as long as you can purchase is also a great help when you do not want to get wet and want to see what you are doing. So is a long stick with a needle inserted in one end. This helps to feed anemones or to impale unwanted debris or a dead fish.

To catch fishes, two reasonably sized nets are needed, unless you want to risk wrecking the scenery. Nets about half the tank width, dull green or brown in color, manipulated together, can catch a fish far more easily than chasing it around with a single net. Even so, it can be difficult in a reef tank and an inadvisable project unless absolutely necessary.

There are several ways of cleaning the inside of the

Test kits are necessary to monitor the condition of your water. This is a kit for determining the amounts of ammonia, nitrite, and nitrate.

glass. Soft algae can be wiped off with a piece of cloth, a sponge, or plastic pot scraper. Harder, calcareous types need a razor blade scraper or one of the magnetic scrapers that have a mass of tiny teeth that do not scratch the glass. Beware of picking up coral sand or other abrasive material when cleaning near the bottom. This can inflict bad scratches on the glass. Although many aquarists only clean the front panel, I like to leave at least a window in the side panels so that it is possible to see into the tank from a different angle. I can look along the length of the tank to see if the water is becoming discolored and needs a change of carbon or other filter material.

Marine algae are seen more commonly in the miniature reef aquarium. *Caulerpa* spp. are the most popular types and at least two species can be seen in this photo.

SALT WATER

Ocean water, away from river mouths or shallow shores, is fairly constant in the composition of at least its major constituents. The surface may be diluted temporarily in storms. At a few fathoms down, though, storms have no effect. Schooling fishes, like herring or tuna, have no great changes to encounter, except perhaps in temperature. There is a phenomenon called the thermocline, a zone around 50-100 feet down. Here the temperature and hence density change abruptly as you descend. Fishes can choose to penetrate it, but they tend not to do so. Thus, pelagic fishes, like the herring, are difficult to keep in aquaria, even large ones. These fishes are accustomed to very constant conditions, not to mention plenty of space. Only oceanaria provide them with suitable surroundings.

Our marine aquarium fishes nearly all come from very different localities— from tidal zones, shallow reefs, even estuaries— where things are far from constant. They are accustomed to noticeable changes in temperature and salinity. Sometimes the changes are quite abrupt but also seasonal. Many of them can stand a much cooler incoming tide after having been heated for several hours in a shallow lagoon or tide pool. This does not mean that we should subject them to such changes on purpose. It does mean that they are tougher than most writers make out. What they normally do have is excellent aeration and water movement. Both are more essential in the marine aquarium than absolute temperature or salinity control. They also have pure water in the regions in which most of them flourish, although many fishes can stand insults much better than many invertebrates.

Sea Water

The oceans and the rivers running into them have had billions of years to dissolve salts from the earth's solid constituents. It is perhaps surprising that they are not more saline than is the case. Even more surprising is that recent evidence leads to the conclusion that over the past half billion or so years they have not gotten any saltier. The blood of vertebrates has only about 1% salts yet the seas have about 3.5%. It was thought that the lower concentration in blood and cells reflected that of early seas, but such would not appear to be so. Just when the oceans ceased to become more saline is an open question.

Sea water contains almost all of the natural elements. Many of them are in minute quantities, however. Not all appear to be needed by any known life forms, but a lot of them are. Some are concentrated to a remarkable extent in the tissues of various invertebrates and algae. The substances dissolved or suspended in sea water are usually divided into the major constituents, dissolved salts, and the minor constituents, other dissolved salts (trace elements), dissolved gases, organic substances in solution and in suspension, and inorganic particles in suspension.

Major Constituents

These are ten in number. They vary little in concentration or relative proportions except in areas like the Red Sea that have only slow exchange with other parts of the ocean and a high evaporation rate, or in coastal areas, where dilution by runoff may regularly occur. The normal extremes are totals of 3.3-3.8%, but the Red Sea is 4% and some coasts have 3% or even less. In terms of specific gravity, this translates to a normal variation of 1.024-1.028, 1.030 and 1.022 respectively at 60°F, or 1.022-1.026, 1.028 and 1.020 at 75°F.

Although I have talked of salts as though it is possible to define sea water as having so much sodium chloride, magnesium sulfate, and so on, this is not possible. Most of the individual constituents are in the form of charged atoms or parts of molecules (ions). Most of the sodium chloride ($NaCl$), is in the form of $Na^+ + Cl^-$, and most of the magnesium sulfate, $MgSO_4$ in the form of $Mg^{++}SO_4^{--}$. It would be just as correct to refer to sodium sulfate and magnesium chloride, since both would be present as well as many other possible combinations of sodium and magnesium with other acids. In fact, about 80% of each constituent is in ionic, charged form. The rest is in temporary molecular form. This is why various mixes for making up artificial sea water differ in the amounts of each salt present, although they all add up to

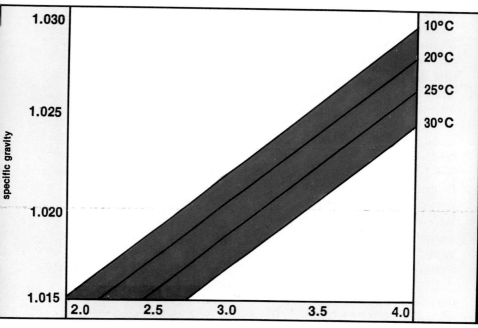

A graph showing the relationship between specific gravity, temperature, and percentage of salts by weight.

much the same when dissolved. The various ions are referred to as sodium ions, chloride ions, etc.

The ionic composition of sea water is thus sodium 1.08%, magnesium 0.13%, calcium 0.041%, potassium 0.039%, strontium 0.001%, chloride 1.94%, sulfate 0.27%, bromide 0.007%, borate 0.003%, and a mix of carbonates, bicarbonates, and carbon dioxide 0.003% approximately. The minor constituents altogether only add a further 0.0005% to the total, plus dissolved oxygen 0.0008%, nitrogen 0.0013%, and organic matter 0.0001-0.0025% depending on locality. Suspended particles in clear water are at about 0.0001%, and up to much higher percentages in turbid water.

Minor Constituents

In nature, most of the minor constituents of sea water that are needed at all are essential to algae and invertebrates. The fishes get them by eating these plants and/or animals, or by eating other fishes that have eaten them. Some may be absorbed directly from the water, but it seems likely that they do not have to be. The trace elements known to be essential to all algae, in addition to the major constituents, are nitrogen, phosphorus, iron, copper, manganese, zinc, and molybdenum. Various individual species of algae need one or more other trace elements in addition. Most need vitamin B_{12} and thiamine. Different invertebrates need those listed for algae and, in addition, iodine, cobalt, vanadium, copper, and arsenic.

Dissolved organic (carbon containing) substances come from living creatures. In some oceanographic studies, mainly northern ones, they amount to 5-6 ppm (0.0005-0.0006%). These compounds are proteins, parts of proteins such as polypeptides and amino acids, and very small amounts of the vitamins thiamine, biotin, and B_{12}. Particles in suspension amount to 0.2-2.0 ppm (0.00002-0.0002%). In northern oceans, half of them are clay-like material containing iron. Inshore, turbid waters have particles containing phosphates as well. They are regarded as important substrates for bacterial growth that in turn influences the organic content of the water.

Inshore waters and even more remote ones may nowadays carry many other, man-made substances. These come from sewage, agricultural fertilizers and sprays, industrial wastes, the bilges of ships, and spillage from ships. Silt may also clog marine life and kill much of it, especially corals. Most of the above are toxic. Others cause the water to become eutrophic (containing too much fertilizer) and promote the excess growth of algae. In the sea, this takes the form of algal blooms. These are masses of floating unicellular algae that color the water. In the end they die off and poison it, killing fishes and other life. Red tides, made up of red-colored blue-green algae, are a particular menace. In the aquarium, eutrophic conditions are usually followed by an overgrowth of hair algae and red or blue-green encrusting algae. It is becoming more and more hazardous to collect natural sea water for aquarium use, even if it is gathered well away from the shore. Your best bet, if feasible, is to obtain filtered and purified water

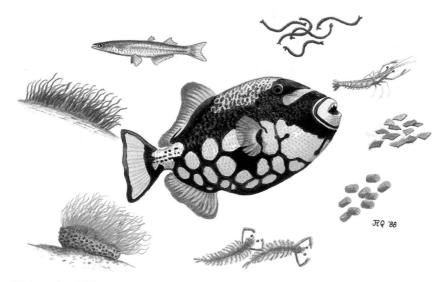

Fishes should be provided with a variety of foods. Shown here are examples of the different foods that can be given to a triggerfish, *Balistoides conspicillum*.

from a public aquarium. Some public aquaria are obliging enough to make it available.

Plankton

Plankton refers to the floating life in the sea that is carried around by currents and tides. Although it is often capable of movement, it is largely transported from one area to another by the movements of the water. Phytoplankton is the algal content of the plankton; zooplankton is its animal content. Some plankton members are intermediate between the two and classed by some as plants and others as animals. Phytoplankton depends on sunlight and is found near the surface. Zooplankton tends to rise and fall with the daily rhythm of daylight and darkness. It rises at night, or much of it does, and descends during the day. The phytoplankton is mostly one-celled plants or the spores of multicellular ones. Zooplankton is much more of a mixture. It includes one-celled animals as well as the juvenile stages of many phyla of higher forms—crustaceans, worms, echinoderms, cnidaria, etc. A quite large number of adult forms pass their whole life in the plankton. Bacteria are common near the edges of the sea, but sparse in the upper layers of the open oceans.

There is a food chain in the ocean that is largely missing in the aquarium. It resembles the food chain on land, depending ultimately on the phytoplankton. This is eaten by many creatures of the zooplankton that in turn are eaten by others. The predators of course get bigger and bigger as we pass up the chain. It ends with planktonic animals, such as krill, a shrimp found in vast numbers that is eaten by fishes, birds, and whales. Although occasional bursts of larvae are seen in the aquarium, this planktonic chain is not maintained. It is imitated by supplying finely divided invertebrate food, newly hatched brine shrimp, and other such preparations. It is not needed by other than larval fishes since we feed aquarium fishes by other means. An imitation of it is necessary with many invertebrates.

Collection and Storage

If you decide to collect natural sea water, take great care to avoid potentially contaminated areas. Out to sea as far as possible is best. Keep an eye out for oil slicks or floating rubbish. Otherwise, collect well away from outfalls, popular beaches, and estuaries. Choose a place where algae, anemones, and other creatures are flourishing. Everything should look clean and clear. Store the water in plastic containers known to be non-toxic (some are not) or in containers lined with safe plastic bags. Plastic jerry cans or small garbage cans are convenient for transport. Avoid large containers as they are going to be too heavy for convenience when full. They may be decanted into larger vessels for longer storage once you are home. This is advisable for reasons about to be discussed.

Stored sea water undergoes rapid changes. Plankton dies, releasing organic materials on which bacteria feed and proliferate. A silt of dead material collects at the bottom of the vessels. Dark storage is recommended so that an algal bloom is not encouraged. Depending on the temperature, oxygen, and nutrient availability,

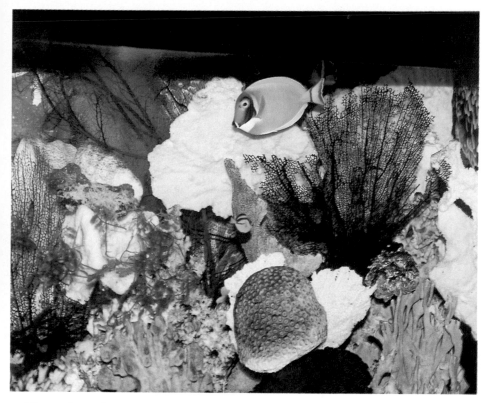

The variety of non-living decorations is relatively extensive. Even properly cured sea fans can be added to the different species of corals available.

and also on the size of the vessel, bacterial proliferation may be mild to massive. The smaller the vessel, the more they multiply. This is because they largely depend on the walls of the vessel to act as a substrate. A peak of bacterial growth occurs between two and ten days in most circumstances. It rises from a few hundred per ml to many thousands in large vessels, or even half a million or more in small ones of a few gallons. After the peak, most of them die off and join the sludge at the bottom. In two or three weeks the count has fallen to acceptable levels, but rarely to as low as it was originally. The water is now oxygen depleted. It is also disease-free or very nearly so, as even disease-causing bacteria will likely have died off in the absence of a host.

You may decide to take the trouble to filter the water before storage. Use anything from clean cloth to a well set-up filter with plastic pads and carbon. This certainly helps, but it is still advisable to follow with dark storage unless the water is used immediately. You can even sterilize the water completely by running it over UV tubes as described earlier. Do this after filtration or you will be introducing dead plankton to the aquarium. This may or may not be a good thing depending on what is there to eat it. Whatever the initial treatment, after dark storage the water must be carefully siphoned off so as to avoid sucking up the sludge at the bottom. If any but a small amount is to be used for a water change, aerate the water before using it to restore the oxygen content. This can be done by pouring it vigorously from bucket to bucket. The oxygen will be rapidly taken up from the air.

There is a small risk in using newly collected sea water. Few reports of any damage from doing so have come to hand, though. It has the advantage of introducing new life to the aquarium and of giving it the genuine article. If it is not used immediately, it should not be used for at least two weeks of dark storage. This avoids heavy bacterial counts and also gives pollutants a chance to biodegrade, if they will. Should there be any doubts about purity, it is a good idea to test for ammonia and nitrites, and even nitrates, before using the water. It is also much safer to add new water to a tank containing filter feeders (clams, corals, tube worms, etc.). They will clean it up and also benefit from it. Do not forget to bring the water to approximately the same temperature as the aquarium if any quantity is to be added. It is easier to stand jerry cans or small receptacles in a bath of warm water than to insert heaters. Twenty minutes in a 120°F bath will raise the contents from 60° to 80°F. If you keep the aquarium water diluted, as many do, adding hot distilled or demineralized water will do the trick. If you use tap water, make sure that it is suitable. Never draw it from a hot water system. The danger of metallic contamination is too great.

Synthetics

Many aquarists have no choice but to use an artificial mix for making up their marine aquarium water. In some ways they are lucky. First, because good modern mixes are excellent. Second, because they are sterile, free from contaminants, and usually have a few goodies added to protect fishes or to stabilize the product. We always learn the hard way. Earlier mixes were usually unsatisfactory, even for keeping fishes. They were made from a few basic major components. Such trace elements as they contained were often just the contaminants present in commercial grade salts. Now, we have what are probably unnecessarily elaborate mixtures of up to 70 elements or so. They contain everything known to be needed to support marine life plus a few extras just in case.

A good mix dissolves easily and is homogeneous. This means that all the components have been so thoroughly blended that it is safe to take a portion from the package without having to use all of it at once. There may be a separate package or vial of trace elements, but even that is absent from some.

The chemists concerned must be congratulated. Earlier workers found it impossible to compound a mixture that could be dissolved all at one go and contained everything. When making up water to be added to an established aquarium, the mix must be dissolved in tap water (if suitable). Demineralized water can be used if your tap water is as bad as much of it regrettably is. The demineralized water can sometimes be obtained from mineral water or soft drink manufacturers at a small charge, from universities, or made yourself by purchasing the necessary equipment. Tap water contaminated by metals can be rendered safe enough by treatment with commercial preparations. However, the metal is still there and eventually accumulates and may be dangerous.

Do not assume that tap water suitable for human consumption is safe in the aquarium, either freshwater or marine. Apart from metals, it is likely to be contaminated with chlorine, chloramines, and nitrates. There may even be industrial wastes. Hardness and pH do not matter to the marine aquarist. Both will be swamped by the addition of the mix and the buffers it contains. Chlorine does not matter much either. It will dissipate as the water is made up and aerated before use. Chloramines are not so easily removed. However, there are commercial neutralizers available from pet shops, or you can use 1 grain (65 mg) per gallon of sodium thiosulphate (photographers' hypo). This also gets rid of chlorine itself. Chloramines are mixtures of ammonia and chlorine. They are added to the water because they last longer than chlorine alone. They also neutralize trihalomethanes, suspected of causing cancer. They are toxic to fishes since they pass through the gills. They act like nitrites by combining with the hemoglobin in the blood and stop it from carrying oxygen. They are also toxic to various invertebrates for diverse other reasons. Nitrates may be present at up to 50 ppm. This is higher than wanted even if only fishes are being kept. There is no simple way to get rid of them. Denitra equipment will do it if included in the aquarium set-up. Otherwise, it would be best to find an alternative source of water if at all possible.

So how do we proceed to make up replacement aquarium water? First, test the tap water for copper and nitrogen cycle components. Enquire of your local water suppliers about chloramines and any other contaminants. Zinc is a danger. However, kits for testing zinc levels are not available despite the fact that it is highly toxic to sea life. Since it is only mildly toxic to humans, it may well be present in dangerous amounts. Then, take all necessary measures to purify the water or find another source if its condition is too bad. It is probably wise to routinely add a conditioner. This should neutralize metals and some other toxins. Next, warm the water to well above aquarium temperatures to aid dissolving the salts—about 100°F will do. Add 4.5 oz (126 gm) of the mix to each U.S. gallon. Stir well until nearly all is dissolved. Some small residue may take time to go into solution, but aeration will take care of that. Then put in an airstone. Cover the receptacle and leave it for some hours, preferably overnight, to stabilize.

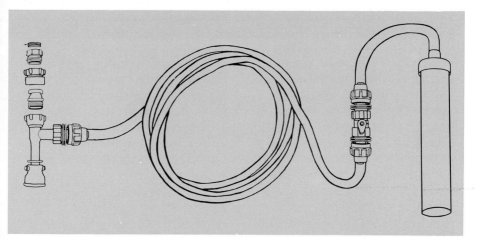

Water changes can easily be made by a commercial water exchanger in fresh-water tanks. This can also be done in the marine aquarium if there is a reservoir of already mixed marine water of proper chemical makeup available.

Check the temperature. Correct it if necessary before adding the water to the aquarium.

Water Changes

Once in the aquarium, sea water or artificial mixes begin to undergo modification due mainly to the actions of animals and algae. The buffer capacity or carbonate hardness tends to fall. It should be checked and kept to a suitable level, say between 7 and 11 as suggested earlier. The pH that tends to fall with it should also be checked. The periodic addition of sodium bicarbonate takes care of both of them. These precautions are necessary whether water changes are made or not. There is no way, without the services of a scientific laboratory, to check on the many other changes that are taking place: the uptake of calcium by corals and shell-building animals and fishes or its release from some of them; the uptake of trace elements and vitamins; or the accumulation of various toxins. Nitrates can be checked. They will usually be found to rise consistently or until a steady level is reached that may be near to zero with denitra equipment.

Many of these changes can be kept under adequate control by water changes, aided by the action of the various filters discussed. In a fish-only set-up, nobody seems to worry very much as long as regular water changes are made. The changes are usually suggested as 20-25% per month or an equivalent more frequent change. Nitrates characteristically rise to 40 or 50 ppm or even higher, but fishes do not mind that. Vitamins and trace minerals, etc., may be depleted despite the changes of water. Fishes do not mind that either since they get what they need from their food. Even an imbalance of perhaps calcium and magnesium would probably not worry fishes unless very marked.

In aquaria housing invertebrates and macroalgae as well as fishes, we find very big differences in opinion. Many aquarists treat them much the same as fish-only set-ups. Others assert that corals and some other invertebrates, even some macroalgae, are too sensitive to sudden water changes. They advocate very gradual, small alterations in the aquarium water. This is achieved by daily changes of 0.1 or 0.2%, or at most weekly changes of 1 or 2%. Yet others make no changes at all. They assert that at least in a reef aquarium they are not desirable. They do, however, add trace elements, vitamins, and in some cases calcium salts in an attempt to make up for losses. As nobody has published analyses of aquarium water maintained under these different systems, it is impossible to say who is right. Possibly they all are as long as their particular regimes are adhered to. It may make a big difference whether a mix or real sea water is used. A newly made-up mix is still raw. It is known that corals do not fare well in such water. Sea water is not raw, whether used immediately after collection or after storage. It is perfectly conceivable that big changes with real ocean water are beneficial, whereas the same changes with a mix are not. Just what is meant by raw could be better explained, but I cannot offer a suggestion myself!

There is also the assertion that fishes do better in reef aquaria than in others. That I believe to be true. I think the reasons are twofold. First, fishes do better and look better in aquaria served by wet/dry or dry filters, even if not otherwise reef aquaria in construction. This is presumably due to better purification and oxygenation of the water. Second, in reef aquaria, fewer fishes are normally kept. Also, they are kept together with a lot of filter feeders. This reduces the chance of disease. The filters mop up disease organisms when they are present. The assertion that diseased fishes are cured when placed in reef aquaria may even have an element of truth if their health and resistance improves and they are not subjected to heavy reinfection.

As long as circulation and filtration rates are adequate, it seems that the bigger the aquarium, the higher the chances of success. This is probably mainly because the water/biomass (volume of living material) ratio rises as aquarium size rises. A survey of the literature reveals that people who assert that their particular system works without water changes, for example, appear generally to have large tanks. There is a factor, too, that very large tanks are expensive to service with water changes

and so they do not get done. (I must be careful of the logic in this argument!) Perhaps those people with large tanks who happen to get away with no or few water changes are the ones that write about them most frequently and so...!

Miniature reefs commonly include only a few fishes. Those selected must be of the type that will not do serious damage to the invertebrates or algae in the tank.

A reef aquarium that includes a denitrification filter in the setup. Delicate invertebrates are doing very well indeed!

SETTING-UP AND MAINTENANCE

Six Steps

Your aquarium is in position and, depending in part on what kind of set-up you intend to have, there is a logical way to proceed to get it going. Let us assume, for the first series of steps, that you are aiming for a "standard" undergravel filter model. This is now beginning to be outmoded, but likely to continue in use by a large number of marine aquarists. This is because it is much cheaper and easier to assemble than reef types. It also gets you into fishkeeping much quicker with no waiting around for maturation of the reef. In the end, though, the reef aquarium is likely to be more attractive and efficient.

So to make the job as quick and easy as possible, the following steps are recommended:

1. If you have not already done so, wash the tank thoroughly with warm, fresh water. Siphon off a couple of rinses. Do not use any bleach, detergent, or anything—just plain water. Wash the gravel very thoroughly a ¼ bucketful at a time. Each ¼ bucketful must be so clean that the water runs off clear. This may take quite a long time with up to a dozen or more rinses per bucket. Even if the gravel or coral sand, etc., is sold as washed, this must still be done. Wash all decorative coral skeletons, rocks, etc., until they look, smell, and feel clean. If you are in any doubt about a coral head, leave it in a bucket of fresh water for a day or two and you will soon find out.

2. Place the undergravel filter in position. See that it fits quite tightly against the sides and back, but preferably not in front unless you have it hidden by a sunken stand-top. Otherwise leave about a ½ inch margin so that the coral sand hides the actual filter. Connect up any airlifts or power-head connections. Then cover the whole filter with a double thickness of plastic fly screen. Tuck it over the exposed front or the sand will get under the filter, unless you have managed to get an uncut edge to the front. Tuck the fly screen neatly around uplifts. If you do not wish to go to this trouble and do not mind seeing a filter edge, get a filter that exactly fits your aquarium. Most standard sizes are available.

In a divided tank, there are no uplifts and the job is easier. All that has to be done is to see that the fly screen, if you use it, fits neatly and does not impede the flow into the rear compartment.

3. Place the coral sand, still wet, over the filter to form the filter bed. Form it into a half-basin so that it is deeper at the back and sides and slopes gently toward center front. It should not be less than 2-3 inches deep at its shallowest point. This not only looks pleasant, but it assists the mulm to collect in center front to be siphoned off readily. It also helps to keep the flow through the coral sand as even as possible. This is because the back near the uplifts has a deeper layer to contend with. I know: the fishes will upset this ideal arrangement to some extent, but you may as well start off right!

You can now seed the filter bed with the bacteria that will start up the nitrogen cycle conversions. These may be purchased from your pet shop. Sprinkle them over the surface or add as directed. This is a safe method. It is preferable to the older techniques of adding some gravel from another aquarium or some garden soil. But if you are out in the sticks and do not have a preparation available, a few teaspoons of soil is best, or some gravel from a freshwater tank. Both will be free of marine diseases. Yet they contain the same bugs that will adjust to the new conditions after a short period.

4. Arrange any dead rocks and coral where you have planned. It pays to make diagrams of front and top views of your proposed arrangement unless you are a natural at the job. Place any other equipment, such as an inside filter, heater, and thermostat, in position so that they will be as inconspicuous as possible. Their lead or airlines can be held in position if necessary by temporary stones or by the other decorations. See that the heater is clear of any obstruction if, as is preferable, it sits near the bottom. Make sure that it cannot roll onto the glass or get buried. Then it could overheat and fuse. I do not like suckers that hold heaters, etc., in position. They too often come unstuck. If you have decided to use them, it will be necessary to make final adjustments after filling the tank.

In a divided tank, you do

not have to deal with the auxiliary equipment at this stage. It is all going in the back. It may as well be arranged roughly in position so as to make sure that everything fits. As there is nothing on the bottom of the heater compartment, the heater is best supported by a cradle of some kind or, in this particular case, by suckers. I find it best to let the heater dangle by its cord so that it cannot touch the glass or fall to the bottom. If necessary, place a collar made of a plastic jar top with a hole in the middle that just fits the heater around the base. You may wish to place an airstone in position in the front section. This is not so much in anticipation of using it as to have it there as a precaution. Otherwise, an airstone in the back compartment will do a good job. Also, it will not splash onto the covers or lights. Contrary to common practice, marine aquaria without airstones in the display are nicer to look at. Airstones are quite unnecessary if the other arrangements are satisfactory.

5. The aquarium can now be filled. As a super-precaution, half fill it with suitable fresh water. Leave it for a few hours or overnight to check for leaks. A leak is very unlikely but possible. Run the water gently into a shallow container placed in center front. The container will overflow equally gently with minimal disturbance of the coral sand. If using tap water, run it for a few minutes beforehand. This will empty the pipes of standing water that may be contaminated. Remember all the precautions about contaminants already discussed. If you are using sea water, omit the fresh water part above. Half fill the tank as before and then proceed to fill up completely. Pour the water into the front container with a plastic watering can or a jug.

The salt can be added to the tank half-filled with fresh water after calculating how much to use. It may be added to an empty tank if you intend to omit the precaution against leaks. Take 80% of the theoretical tank volume as a rough guide to how much mix to put in. Fine adjustments can be made later. Then dump in 4½ oz per gallon. Use a little less if you intend to use somewhat dilute water for

Live corals are favorites for reef aquaria. Particularly in demand are the orange cave corals *Tubastrea* and *Dendrophyllia*.

fishes only. When circulation is started up, the salt will dissolve. Although it may clog the undergravel filter a bit at first, it will soon all disappear into solution. Some of the calcium salts may take a few hours to dissolve. Do not worry. You have plenty of time before putting living things into the tank.

In a divided tank, the water is run into the central back compartment. It will flow up through the filter bed without disturbing it or causing clogging as it dissolves the salt. If you are using sea water, the same technique applies. In both cases, you do not have to worry about being quite so careful over introducing the water.

6. The aquarium can now be put into action. Start everything up. Check to see that the air flow to filters and any airstones is correct, the water pump, if any, is working well, and see that everything stays in position. Next day, make another check. See that the temperature is between 76° and 80°F. Test the specific gravity. Correct it if necessary to between 1.022 and 1.023. It should not be less than 1.018 if you intend to keep fishes, but only fishes, in a more dilute water. At less than 1.018 you will have trouble with

introducing fishes that have been accustomed to denser water. When the tank has settled down a day or two later, start the maturation process described later on in this chapter.

Reef Aquaria

A reef aquarium does not have an internal filter or a deep coral sand bed. All the action happens outside, except for the pre-filter. Naturally, the situating and cleaning of the tank is the same as before. However, the main installation of living rock and other life follows a maturation process that, if the set-up allows it, may be applied only to the dry or wet/dry external filter. Otherwise it resembles the maturation of an aquarium with an undergravel filter. A reef aquarium is not set up with dead coral skeletons and rock as an ordinary aquarium often is. It is filled with living rock, old coral that has been colonized by all kinds of invertebrates and algae. The same applies to a natural system aquarium, from which the reef aquarium is in reality derived. Of course, any aquarium can have pieces of living rock or corals and other invertebrates added to it. But a natural system and reef aquaria have only living rock or at most a few dead rocks or coral skeletons hidden at the back behind the living stuff.

Maturation

The bacteria in the biological filter will not multiply unless they are fed. Then they will increase in numbers to a sufficient level to deal with the food supplied. If the supply decreases, their numbers will also decrease. If it increases, they will increase in numbers to deal with it, up to the limit allowed by the filter bed. As soon as fishes are placed in an aquarium, they start to release ammonia and other waste products. The bacteria start to feed on this. The species of bacteria concerned can use various substances, but if ammonia is predominant, they will adapt to it and process it efficiently. Other bacteria may utilize different wastes. Therefore, competition for space may occur. The ammonia-processing ones may lose out on some of the space available.

When fishes or any other living animals are first introduced to a new aquarium that has not been matured beforehand, the ammonia they release accumulates faster than the bacteria can deal with it. If too many creatures are added at once, this causes the development of the "new tank syndrome." After a few days or a week or so, in the words of one sufferer, "all hell breaks loose." The fishes show signs of distress, break out with diseases, and you are lucky not to lose the lot. All this happens because they are poisoned by the excess of ammonia not yet mopped up by the biological filter.

This is because bacteria of the right sort happen to be rather slow to grow compared with most others. They need time to catch up.

The old solutions to this problem were either to introduce one or two hardy fishes at first and then increase the numbers gradually from then on (it took many weeks to build

Several species of algae do well in reef tanks. This flat-bladed form of *Caulerpa* is *C. prolifera*.

Another species of *Caulerpa* that is commonly kept in reef aquaria is *C. racemosa*, easily identified by its "berry-like" form.

up to a full tank), or by putting a turtle or two in the tank to do the same initial job. A less attractive method was to put some decaying fish or meat into the tank. An increasing load of ammonia (plus other wastes) was allowed to build up. Then someone a bit brighter than most of us reasoned that we might as well use ammonia itself from the outset. This is what is best to do. There are advantages to this method, too: no risk to fishes or other life, a minimum period of waiting to build up the filter, and a better development of the right bacteria since a good supply of ammonia can be fed from the start. Some aquarists claim 300% improvement in filter performance compared with older methods. This is despite the fact that most of them add less ammonia than is optimal.

Ammonia can be added to the aquarium. Alternatively, it can be added separately to an external filter in the form of ammonium chloride or sulfate. The chloride is preferable since less is needed. Also, chloride ions are already the predominant ones in the aquarium. Hence, the balance will be least disturbed by the addition of more of them. Whatever the dosage used, you will get an ammonia peak followed by a nitrite peak. Their occurrence does not guarantee that the filter is fully conditioned. Whatever method of conditioning used, a peak of ammonia can be expected within a few days to a few weeks. A peak of nitrites will occur within about ten days to several weeks, or even months if a slow method such as a fish or two is used. A peak of ammonia is masked by the ammonia you are adding when maturing the tank by that method. Therefore, look for a nitrite peak from day ten onward of around 10-20 ppm. Continue until it falls to less than 0.25 ppm before considering adding fishes. It is best not to add invertebrates at so early a stage.

To mature the filter to as high a capacity as you can reasonably expect it to achieve, proceed as follows. Make up a solution of 10% ammonium chloride or 15% ammonium sulfate in a glass of distilled water. These will contain approximately equal amounts of ammonia. Add 2 ml per 25 gallons of aquarium water (100 liters) on each of the first two days. Add 4 ml per 25 gallons on days three and four. Add 6 ml per 25 gallons on days five and six. Continue so on until 10 ml per day is reached on days nine and ten. Continue with this dosage of 10 ml per 25 gallons per day (two standard teaspoons per day) until the nitrite peak has passed. The nitrite level should be below 0.25 ppm. Then you can add fishes to a fish tank, but not any delicate invertebrates yet, or living rock to a reef tank. Why? Because you can add a full or nearly full load of either fishes or living rock, but not both at once. Some invertebrates, particularly corals, do not do well in new tank water even after maturation. Do not stop adding ammonia until at most a couple of days before adding the fishes or rock. Of course, do not continue to add it on the day you put them in or thereafter. If you wait too long before adding the

A third species of *Caulerpa* can be seen in the center of this photo. It is *C. crassifolia* or *C. mexicana,* and can be recognized by its feather-like appearance.

living material, the filter will decline in capacity. You will have to add fewer creatures or rocks and build up the rest slowly. A 50% change of water to reduce the nitrate level is beneficial.

In some aquarium designs, the external biological filter can be matured separately. You can save on ammonium chloride by circulating the water only through the filter. This also reduces the total nitrate level and saves on water changes at this early stage. When the filter has matured, fill the tank with sea water or synthetic mix properly matured. You cannot dump the salts for a synthetic mixture in and wait for them to dissolve as in an aquarium with an undergravel filter. Connect the filter to the tank. When everything is running smoothly and the temperature is right, in go the fishes or living rock.

Building a Reef

There are two ways of building up the reef aquarium. The first is presented above. As soon as the tank is ready, a complete reef is constructed and is allowed to settle down. Invertebrates can be added in stages so that the filter keeps up with them. It is generally best to leave corals, large anemones, and most fishes for a month or two, even if the reef looks fine and ready to accept them. How long the reef takes to be in a state of readiness depends, however, on how the living rock was treated before you received it.

If it was received within a few days of leaving the ocean and still had plenty of flourishing life on it, it could recover immediately. It will present the appearance of a well-developed reef straight away. If it had been packed dry for a period, much of the life would be dead. The rock must then be cured by placing it in filtered tanks to remove the dead and decaying material. This allows the residue to recover and build up again. Your dealer usually does this prior to selling it, or at least cleans it up. Some collectors brush the living rock free of nearly all its surface content. It can be used more readily on reception, but it then takes months to recover. It is surprising that such seemingly near-dead rocks can eventually produce a mass of regrown algae and a population of invertebrates that has arisen from larvae and spores, etc., hidden away in crevices. This does take time, though.

Each of these methods has its advantages and disadvantages. Newly collected rock is liable to be harboring unwanted crabs, mantis shrimp juveniles, and other nasties. Each of these can be a terrible nuisance and very difficult to eradicate. Yet it will also have a greater variety of creatures and algae than the other types. The risk may be worth the taking. Dried or brushed rock will have fewer nasties, but also fewer nice things as well. It will never present the range of invertebrates that fresh rock contains. It pays, by the way, to get samples of rock from different sources or localities so as to increase the variety of organisms eventually appearing.

Zooxanthellae may be found in the expanded mantle of this clam (*Hippopus* sp.). Bright lights of the proper spectrum are needed to keep these healthy.

You may not wish or be able to acquire all your living rock at once. Do not be misled by statements appearing in the literature that a reef aquarium must receive all its rock in one go or for some unstated reason it will fail. It will not. It is perfectly possible to build up the reef in stages. If this is done, there is no need to mature the filter beforehand beyond seeding it with the necessary bacteria. As long as not more than, say, a quarter of the eventual reef is added at a time, the seeded filter will keep up with it. It is best to check ammonia or at least nitrite levels to make sure that all is well. There should be no more than initial small peaks. Often there are none at all if dried or brushed rock is used. New additions can be made at weekly or longer intervals.

In the absence of fishes and delicate invertebrates (not to be added until later) these small ammonia or nitrite peaks are relatively harmless. They even help to feed algae. If you want to add more than a quarter tankful to start with, condition the filter with half or a third of the ammonia dosages recommended above. There is no point in building up the filter only to have it decline again.

The amount of living rock usually recommended for a reef or a natural system aquarium is 2 lbs per gallon. This is purely optional. Such an amount builds a solid reef reaching from near front to near top back of a typical tank. Indeed, it gives a very good effect. You do not have to do that if you do not want to. Articles on the subject often make it sound as though such a build-up is essential for success. Far from it. An underpopulated aquarium is always safer than a crowded one. The point about the reef tank is that its super-efficient filter and other equipment allow overcrowding, but they do not demand it.

When you do build up a full reef in stages, build it in vertical sections across the tank. Do not build it from the bottom upward, or you will kill off the creatures at the lower levels, or some of them, by placing new materials on top of them. Although some disagree, I find it helpful to use some old bleached coral at the base to support the earlier living rock. It also keeps the rocks nearer to the lights. Eventually it is covered over by later batches—but still built-in vertical sections initially. This way you do not disturb the structure you are creating more than necessary. It becomes less and less beneficial to move creatures around once they have settled in.

The Natural System

After all I have said about biological and other filters, about the circulation of water though them, and of course the aquarium, it must seem odd to the reader to be told that you can do without them. Lee Chin Eng of Jakarta introduced the natural system. In his hands it underwent a degree of development. It the hands of others, though, it had very variable success. This was in part because Lee died without fully discussing his methods. Also, he had access to the tropical oceans. His specimens went straight from it into his tanks. Most of his followers were not so lucky. They had to depend on freighted materials and, at that time, on less than perfect salt mixes. I have

Small mandarinfishes are excellent fishes for the reef tank. This *Synchiropus* sp. is hard to detect at first.

described this system and ways of setting it up in several previous TFH books. I shall only summarize the details here.

A natural system aquarium is best built up with fresh living rock in stages, much as described for a reef tank. However, there is only aeration added, about one airstone per two square feet of surface area. The products of the nitrogen cycle are dealt with by bacteria living in the rock and by algae, the algae if both free-living and dwelling in the tissues of invertebrates, such as anemones, sponges, and eventually corals. This is a frail system compared with a biological filter. It can be made to work, though, and it develops more robustly as time passes. Natural sea water with its plankton, fresh from the ocean, works best, but synthetic mix of high quality can be used instead. Very great care is needed, however, to see that the water from which a mix is made up is really pure. It must be filtered through carbon before use and treated to eliminate chloramines. Better still, use glass distilled or demineralized water.

As again with a reef tank, you can use dried or brushed living rock and have to wait for a long time for it to recover. In some ways in the natural system, this is a safer way to start. The aquarium slowly develops its own bacterial population. It keeps pace with the equally slowly developing living rock— really half-dead rock at first. By whichever way you have proceeded, when the living rock is flourishing and all seems well, start adding invertebrates quite slowly. Add one piece of coral or one anemone or clam per week until you have built up a respectable population. Go mainly for organisms that are filterfeeders or have *Zooxanthellae*, the indwelling algae that mop up waste products and feed their hosts. Healthy corals and anemones, etc., actually help to cleanse the aquarium and remove more wastes than they create. This is only in adequate light, though. This must be provided just as over a reef tank or things will go sadly wrong.

You can add something to the natural system that gives it a much greater stability and starts it on the way to resembling a reef aquarium—an undergravel filter. In fact, you can get very close to a miniature reef aquarium with such a filter, but can only rarely equal it. Still, it is a very much cheaper way of doing things and it works well with a divided tank. There is also the advantage that plankton is not filtered off as it tends to be in a reef system with its external filters and pre-filter. The rate of flow through an undergravel filter is quite slow. It is typically less than one inch per minute through the surface of the filter bed, from which zooplankton can easily escape, even if phytoplankton does not.

The fish capacity of a natural system aquarium is limited as fishes contribute nothing but pollution. This naturally limits its interest to aquarists mostly concerned with invertebrates. Fishes do well. This is if not too many are introduced and are not of the wrong sort that would destroy the scenery. Lee had up to a dozen small fishes around an inch in

Cleaner shrimp (*Stenopus hispidus*) add some color to the reef tank. Although they remain hidden much of the time, they can usually easily be located by looking for their long white antennae.

body length in a 36″ tank. Larger fishes would naturally have to be fewer in number. Feeding must be kept to a minimum to limit ammonia production and general waste. Those interested in breeding or in larger fishes must look elsewhere, to a reef tank or back to older methods that are adequate for fish-only aquaria.

Periodic Maintenance

All aquaria need regular attention. They need a daily look-see to check that all is well (just a minute while feeding usually). Note the temperature and observe whether all fishes are feeding and looking good. The equipment should be functioning properly. The practiced aquarist can tell at a glance if anything is different in a tank and will immediately seek the cause. Other than that, there are weekly, monthly, and less frequent checks to be made to ensure the continued functioning of the set-up. The exact timing of these may be varied according to circumstances, such as the size of the aquarium (small tanks need more frequent attention), its biological load, and the type of filtration it has. The following is just a guide.

Weekly:

Clean all glass covers. If necessary, clean the fluorescent tubes and the inside front and side glass panels. Algae can be removed with a sponge or ball of material on a stick. Resistant algae can be loosened with a razor blade or magnetic scraper. The latter has lots of tiny, non-scratch hooks on the cleaning surface that follow an exterior magnetic hand piece. After the first month, add ½-1 teaspoon of sodium bicarbonate per 25 gallons of tank water. In a reef tank, check the carbonate hardness and act accordingly. Also in a reef tank you may change up to 5% of the water rather than making larger and less frequent changes. The pre-filter may also need attention.

Monthly:

Siphon off 20-25% of the water in tanks under 40 gallons. Siphon off at least 10-15% in larger tanks, depending on the biological load. Use the siphon to remove as much mulm and debris as possible. Lift dead rocks slightly and stir up the gravel somewhat. The siphon can also strip off some unwanted hair algae. The rubber tip of the siphon recommended in Chapter 3 does this admirably. Renew any clogging filter pads in carbon filter or pre-filters. Check pH and nitrite levels in all tanks. Check nitrates in a reef tank or natural system. Check and correct the specific gravity if necessary.

Quarterly:

Renew the activated carbon, first washing it thoroughly. If necessary, change some of the decorative dead coral, or take some out and scrub off much of the algae. A mixture of algae-covered, semi-covered, and newer coral can look very attractive. Bleach any coral removed in the sun if possible. If not, household bleach may be used. However, very thorough rinsing is required before the coral can be returned to the aquarium. Check all equipment. Replace airstones and dry out old ones for reuse.

Up to Yearly:

In an aquarium with an undergravel filter, siphon

Some algae are brought in with the rocks and seem to do just fine without any extra attention. This bubble alga is very common.

off about a third of the gravel. Do not remove it all from the top or you will remove too many bacteria. If necessary, additional thirds can be removed at weekly intervals. Wash the gravel gently in salt water. Return it to the aquarium preserving as much bacterial coating as possible. In a divided tank you probably need to do this less frequently, even up to several years.

Indications that it is necessary are gurgling airlifts, much swirling debris on disturbing the gravel, or an inadequate flow to back compartments. Feed lightly after such a procedure as you have weakened the capacity of the biological filter. You must wait for it to build up again. The top layers of the filter bed do most of the work. That is why they must never be severely depleted at any one time. The biological filter of a reef aquarium rarely needs frequent attention but it must be handled with care. Replace only part at any one time and cut down on feeding for a week or so at that time. DLS filters can be washed gently without losing too many bacteria.

All mechanical equipment needs an annual overhaul. Preferably this includes replacement of diaphragms in air pumps and a thorough cleaning of water pumps or power heads. Some types of fluorescent tubes lose output fairly quickly and may need annual renewal, others have longer useful lives and can be left in service for as long as three years. Airlines, gang valves, and all connections need checking. This is because plastics harden and may start to give trouble. Equipment, such as ozonizers and protein skimmers, that has not been mentioned need frequent but less predictable servicing depending on circumstances. Ozonizers need renewal of the dehumidifying crystals and frequent cleaning of the electrodes, as do redox potential meters. Protein skimmers need frequent emptying and cleaning to keep them efficient. Ultraviolet tubes fall off in output quite rapidly and may need replacing more often than yearly. The more gadgets you have, the more it is going to cost you!

Many of the small fishes obtain a portion of their food from the invertebrates in the tank and need very little additional food from the aquarist.

HANDLING FISHES

One famous old aquarist, on being asked the secret of his success, replied that you have to think like a fish. Perhaps he should have said think for your fishes, for that is a necessary step. I was once telephoned by a fisheries official who asked me why his trout died when they were shifted from fresh to salt water (where they grow quicker and larger). After a few questions, it emerged that the unfortunate fishes were moved straight into full strength sea water all in one go. That gentleman certainly did not think for his fishes. He appeared to lack any knowledge of their physiology as well. The "secret" is, of course, adequate knowledge and care.

Collecting Fishes

Collectors and exporters of fishes vary tremendously in their knowledge and skill. They differ in the care they give the catch, and also in how they collect. These are factors over which we, the aquarists, have no control unless we do our own collecting. It is only by experience that importers, rather than individual aquarists, learn to discriminate between good and bad suppliers. Sometimes importers even manage to educate the suppliers a little. But then importers themselves vary. In some cases, this is another hazard in the history of the fishes we eventually purchase. So do our immediate contacts, the retail pet shops. But at least we can see something of what goes on there. It is fortunate that a good deal of effort is now going into educating all concerned. This minimizes losses and present the public with healthy, robust fishes.

Practically all marine fishes are caught in the wild. This is in contrast to the freshwater trade in which many are pond or tank-bred. The best way to catch them is as gently as possible with hand nets or similar equipment. With many species, this is fairly easy to do at night. This is not to the liking of most collectors. The worst way to catch them is with cyanide or some other drugs. The drugs not only harm the fishes and the fishermen but the environment as well.

Cyanide is a poison. It has been used extensively in the Far East by individual collectors as a cheap way of catching

Many fishes are collected by hand with the aid of SCUBA. Some divers not only collect fishes but take along a camera and take a few photos on the same dive.

masses of fishes quickly. It paralyzes without killing if used effectively. However, there is a range of possible consequences. A mild dose paralyzes and the fishes can recover relatively unharmed. A somewhat stronger dose does not kill the fish, but destroys its powers of digestion. Eventually the fish gradually starves—even with a full belly. A stronger dose yet kills, quickly or more slowly according to just how strong it is. The usual result is a mixture of effects, depending on where a fish was when the cyanide was released. The survivors may look perfectly healthy and behave as if all is well. The unlucky ones will eat to no avail and die in a few weeks or even longer time. Their purchaser cannot diagnose the condition just by looking at them. He must try to find out how they were caught. Regrettably, this is often impossible, although some shops do advertise cyanide-free fishes.

If handled really carefully, a newly caught fish, even one like a marine angel, can be eating in an aquarium within a day or two. Most often they do not. The fishes are held in unnatural surroundings in holding tanks without any attempt to feed them if they are to be shipped off soon. Then they do not foul the shipping water as readily as if they had been fed. Fishes that must be held for some time should be fed. But they should be starved for a few days before shipment. Dealers will often withhold food for a long time since most fishes can tolerate such treatment until they are in someone else's hands. Then their new owner may often stress them by crowding them into his tanks for retail or wholesale distribution. By the time you get your selection it is no wonder that losses occur.

Marine fishes should be packed individually in plastic bags for shipment from dealer to dealer, or dealer to you. This is so even for a short journey. About ⅓ water and ⅔ air enriched with oxygen is about right for the average fish. The oxygen should be almost pure for long journeys. During a journey, the shipping water becomes charged with ammonia and carbon dioxide. This has the useful action of lowering the pH which in turn renders the ammonia less toxic. On arrival, the fishes are gradually introduced into the new tank water by floating the bag on the surface after opening it. The water in the bag is exchanged for the water in the aquarium (or quarantine tank) in stages. This takes at least half an hour—longer if pH and temperature differences are large. This also takes care of any difference in specific gravity or other aspects of water quality. Most important, it dilutes the ammonia as the pH rises and helps to prevent any harmful effects. These could be avoided by plunging the fishes straight into the aquarium, but then temperature, osmotic, or other shock could occur.

This step into your own aquarium may be the third or fourth a fish has had over the last few days or weeks. You are handling a stressed animal. Try to avoid further stresses, such as bullying by tank mates or overcrowding. Feed lightly for a period as soon as the fish will eat. Recovery should be quite rapid.

Buying Fishes

Be very critical when buying a fish. Look at the entire tank of fishes in which your selection is swimming. Make no purchase if any of them is in trouble. Ignore the odd torn fin or tail as long as it is not inflamed. Marine fishes are belligerent and often do a little damage to each other but the damage will soon heal. Do not buy a fish with a chunk out of a fin since it may never grow back perfectly, even though it usually does. Sometimes the color does not return, for instance, and you have a permanent blemish.

Look carefully at your chosen fish. It should be plump, the belly should not be hollow, and there should be no wasted musculature on the head or body, such as the well-known razorback appearance of a starved fish. This is characteristic of cyanide-caught specimens that have full bellies and wasted bodies. If the fish looks a little thin and you are still keen to purchase it, ask the dealer to feed it. Make sure that it is eating. It is not a bad idea to ask for any proposed purchase to be given a few flakes or other suitable food just to make sure.

The eyes should be clear. The fins should be held out naturally, neither being

One of the collecting sites for an aquarium favorite, *Holacanthus clarionensis*, in the Revillagigedo Islands.

clamped nor rigid, and moving in unison. The mouth and gills should be undamaged. There should be no sign of infection or injury. The fish should be breathing at below 100 opercular movements per minute if small, even slower if a relatively large fish. Infection or parasites often strike the gills first, so a healthy looking pair is important. There should be no undue redness or brown coloration, and no excessive opening and shutting even if at a normal rate.

Next look for symptoms of disease. There should be no blemishes, torn scales, or signs of the two common diseases, velvet (*Amyloodinium*) and white spot (*Cryptocaryon*). White spot is easy to see. There are obvious pinhead or smaller white cysts embedded in the skin or gills. To detect velvet, look obliquely along the sides of the fish for a white, powdery appearance. Damselfishes in particular can be quite heavily infected and yet show no distress. If there is a particular fish infected with velvet that you are nevertheless keen to have, it is not that hard to cure. Keep the infected fish in quarantine as outlined below. Any other fishes bought from the same shop at the same time might as well go in with it. They will probably have been dipped out with the same net. Therefore, it is likely they are infected even if they

are from a different tank.

Some fishes behave differently from the general run. Wrasses, blennies, gobies, and a few others do not necessarily swim up into the water column. Wrasses in particular are liable to be buried in the sand. Make allowances for particular genera, but otherwise expect to see active fishes swimming around freely.

Collecting a *Dascyllus* sp. by hand net in the Philippines. Notice the use of the larger net to surround the area being worked so that the fishes cannot flee.

job of it and does not stress the fish. Do not expect him to select an individual fish from a tankful of the same species but do expect him to show you what he has caught for you to approve.

Compatible Tank mates

Many marine fishes are territorial, particularly those from a coral reef. They typically live in a restricted area, sometimes of only a few square feet. Anemonefishes rarely go far from their home anemone. Fishes guard their home and chase others off to a variable extent, tending to be especially vicious toward members of their own or related species. Some nevertheless live in communities, like anemonefishes. They tolerate their own species while being hostile toward others (or at least some others). Mated pairs are the most likely to chase everyone else away guarding their nesting site if they have one. Yet a territory may be shared by several species as long as they do not compete for food or shelter. Angelfishes, for example, are most unlikely to tolerate each other if both are large adults. They do

Watch them being caught, looking out for any ineptness on the part of the person doing it. If he chases them fruitlessly around the tank with a small net they may eventually become damaged and should be rejected. A good catcher has nearly empty tanks, as far as decorations go, and uses two reasonably large nets. This makes for a quick

not mind a swarm of *Dascyllus* around them, and sometimes they tolerate small juveniles of even their own species.

What all this means is that you do not in general follow the freshwater aquarist's habit of buying pairs of marine fishes. Either buy single specimens or a small school of half-a-dozen or so. Which is appropriate you can determine from the literature or ask your dealer. A true mated pair, rarely available, is a different proposition. You can also buy a number of juvenile fishes, such as angels, if you are lucky enough to find a collection of the same species. Eventually you can select a mated pair from them. Expect bickering as they grow up and be ready to separate the combatants. Less stressful on the aquarist is a similar project with damselfishes, including the anemonefishes.

Predatory fishes, other than most triggers, can often be kept together. This is as long as they are of a similar size and are kept with other tough, large fishes. Remember that good swallowers, such as anglerfishes and lionfishes, can engulf others nearly as large as themselves. Be sure that they have only bigger companions. Slow feeders, such as pipefishes,

Adult Clarion Angelfish may move up in the water column without hesitation. Perhaps some food item is floating by.

The Leather Bass, *Epinephelus dermatolepis*, heads for refuge as a diver approaches.

seahorses, and dragonets, cannot compete with the general run of other species. Such fishes should be kept together. If kept with fast moving fishes, they will miss out on practically all of the food and gradually starve. However, in a reef aquarium fishes like mandarins can usually make a living by wandering around all day picking up tiny creatures. Unfortunately, this sometimes includes favorite tubeworms.

A great advantage of maturing a tank with ammonium salts is that many fishes may be put in together from the start. This avoids attacks that are often made by old inhabitants on newcomers. The group has a good chance of settling in peacefully together since nobody has an established territory. There may be an amount of peck-order bickering that does not usually develop into anything serious. Further additions may give trouble. It pays to start with the smaller, most timid and peaceful of your intended purchases. They will then be the established owners of the aquarium and its furnishings. The normally more belligerent newcomers will be less likely to cause trouble. Sometimes the difference such a strategy makes is

A fine specimen of the Maroon Clownfish, *Premnas biaculeatus*. These are popular with aquarists as they come in several different shades of color.

spectacular. The anemonefish, *Premnas biaculeatus*, for example, must be a dominant individual to remain intact. Put him in the aquarium with the first batch and he settles in. He even tends to bully the others if he is fairly big. Put him in later and he gets chased around the tank and his fins torn.

If you expect trouble or experience it unexpectedly, rearrange the scenery. This is not usually feasible in a reef tank, but it is possible in most others. A rearrangement puts everybody back to square one. It may upset some of the older inhabitants, but rarely seems to cause serious setbacks. It is often much easier than trying to fit in a partition of some kind to separate the newcomers from the others. Not only does the partition demand a rearrangement anyway, but it cuts into water circulation. It may demand considerable ingenuity if it is not going to cause trouble of its own.

Quarantine and Treatment

If you know that the pet shop from which your purchases come quarantines its fishes and treats for any suspected diseases, you may safely put them into your aquarium without treatment of your own. But how often do we know this? It pays to quarantine all incoming fishes for absolute safety. But how many aquarium keepers have such facilities? Not very many. In an aquarium without invertebrates, it is feasible to dose the water with copper. This gets rid of velvet, white spot, and perhaps a few other surface infestations each time there is a new introduction. But copper must not be added too often. Some dealers keep a low level of copper in their tanks for the same purpose. Ask about that and omit the copper from your tank if the shop does. Some use antibiotics with a similar purpose, although they are not a cure for the diseases just mentioned.

There are several popular quarantine methods. The mildest is to put all new fishes into an almost bare tank. The tank should have a heater and an airstone. It is usually 5-15 gallons in capacity, according to needs. The water is changed partially every few days. The quarantine is maintained for 2-3 weeks. If no untoward symptoms are seen, the fishes are then moved into their permanent home. Most aquarists prefer to add some kind of treatment. The most common is copper sulfate at around 0.15 ppm metallic copper. This may perhaps be combined with the addition of formalin, but I would not recommend it. A one hour formalin bath prior to transfer to the quarantine tank is better. This should contain 1 standard teaspoon (5 ml) of concentrated formalin per 5 gallons. This helps to kill off flukes and other external parasites. Some aquarists dunk their new fishes in fresh water at the right pH and temperature for a few minutes only. Care must be used to remove them at any sign of shock. This rather drastic treatment kills off many parasites and bursts cysts of protozoans. My feeling is that already stressed fishes are better off without further shock, which undoubtedly occurs whether it is manifested or not.

A more elegant form of quarantine is the permanent quarantine tank. This has undergravel and carbon filters. The latter is turned off when any medication is given. Most medications except antibiotics do little harm to the undergravel filter. The filter in turn does not remove most of them. The tank is then treated as a normal aquarium and medication is given, if at all, as above. The fishes are happier in such a tank. They get accustomed to your usual routines of feeding, lighting, etc. The drawback to such a system is a tendency for the quarantine tank to become a second aquarium. Then a new one is added and so on. On the other hand, this may not be a drawback. Perhaps it is the nicest way for a hobby to grow.

Despite everything, no aquarium is sterile. The main defense against disease is a healthy tank full of healthy fishes that are in fact infected mildly with several diseases that do not become serious. The fishes' immune systems keep the diseases in check and the very mild infections keep their immunity up. White blood cell activity and antibody production, although not quite like our own, still exists. As long as conditions are good, nothing is going to happen. That is why quarantine is

Two freshly caught Clarion Angelfish. These are subadults as indicated by the rather indistinct body striping.

recommended. The sudden introduction of a new, infected fish can upset the balance, either with a massive assault from a common disease already present but dormant, or with the introduction of a new disease.

If a new fish is not in good condition it can actually catch a disease from the aquarium. This is another reason for a quiet quarantine period in which the fish can adjust. Otherwise, in goes the new fish, it catches white spot already in the tank, and the pet shop gets blamed for selling you an infected fish. Meanwhile the new fish breaks out in a severe attack. It releases cysts that give rise to swarms of tomites, the free swimming infective stage. The other fishes succumb because this is more than they can handle and you have a problem. The same thing can happen when an existing fish becomes weakened for some reason—bullying or a hunger strike for example. It then falls victim to something and infects the others. This is a good reason for removing such a fish when you notice that all is not well. A sudden chill or a wave of ammonia can hit the whole community with instantaneous effects and all the fishes get sick. It is a wonder we can keep fishes at all! In fact, these are very rare events in the hands of seasoned aquarists. This does explain why beginners get into trouble sometimes.

Fish Physiology

To help you to "think like a fish" a knowledge of its physiology is a great advantage. Fishes differ from ourselves in a large number of ways. Some of the ways are difficult to appreciate. For instance, we get hungrier when the weather is cold and our appetite drops off in hot weather. Not so with a fish. He gets hungrier as the temperature rises. Coldwater fishes barely eat when the temperature is near freezing. Why so? Because we are warm-blooded. Much of our food intake goes toward keeping warm. The colder it gets the more we need. A fish is cold-blooded. Its activity falls as the temperature falls. The colder it gets the less it needs. At very high temperatures its appetite eventually also falls. That is because it is suffering from abnormal conditions, including oxygen shortage and a complete upset of its physiology. Failure to appreciate these facts can lead to inappropriate feeding and fouling of the water or semistarvation.

Living in Water

Air offers little resistance to everyday movements. It is also charged with plenty of oxygen and little carbon dioxide. It does not make much difference to the effects of gravity on our bodily functions. Water offers a good deal of resistance, holds very little oxygen, and plenty of carbon dioxide if the gas is available. Water virtually abolishes the effect of gravity on a fish swimming in it. Being a fish is more like being a bird than a mammal—living in three dimensions rather than the near-to-two of most of us. Even then, though, gravity effects are different.

Instead of the easily-managed gas containing 20% oxygen with which land animals deal, a fish has to extract its oxygen from a heavy fluid. The fluid contains only 8-10 ppm oxygen in fresh water and 7-8.5 ppm in sea water in the temperature range of 60-80°F. The lower figures apply to the higher temperature—the warmer the water, the less oxygen it holds. Thus the fish manages rather effectively with its gills. At the same time it has to control its salt and water balance. Freshwater fishes take in a lot of water through their gills, drinking almost none and excreting a great deal

An attractive land crab collected at Socorro Island.

via their kidneys. Marine fishes lose water from their gills and other surfaces. The saltier water surrounding them actively extracts it from their blood and tissues. This is a process known as osmosis. So the fishes drink a lot of water with the gills and the gut excreting a lot of the salts it contains so as to leave a more dilute solution of the right salts in the blood. Sea water is approximately a 3.5% solution of salts whereas a fish's blood is only about 1%. The whole process is known as osmoregulation.

Most fishes are stenohaline. This means that some can live only in fresh water, some only in sea water, but not both. Some, however, are euryhaline. This means that they can live either in fresh or salt water. Take salmon for example. In particular, their kidneys are different in structure from those of stenohaline fishes and enable them to adapt gradually. Even so, juvenile salmon can take a sudden change without obvious distress, although trout cannot. Elasmobranchs differ from teleosts (bony fishes) in that they secrete urea into their blood. This has an osmotic effect enabling them to absorb water instead of losing it. Some of them, particularly sharks, can swim into brackish or

Scats (*Scatophagus argus*) are among the few fishes that can claim to be favorites with both marine and fresh water aquarists. They do best in somewhat diluted marine water or brackish water.

even fresh water and survive. They lose some of the urea, thus becoming euryhaline or partially so.

Although stenohaline, marine fishes can take considerable changes in specific gravity as long as they are gradual. Otherwise, as already mentioned, they are liable to osmotic shock. The general lower limit appears to be about 0.013; the upward limit has not been well investigated but is certainly somewhere over 0.040 for the majority. Aquarium favorites such as scats (*Scatophagus* spp.) and moonfishes (*Monodactylus* spp.), are fully euryhaline. In most cases, they depend on the presence of calcium when in completely fresh water for calcium is important for osmoregulation. The process fails in the absence of calcium. Calcium is also important in membrane physiology.

Breathing in Water

The gills are richly supplied with blood vessels that are connected directly to the heart. All of the blood from the heart flows through them. This is scheme no.1 for taking up as much oxygen as possible from the water. The blood then flows through the gills in masses of small vessels in the opposite direction to the inflowing water that bathes them. This is scheme no. 2 for taking up as much oxygen as possible. This is a countercurrent arrangement such as has been seen in a protein skimmer. If the blood and the water flowed in the same direction, at most 50% of the oxygen in the water could enter the blood. By the countercurrent arrangement, up to 80 or 90% can be taken up. After the blood leaves the gills, it is mainly collected into an aorta (large blood vessel) and distributed to the various organs as in our own circulatory system. Veins from the various parts of the body then return it to the heart. This is done mostly via two large collecting veins, one from the liver collecting blood from the gut, the other collecting blood from the rest of the body. A fish's heart is a simple 2-chambered pump, not 4-chambered with separate lung and bodily circulations like our own.

Fishes have red blood corpuscles containing hemoglobin similar to higher vertebrates, with the exception of some fishes that live in very cold waters. These take advantage of the higher solubility of oxygen in colder water and can get away with no corpuscles. Sufficient oxygen is carried in the serum, the liquid part of the blood. They also have white corpuscles performing the same functions as ours. They mop up bacteria, inactivate viruses, and produce antibodies.

The Swim bladder

Without a swim bladder, a fish is somewhat heavier than either fresh or salt water. This would make it laborious to swim since a good deal of energy has to go into maintaining its position in the water. With it, a fish can adjust its weight to equal that of its surroundings saving a lot of fuel and effort. The swim bladder (or gas bladder as it is commonly called) is filled with gases secreted into it from the blood stream or is directly connected to the throat via a tube, the pneumatic duct, through which it is initially filled. Some fishes have both mechanisms. Clearly, the air entering via the pneumatic duct must be gulped in at the surface. It also provides a method for rapid changes. The secretion of gases from the blood is a slower method of adjustment. Special glands control the gas exchange between the blood and the swim bladder. Adjustments may take several hours or longer.

It appears that the swim bladder was originally a primitive lung. It became the lungs of amphibians as they evolved from fishes. In other fishes it became what it is today and has lost its lung-like role. Many species of fish breathe air in addition to using their gills. They extract the oxygen from it by one means or another. The ones best known to aquarists are the freshwater labyrinthine fishes, such as Siamese fighters (*Betta* spp.). They are now so dependent on their air-breathing equipment that they drown if prevented from gulping air at the surface of the water.

Special Senses

The eyes of fishes work much like our own. They are more like ours than those of most mammals, which are color blind. Such studies as have been reported suggest that many fishes see colors. Almost certainly the typical run of

These Clarion Angelfish were attracted to the photographer by a freshly crushed sea urchin.

coral fishes do. Their eyes contain many different visual pigments. It seems likely that they see more colors than we do, like many insects that see into the ultraviolet as well as our "visible" spectrum. Insects like bees that visit flowers may be presumed to have developed color vision in parallel with the flowers' development of color. What caused tropical fishes in particular to do so? What caused tropical creatures in general to become so colorful?

The chemical senses of fishes may be supposed to resemble our own senses of taste and smell. They are not confined to the nostrils or mouth, though. Chemoreceptors are found in various areas of the body in some fishes. Some are even associated with the lateral line. External parts of the head, such as the snout or barbels, commonly carry them. Species like the goatfishes actively seek food with their aid. Whether they respond to exactly the same stimuli as our own receptors is only partially known. In some cases they do not. An interesting example is the drug, chloromycetin, that tastes horribly bitter to us yet is unnoticed by fishes when mixed with their food.

Receptors for pain, heat, and cold have been little studied in fishes. We are uncertain about them.

A newly caught Clarion Angelfish (*Holacanthus clarionensis*) being brought to the surface.

General evidence suggests that they exist and that fishes probably feel pain. But it is possible that they do not and merely have avoidance reflexes that makes it look as if they do. We just do not know enough about the fish brain to know how it interprets signals from such receptors.

Fishes do not hear by the same mechanism as mammals. They can hear relatively low-pitched sounds, though. There is no cochlea. This is the curved organ containing graduated hair cells that functions in vertebrates. Otoliths, little bones that lie over sensory hairs in the inner ear, are present. They vibrate to sound and are the mechanism of hearing.

The maintenance of balance in vertebrates in general depends on a number of factors. The semicircular canals in the ear are of prime importance. Input from the eyes and from the muscles also contributes. Fishes have the standard equipment—three canals looped at right angles to one another. They give information about the position of the body in all three dimensions. The whole organ is called the labyrinth and is comprised of the canals plus various chambers and otoliths. The pressure of the otoliths on various parts of the organ informs the fish of its position in the water. In our own case, they tell us about posture only. They do not function in hearing. In both cases, various "righting" reflexes ensure that we and fishes take the actions necessary to remain upright. Oblique light, though, can confuse a fish and make it swim at an angle.

The Lateral Line

The hair cells in the semicircular canals of a fish are imitated in the lateral line complex. This is a system of canals that branch around the head and run down the sides of the body. Either in grooves or in tubular canals with openings at intervals to the surface, organs called neuromasts, modified hair cells, are the functional parts of the lateral line.

They respond to pressure changes and are probably important in blind swimming at night or in keeping station in a school. Some neuromasts are sensitive to electric fields. They act as a kind of radar that may back up the others by providing additional information and may serve as a means of communication as well. Fishes produce electric fields when they use their muscles. Thus they radiate information continually at a level of at most a few volts, more usually a fraction of a volt. This phenomenon has been adapted in some fishes to produce electric shocks (up to 10 kw in the electric eel, at a pressure of 500 volts). The shocks are used in defense and to stun prey. Layers of special muscle, like the plates of a battery, are connected in series to build up such currents.

A good catch of Clarion Angelfish. These do not seem to be spooked by a hand placed in among them.

Feeding must be done carefully in a tank so decorated because of the possibility of food items falling into a crevice and being hidden from view. The fishes may not be able to get to such food items, which may eventually cause pollution.

FOODS AND FEEDING

Much more is known about the dietary needs of salmon, trout, carp, channel catfishes, and eels than those of most aquarium fishes. Perhaps the least is known about tropical reef fishes. We can only extrapolate from the results of work with the food fishes, sport fishes, and manufacturers of aquarium fish foods and hope that our general statements are near enough to the truth. Certainly, most modern prepared fish foods can form a predominant part of the diet with good success. Many fishes live on them entirely for long periods.

Proteins

Proteins are made up of amino acids. Long chains of them go into the making of almost innumerable natural proteins. Yet there are only a few essential amino acids—ten of them in the species tested. A protein mix containing these necessary amino acids is a first class protein. Others lack one or more of them, eventually causing the appearance of deficiency diseases if they are the sole source of protein. The total protein content of a diet, therefore, is only an indication of its value. The diet must contain all of the essential amino acids to be satisfactory.

Various vertebrates differ a great deal in their protein needs. Cats, for example, need a high percentage of protein in their diet. So do fishes. Figures about to be discussed refer to the dry weight of a diet. They do not refer to the percentages in meat or vegetables, for example, that may have a high water content. Percentages given for flakes and other dry foods should conform to them. This is not so for frozen or live foods. Young tropical marine fishes just about top the bill for protein requirements. Older or colder water fishes need less, but not much less. As an example, young chinook salmon need 47-56% of first class protein. Half-grown ones need 43-47%, and adults 40-42%. Consider, however, that these are not tropicals!

A satisfactory dry food should contain not less than 45% protein that we hope is first class. This is rarely stated. If it is not, supplements of the missing amino acids should have been added. Foods like frozen brine shrimp may well have very much less because they are mostly water. We feed them to the fishes for other reasons, though.

Fats

Just as there are essential amino acids, so are there essential fatty acids. Insufficient research has been done in fishes for us to be sure of all of them. Linolenic and linoleic acids are needed by carp, but only linolenic by trout. These are polyunsaturated fatty acids—oily, not solid—now famous in human dietary schedules. Liquid fats are much better digested by fishes than solid ones. Their level in the diet does not have to be high, at most 15%. Tropical fishes can take hard fats better than coldwater ones. Beware that they may not contain enough of the essential fatty acids. Hence feeding beef or ham to tropicals may be dangerous except as a treat. Much of the energy produced by a fish may be derived from fats, since they yield more than twice as much per gram as proteins or carbohydrates.

Carbohydrates

Starchy foods do not figure in fish metabolism in the way they do in that of higher vertebrates. Natural fish foods tend to be low in carbohydrate content. Too much in the diet can cause trouble, such as liver damage. A suggested maximum dry weight is less than 20%. Many diets contain half that amount or less and still prove to be satisfactory.

Vitamins

For the most part, vitamins perform the same functions in fishes as in higher mammals and must be present in the diet. As usual, detailed requirements have been studied only for the popular food and sport fishes. It is assumed that others have similar needs. Some fish livers are loaded with vitamins A and D, the fat-soluble vitamins. It is not clear whether they are needed by the fishes or are stored as unwanted byproducts. Indeed, studies have shown that excess of both in the diet is harmful. D_3 (cholecalciferol) is needed in the trout for calcium uptake. Of the B vitamins, B_1 (thiamine) is needed in fishes for the normal utilization of carbohydrates, B_6 (pyridoxine) is needed for protein utilization, B_2

(riboflavine) is associated with normal eye function, B_{12} with blood development, and B_3 (niacin) with gut and skin development. Various other water soluble vitamins have been allotted different functions resembling those in higher vertebrates. Vitamin E (tocopherol) has to do with fat utilization. Vitamin C (ascorbic acid) is required for cartilage and bone development and normal healing. Biotin, choline, folic, and pantothenic acids have to do with normal growth, a rather vague concept. Vitamin C is only needed by some fishes since, as with mammals, the others can manufacture their own. Very few vitamins are found in sea water.

Minerals

Sea water contains lots of minerals. Their availability to fishes depends on how readily they pass into its body via the gills and perhaps the skin. The mineral content of normal foods is thought to be adequate in most circumstances. Manufacturers of flakes and other dry foods sometimes add small amounts of them to make sure. Test diets used in establishing the need for various food constituents are made up with mineral supplements covering all likely needs for the same reason. This includes copper, cobalt, iodine, phosphorus, manganese, etc., all in trace or conservative amounts. Concentrations of minerals like copper and zinc in the food can be much higher than is tolerable in the water. Safe levels of the two mentioned are respectively less than 0.015 ppm and 0.05 ppm in the water but they have been used at 12.5 ppm and 14 ppm in long-term diets.

Assimilation

The term above means the ability to turn food into body tissues. It varies considerably from one vertebrate group to another. We often read that only 10% of the food we eat becomes part of our bodies, and that only 10% of the vegetation that herbivorous animals eat becomes a part of their bodies, from which we obtain our meat and meat products. Much of this is rubbish. Naturally, a grown man or cow does not turn much of what is eaten into flesh, but growing creatures may. Thus, while some wild animals and man do not much exceed the stated 10%, domestic animals, bred for thriftiness, convert up to 30% and chickens 35%. Brown trout do better than 50% with a good diet. We may assume that other fishes do as well. This is why they need such a high percentage of protein in the diet, and what seems to us such a meager diet to remain healthy. Fishes are remarkably accommodating in contrast to birds or mammals. These latter creatures grow skinny and miserable on restricted rations. Not so fishes. They just do not grow much. In many cases they can mature and breed at a fraction of the normal weight for the species. Naturally, if you starve an adult fish it will get thin because its bones cannot shrink. If you simply restrict the diet as it grows it is a different story. The restricted diet must, however, be nutritious and high in protein.

Natural Foods

Fishes may be herbivorous (vegetarian), omnivorous (eating almost anything), or carnivorous (eating meat). So-called vegetarians are rarely strictly so. They will eat other foods as well. Some carnivores do in fact eat nothing but flesh. Examples of herbivores are surgeonfishes and some angelfishes that eat mainly algae. Examples of carnivores are triggerfishes and scorpionfishes that eat various other animals. The majority of fishes are omnivores. Young fishes and some small adults are plankton feeders. They eat mainly the zooplankton (small drifting animals and larvae) that in turn feed on the phytoplankton (one-celled floating plants). As they grow up most species leave the plankton, of which they were part, and change diets as well.

Many young fishes are cleaners as well as plankton feeders for a period. As they grow up, except in the case of some very small species, they feed on other materials. They may continue to eat plankton as well. For example, some juvenile angels and chaetodons eat plankton, algae, and pick parasites from other fishes. Later they transfer to sponges or corals and larger foods. In the aquarium most of them accept different foods from that which they obtain in the wild—some awkward customers will not.

A young, but not too young, fish is more likely to adapt to conventional foods than a wild-caught adult. Very young chaetodons are hard to feed because they need almost constant supplies of tiny live food. At around 2 inches in length they can get along with sporadic feeding on preferably live or frozen foods. If caught at a large size they may be hard to adapt to aquarium foods. They will not eat enough to thrive if they eat anything at all. Luckily, fishes such as damsels, wrasses, blennies, and gobies usually wolf down anything offered and give no trouble.

Fishes should be offered as large a choice of foods as you can provide. It is a great mistake, even though they seem to be going along okay, to restrict them to prepared flakes or frozen brine shrimp. This is true for two reasons. First, they may be getting insufficient amounts of one or more constituents, such as first class proteins or vitamins. Second, they may become addicted to a particular food and refuse others, preferring rather to starve. So keep providing a good mixture. Rotate or change

foods from time to time. You can even make up your own mix with a lot of substances in them. Flakes, for example, have many different constituents. There is no guarantee that any mix is adequate or that your own is constant in composition. The best and easiest way is to use flakes today, frozen food tomorrow, and live food next. A good aquarium will have its own growth of algae for those that will eat them. A miniature reef will offer all sorts of delights.

Live Plants as Food
The algae just referred to may be encrusting, filamentous, or macroalgae (those that have a leaf-like thallus and a holdfast). Herbivorous fishes crop away at all of these so if you wish to keep a good display of the macroalgae you may have to choose the fishes carefully so that it is not destroyed. The algae are doing a good service. They purify the water and provide food for the fishes at the same time.

You may choose to go in for a "sterile" aquarium. This contains bleached dead coral skeletons that are taken out and cleaned as soon as a lot of algae grow on them. This is going out of fashion, though. The presence of some algae makes a decorative addition to an aquarium and performs useful functions. To keep a nice balance, there must be adequate light. There must be few or no eaters of the higher algae, although some fishes are needed to crop the lower types, hair and encrusting algae. Perhaps a periodic algal food is a good idea. It is sold at your pet shops. It is a mixture of trace elements and vitamins that are needed for good algal growth.

If there is not sufficient algae in a tank to feed herbivores, other foods must be substituted that resemble them. Chopped up vegetables or plants, such as lettuce, spinach, peas, lentils, etc, may be eaten readily or they may not. The secret is to freeze them, particularly the leafy vegetables. They go mushy and are usually more acceptable. Perhaps after freezing they resemble seaweeds. Anyway, many a fish that will not touch them unfrozen will gobble them down after they are frozen. They are more digestible, too, because the plant tissues have been disrupted and the cellular contents made available. You can also try dried seaweed from Oriental stores or health stores. Soak it first in salt water.

The living rocks for a reef tank should be arranged so that they will not fall if jostled by nets or other equipment used in the tank. Falling rocks could kill animals or, worse yet, break the bottom or sides of the tank.

Caulerpa species propagate by means of runners. These can clearly be seen in this healthy stand of *C. mexicana.*

Live Animal Type Foods

Now that frozen and other foods of high quality are available, it is not as essential to feed live foods as it once was. Even so, with freshwater fishes at least, it has been found that they are still needed when breeding. With livebearers, live foods improved litter sizes considerably compared with those from fishes receiving only preserved or even meaty materials. It is still true that a few feeds of live foods increases the growth and survival of young fishes after the early period when, as fry, they normally receive them.

It is best to offer marine fishes live foods from freshwater sources. This is usually easy to obtain and free from marine parasites. If a collection is made from shore pools or plankton, a dunk in fresh water for a few minutes helps to eliminate trouble. It does not matter if the dip harms the creatures concerned. They will be eaten rapidly anyway. Practically all of the foods suitable for freshwater fishes can be fed to marines. Those that breathe at the surface, such as mosquito larvae, can be given freely as they will live until eaten. Others should be given in small quantities so that they do not foul the water when they die in the salt water. Also, most of them are rich in saturated fats.

Whiteworms, Grindal worms, earthworms, and tubificid worms are all fatty and they can be either purchased or cultured.

Insect Larvae

The larvae of insects that live in the water are valuable. They breathe air and so do not compete with the aquarium residents for oxygen. When collecting them for feeding to marine creatures, you can be quite careless. Predators like dragonfly larvae could survive in and be a pest in a freshwater tank. Mosquito larvae are not as easy to collect as they used to be since local councils are apt to eliminate their favorite pools or douse them with insecticides. However, they can still be found, even in brackish or salt pools that form above high tide levels

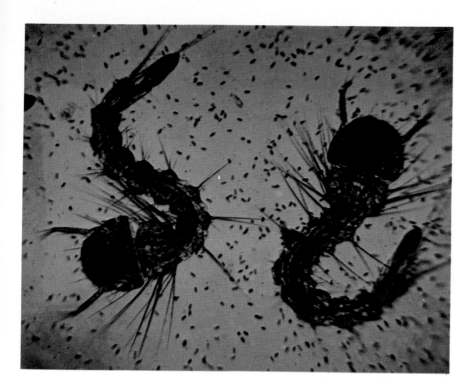

Live mosquito larvae are a good food for marine fishes and, being air breathers, last quite a while if not eaten right away.

Tubificid worms are primarily a freshwater fish food but they can also be fed to many marine fishes if done so with caution.

and last long enough for a hatch to occur. If you collect them from salt pools, store them in fresh water for a short time to get rid of pests.

Mosquitoes lay eggs in the form of little dark brown rafts, each with hundreds of individual eggs. These hatch into tiny larvae that hang under the surface of the water. They grow during the next 8 or 9 days up to ½ inch long, feeding on debris and vegetable matter. They then turn into comma-shaped and hard-bodied pupae. In a few more days

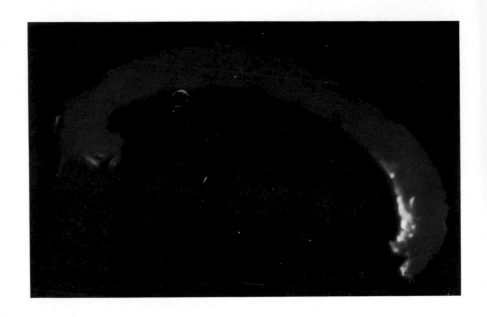

the pupae give rise to adult mosquitoes. The larvae and pupae are collected with a fine net and can be sorted for size with kitchen sieves if necessary. Store them cold, not frozen, in closed jars of water. They will be slowed down in development and available for feeding over the next few days. Most fishes will eat pupae as well as larvae. If adults develop, they can be drowned by shaking the jar and fed to the fishes.

Bloodworms are larger red larvae of the gnat *Chironomus* and are a fine food for fishes. They are found in pools, but less often than mosquito larvae. They also live deeper in the water. Glassworms are the

Bloodworms or chironomid larvae are excellent for both freshwater and marine fishes. These can be fed live or prepared.

larvae of another gnat, *Chaoborus*. These are found in cold fresh water and are equally desirable as food. Mealworms are the larvae of a beetle and are not found in water. They are cultivated commercially and sold as fish bait. They are fine for large fishes and can be chopped up for smaller ones. However, they are fatty compared with the foregoing and should be used with restraint. The same applies to blowfly maggots. They are not usually sold but are easily cultivated. Place some blowflies in an enclosed vessel with some meat. They lay their eggs on the meat and the maggots feed thereon.

Crustaceans

The most famous live fish food is the brine shrimp, *Artemia salina*, that, in its newly hatched state, has been fed to fry and small adults of all kinds for many years. Many freshwater fry can take newly hatched brine shrimp as soon as they are free-swimming themselves. Unfortunately, few of our aquarium

Mealworms are the larvae of a beetle. They can be fed whole to larger fishes but need to be chopped up for the smaller ones.

marines can do so. They must be given smaller organisms. Half-grown and adult brine shrimp are available live in pet shops and are a fine food for juvenile and adult fishes. They are low in fat content and so should not be fed exclusively. They are available frozen or freeze-dried.

The dry "eggs" are really embryos in a state of suspended animation. They accumulate around the edges of salt lakes or the evaporating pans of salt-works. Here they often are blown by the wind into heaps and can be readily collected. They are tiny brown spheres, hundreds per pinch, that can be stored, preferably in a vacuum or in nitrogen, for many years. Even in small

Collecting brine shrimp eggs is difficult work. Imagine how many millions (billions?) of eggs are in this one scoop of the net.

packets in pet shops they survive for a long time and continue to be viable. When placed in sea water or in a sodium chloride solution, they hatch in a few days or hours if the temperature is over 70°F. They do not need brine at that stage. In fact, they often hatch best in rather dilute salt water. This may be so since they typically get washed back into the brine when it rains and the water is diluted. The eggs are also available shelled but these are rather expensive. The shelled eggs can be placed directly into the marine aquarium to be eaten as they are or after they hatch, although not many usually survive to hatch. That does not matter since they are more nutritious as eggs than as nauplii (the early larvae). Both the eggs and the nauplii have sufficient fat content to be a complete food for fishes or their fry. It is only the adult shrimp that have a low fat content.

Using ordinary eggs, a hatching procedure must be followed that avoids getting unwanted shells into the aquarium. The best one depends on how many are to be handled. If only a few hundred thousand eggs are to be hatched, they can be carefully floated in a shallow dish of sea water, a prepared mix, or two heaping tablespoons of butcher's salt per quart of tap water. Do not use table salt as it has unwanted additives. Use no more than one half teaspoon of eggs per gallon of water. This is best measured with one of a set of kitchen spoons rated at ¼, ½, and 1, etc. Cover the solution. The temperature must exceed 70°F or the eggs will not hatch. At 70°F or more they may take 48 hours; at 80°F they may take only 24 hours. Do not let the temperature get any higher than 85-90°F. Spread the eggs evenly and do not sink any of them. As they hatch the nauplii sink to the bottom for about eight hours. Then they swim up into the water as they are attracted to light.

Wait for a day longer than the estimated hatching period to be sure that all that are going to hatch are out. Then harvest the nauplii. To do this, run a flexible siphon tube below the water surface and transfer the swimming nauplii into another vessel or onto a very fine filter cloth if you want to collect them dry. The latter is best as you can reuse the water once or twice. It is best not to pour it into the aquarium. If one end of the hatching tray is brightly lit, nearly all of the nauplii will collect there. You only need to run off part of the water. The egg shells will be left behind to be discarded.

If millions of nauplii are needed, it is easier to use deep vessels with aeration. Fill large fruit juice or similar bottles around a gallon capacity with ¾ volume of the hatching fluid. Put an airstone in each. Use a teaspoon of eggs per gallon. Aerate briskly during the hatching period. When all is ready, turn off the aeration and wait for 15 minutes. The egg shells and nauplii that have been whisked briskly around during hatching will now settle out, the nauplii swimming, some shells floating and some will be on the bottom. The nauplii can now be siphoned off as before. This can be aided by a beam of light directed onto one side of each bottle. This method should be used to hatch shell-less eggs because they sink and do not hatch as well by the shallow pan technique. Use it even with small batches. These eggs come in liquid suspension in small containers. Each drop of liquid usually contains 1500-2000 eggs, or 30,000-40,000 per ml. A small 1 oz-container (28 ml) thus holds around a million eggs.

Brine shrimp can be fed easily with baker's yeast. They grow up quite rapidly to adult size at 6-8 weeks in brine at 70-80°F. The nauplii carry yolk on which they live for three or four days. After that they must be fed. Before that stage, transfer a few, about 500 per gallon if you wish to grow them right up, more if not, to an appropriate brine. For San Francisco eggs, a brine made of 10 oz of butcher's salt (NaCl), 2 oz of Epsom salt ($MgSO_4$), and 1 oz of sodium bicarbonate (baking soda, $NaHCO_3$) per gallon is needed. This is twice the strength of sea water and more alkaline. Other types of shrimp need different brines. Some will grow up in double strength sea water or even sea water itself. Information about their needs is often on the container or can be obtained from the manufacturer.

Many different methods are used for hatching brine shrimp eggs. This is one of the more elaborate ways of obtaining a large number of nauplii.

Newly hatched brine shrimp nauplii. The bright yellow-orange object is a discarded egg shell.

Brine shrimp can be fed easily with baker's yeast. They grow up quite rapidly to adult size at 6-8 weeks in brine at 70-80°F. The nauplii carry yolk on which they live for three or four days. After that they must be fed. Before that stage, transfer a few, about 500 per gallon if you wish to grow them right up, more if not, to an appropriate brine. For San Francisco eggs, a brine made of 10 oz of butcher's salt (NaCl), 2 oz of Epsom salt (MgSO$_4$), and 1 oz of sodium bicarbonate (baking soda, NaHCO$_3$) per gallon is needed. This is twice the strength of sea water and more alkaline.

Other types of shrimp need different brines. Some will grow up in double strength sea water or even sea water itself. Information about their needs is often on the container or can be obtained from the manufacturer.

Feed conservatively with just a few pinches of yeast kept in suspension with brisk aeration. Add more as the water clears. Once adult, the shrimps will breed. They can produce eggs that do not need to be dried before developing. You can have a rich culture of adults plus juveniles from which to collect moderate numbers for feeding to the fishes. Such cultures can be bought from pet shops. They can either be fed to the fishes after straining the water and giving them a wash with salt water for safety, or used as a basis for your own further breeding.

Water fleas, *Daphnia pulex*, are found in freshwater pools in temperate climates. They do not flourish if it is either too cold or too hot. They appear as red, green, or yellow clouds in the water, the color depending on the strain of fleas and their diet. The fleas can be collected with a fine net. They are often available commercially as well. They are a good food for marine fishes and live long enough in the aquarium to be eaten. Well washed in light

Shrimp (both fresh and salt water) are almost always accepted by marine fishes. They may be fed live or prepared.

tap water, they can be kept in cool storage in the refrigerator for several days. Their freshwater origin makes them a safe diet for marines because they will not be carrying parasites. If you catch them in large quantities, they are best transported in layers of soaking wet cloth rather than in vessels of water. Kept wet, they live for hours, but not days. Breeding water fleas is not easy. They only do well in large, cool vessels of 50 gallons or more. Aeration is preferable. They can be fed on liver powder, dried blood, or other similar powders rich in protein.

There are various other freshwater crustaceans welcomed by marine fishes. Most can be bred in large vessels as for *Daphnia.* Few

Daphnia pulex is another freshwater fish food that can be fed to marine fishes. They do not live long in a marine tank, however.

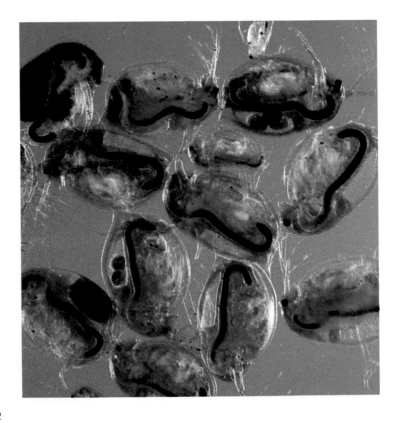

have common names. Smalls one are *Moina, Diaptomus,* and *Cyclops*; larger ones, really coldwater shrimp, are *Gammarus* and *Asellus,* which are found in both ponds and streams. *Myallela* breeds in warm water. Many species of saltwater crustaceans can be collected from tide pools and shallow waters. They can be dunked in fresh water for safety before feeding to the fishes. They fall into several groups. Small specimens of decapods (shrimps, crabs, etc.) make good food. Even larger ones are good if you keep triggerfishes, wrasses, or other large predators. Wrasses are particularly fond of crabs. Copepods other than the parasitic ones are eaten, however they sometimes are able to breed in quantity in the aquarium and become a nuisance. They swim, like *Daphnia,* by rowing with their antennae and are easy to recognize. Amphipods are relatives of the sandhopper. They have upright thin bodies like a human flea. Isopods are flattened the other way and are relatives of woodlice or pillbugs. Phyllopods, or fairy shrimp, are relatives of the brine shrimp. It is

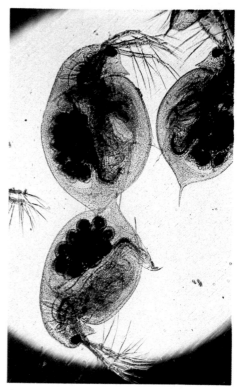

Young *Daphnia pulex* can be seen developing parthenogenetically in the egg chambers of these adults.

quite possible to collect parasites with some of these crustaceans but it is most unlikely if your hauls are made in temperate or cold waters and are intended for tropical fishes.

Fresh *Daphnia* is collected in freshwater ponds. They may transmit diseases to freshwater fishes but like most freshwater live foods they pose little problem to marines.

Copepods are one of the most natural foods of marine fishes. This is a species of *Cyclops*. Like brine shrimp they are positively phototropic.

Worms

Earthworms are good for fishes. They can be purchased and stored in leaf mold. Fed whole to large fishes and chopped up for smaller ones, they make excellent occasional treats. If you collect them from the earth, pull them through the fingers to get rid of the contents of the gut. Avoid the smelly, yellow dung worm. Tubificid worms *Tubifex* come in many species which vary from red to black in color and can be up to several inches long. They make burrows in filth

Tubificid worms are accepted by many marine fishes. Marine relatives of this freshwater puffer will probably eat them just as avidly. They tend to be fatty so feed with caution.

Whiteworms can be cultured and fed to both marine and freshwater fishes. Here they are being harvested with a plastic fork.

and wave their tails in the water to extract as much oxygen as possible. In the right situation they occur in thick masses. This happens usually in sluggish polluted water rich in all kinds of horrors. You can purchase them or collect them yourself if immune to disgusting processes. Once collected, they must be thoroughly cleaned by keeping them for a day or two in a bucket under a dripping tap. They must be kept that way until they no longer smell nor are surrounded by muck. Any solid masses should be broken up because the centers are apt to be a mass of dead worms. Once cleaned, they can be stored in a refrigerator in closed jars until they are rinsed and fed to the fishes. They do not last long in storage, though. They are too fatty for more than occasional use but are often used to tempt finicky fishes to eat— chaetodons may take them and nothing else at first. Stuff the worms into a piece of coral skeleton and fishes accustomed to eating coral may be fooled.

Whiteworms and Grindal worms, *Enchytraeus*

species, are also rather fatty but are good for treats to small fishes. They grow respectively to about 1″ and ½″ long. Roundworms, related to the tubificids, are found in damp soil. They flourish in small containers of soil when fed milky porridge, baby food, or the like. Use an inch or so of soil and put little pockets of food on the top. Add a few of the worms and cover them with a sheet of glass and an opaque cover to keep them dark. There will be rings of worms around the food in a few days. The worms will adhere to the glass as you lift it and can be scraped off and washed for feeding to the fishes. Whiteworms like a cool situation, up to 70°F; Grindal worms (named after Mrs. Morten Grindal who discovered them) like a warmer temperature, 75°F or so.

Microworms, *Anguillula* sp., are too small for any but fish fry. They might be well suited in their tiny younger stages as an early food for marine fry. Even

the adults are only ¹/₁₀ inch long at most. Perhaps they would die too quickly for marine use, but I have not heard of anyone trying them out.

Whiteworms also tend to be rather fatty, but are good for treats to small fishes.

Mollusks

It is perhaps a little too dangerous to feed live or recently chopped-up mollusks to marine fishes, particularly filterfeeders like clams or mussels. Fishes certainly like them, though. In fact, I have not

Earthworms make excellent occasional treats. They can be fed whole to large fishes or chopped up for smaller ones.

come across an example of transmitted disease. I had a royal empress angelfish (*Pygoplites diacanthus*) that would eat only freshly opened live mussels in any quantity. The fish lived for a year on them. It died eventually from bullying by a large Koran angel before I realized what was happening. The fish would not touch frozen mussels or even chopped up live ones. It accepted only a whole creature.

Fishes

Strangely, too much fish muscle is not good for fishes. Occasional feeds, though, are much appreciated. The exception is straight predators that are adapted to eating as many other fishes as they can catch. Since they are eating whole fishes, they get vitamins and other goodies that muscle alone tends to destroy. Unwanted freshwater fishes and fry are fine for marines. They provide a dietary supplement of essential amino acids, vitamins, and so forth. Do not feed predators on fishes that are too large, even if they can readily swallow them, for there are two unfortunate consequences that are liable to happen. Either the prey is regurgitated half digested and fouls the tank, or, especially with lionfishes, the overstuffed predator sits on the bottom panting away and gulping frequently. The fish is seemingly unable to either digest or throw up the meal. In extreme cases I have known the fish to die in this way.

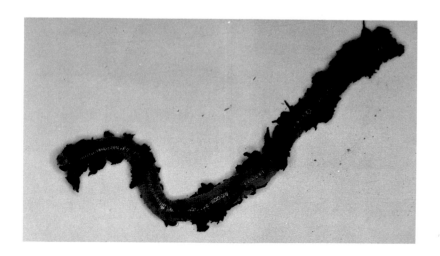

Frozen Foods

Many of the foods mentioned above, plus some others, are offered in frozen form. They come in packets of an ounce or two up to a pound or more. These are the next best thing to the live article. Vast numbers are sold to aquarists, tending to replace live foods. For most purposes, they appear to be an adequate replacement. The desirability of live foods for breeders must be borne in mind, though. Also, do not stick to one variety. This is particularly important when feeding the popular frozen brine shrimp as they are lacking in adequate fats. They are also very soft-bodied and may fail to stimulate the gut. Mysid shrimps are better. They are hard-shelled and readily eaten, even by seahorses. There is a good range of different shrimp available, from tiny planktonic forms to large krill.

Other frozen foods on the market include various mollusks, tubificid worms, small fishes, "plankton" seaweeds, and so forth. You can give the fishes a beautifully varied diet with this type of food alone. Add to these types various others offered for human consumption and you can vary it even further. I make up ice cubes of frozen foods. The food is a good mix of chopped-up shrimp, crab, baby squid, lettuce, and other vegetables. One cube makes a good meal for around thirty 2-4 inch fishes. Alternated with mysids or brine shrimp and an occasional feed of flakes, these provide a safe mix of different and mainly high protein foods. It is best to thaw out the frozen foods before feeding. This is not so much because ice-cold meals are probably bad for fishes, but because a voracious customer can swallow a half-thawed cube and rob the rest of the fishes of the full meal.

Beef heart is sometimes available in frozen packets for fishes but it should be fed with restraint. This is true of all meats, frozen or otherwise. Except for sharks and a few other predators, fishes do not consume red meat. It has a lot of saturated fat and is relatively indigestible. Chicken is better, although I have never seen it frozen for fishes. Whole fishes, even if chopped up prior to feeding, are better still. The fish muscle has vitamin-destroying properties and should not

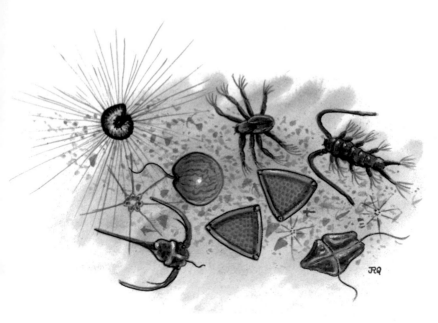

Planktonic animals are necessary for some of the smaller fishes and for most of the filter feeders. Packaged plankton is available from your marine pet dealer.

be offered as food without the rest of the fish.

Be careful when buying frozen foods. Be sure that they have been kept in the frozen state. Be wary of any uneven or rumpled packets as these have thawed at some stage and may contain only a soup of leaked-out nutriments and the shells of shrimps or whatever.

Freeze-dried Foods

The same range of foods that can be frozen may also be available freeze-dried. Freeze-dried bloodworms probably offer the best value for the money. They can contain as high as 65% protein, although I am not sure if that is first-class protein. Freeze-dried tubificids, brine shrimp, *Daphnia*, and other small crustaceans are all good value.

Freeze-dried foods are very concentrated. Do not feed them too freely as the fishes may swallow too much. The food will swell in their stomachs, possibly causing trouble. Some people soak them first. Most fishes seem to prefer the dry food, so just be careful not to dump them into the tank in large amounts. This is true of any food though it is more dangerous with the dry foods. Also, try to purchase all dry foods as fresh as possible. Fish foods do not have a "use by" date on them. Therefore, go to a busy pet shop that has a good turnover. Do not buy more than will last you for a few weeks. It may be cheaper to buy in bulk but once opened though a tin or package will start deteriorating and also lose vitamins.

Other Dry Foods

Many a freshwater fish lives on flakes, pellets, or the like. If these foods are of high quality the fishes do quite well. Not so with marines. Give them occasional feeds of dry food so that they get a change, but do not depend on dry food alone. Dry foods tend to be a nuisance in reef aquaria. Flakes float across into the pre-filter before somebody can eat them.

When buying flakes, choose those intended for marine fishes. There should be a high protein content (45% at least), a good mix suited to your particular fishes, and added vitamins. You can get flakes for herbivores which must be below 45% protein, but buy this type only for that specific purpose. Colors of flakes usually mean nothing, although there are a few brands in which the color indicates content. A good flake is reasonably chunky and does not crumble until you want it to. It floats on the water at first unless dunked. It gradually sinks, if not eaten rapidly, and eventually reaches the bottom dwellers. It should not cloud the water. It should not break up quickly and clog undergravel filters if you happen to feed too much. Remember that they are dry foods and swell on being eaten so do not be too generous when feeding to the fishes.

Flakes only contain very finely ground particles. They are made by drying a gelatinous mixture over rollers. They cannot contain chunky pieces of recognizable material or give a fish something to chew. Granules or pellets can do this and therefore are good for large fishes. They contain bits of insects, crustaceans, roe, etc. Sometimes meant for pond fishes (in most countries this means coldwater fishes) they are apt to contain too little protein for marines. Sometimes they contain too little even for the fishes they are meant to feed. So be careful about buying pellets. Never feed "ant eggs" to your fishes; they are the pupae of ants and of little nutritional value. Not many dry foods are made from dried whole crustaceans, tubificids, etc. They tend to be harder and so soften less readily in water than freeze-dried products. Their food value is dubious.

Homemade Foods

Textbooks give all sorts of mixes to be made up by the aquarist. The mixes are usually intended for freshwater fishes and for the keepers of large numbers of aquaria and may result in cooked, gelatinized, or frozen preparations. None of these are worth making if they take a lot of work time and you have only a few fishes. An easily made mix is the one mentioned earlier, consisting of various constituents frozen in ice cubes. Otherwise, supplement commercial foods with tidbits from your own kitchen, such as bits of prawn, lobster, crab,

mussel, non-oily fishes, frozen vegetables, and so forth. They can be fresh or canned. Even cooked foods will be eaten. Unless your aquarium provides plenty of greens, these are the most likely items to be in short supply in commercial foods. Make sure that tangs, angels, and other vegetarians get enough.

Although I no longer keep large numbers of fishes, I make up a frozen mix. I know what is in it. I also vary it from one batch to another. You can often buy mixtures of seafood in fish shops that contain a good start for your own product: tiny squids, mussels, shrimps, etc. You can add chopped baitfishes, vegetables, whole meal bread, or breakfast cereal. All should be reduced to a size suitable for your fishes and invertebrates. Mix in some frozen mysid shrimp, brine shrimp, or other such foods if you are so inclined. This way you finish up with a wide variety in the mixture. Make enough to last about three weeks. Freeze in ice cube trays with a little water as a filler. A single capsule of human mixed vitamins and minerals (no megadoses) may also be added to 1 lb of mixture.

Get rid of any crustacean shells, apart from tiny ones like mysids, when making up the mix.

Thaw an appropriate quantity for one feed. This is about one cubic inch per 20-30 medium sized fishes. Before feeding a particular tank, a fairly coarse mix may need to be chopped up a bit further with scissors or a razor blade for smaller fishes. Some larger pieces may be left whole for the bigger fishes and anemones or crustaceans. Feeding a tank of mixed inhabitants needs a little ingenuity on occasion if all marine life is to be served adequately. Anemones may need to be trained to take pieces of food. The food must be brushed against the tentacles and then thrust into the mouth before fishes steal it. Alternatively, they may be fed at night while the fishes are resting. This can be difficult if they house anemonefishes. The real problem arises when large crustaceans, immune to anemone stings, learn to thrust a claw down the anemone and extract the food for themselves.

A good general plan for feeding a mixed batch is to swirl the whole meal into the water in two or three places at once. Nobody can monopolize the food and the shyer fishes get a chance to feed. This assumes that you know how much to feed so as not to foul the tank. Do not worry about temporary cloudiness. Even if there are no filterfeeders to take care of it, the filters will do so. The filters cannot continue to deal with heavy excesses, though. Keep your fishes on the hungry side. It does them no harm and ensures healthy fishes and clean tanks. This is not a book about keeping invertebrates, but with the growing interest in reef tanks, a lot of aquarists are going to be doing so. Feed the filterfeeders independently of the fishes and other invertebrates in such tanks. Turn off all filters temporarily for an hour or so. Swirl some commercial invertebrate food (finely divided plankton-like material) as directed on the bottle into the tank. An extra little powerhead that just keeps the water circulating is a good idea. This way the food adequately reaches the filterfeeders. When the water is fairly clear again, turn on the normal circulation.

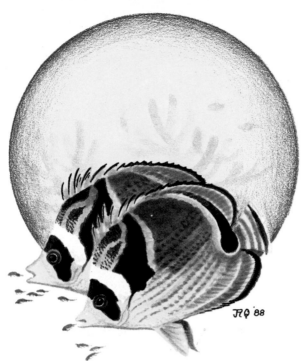

Butterflyfishes have small mouths and should be fed accordingly. Although normally they pick food from the substrate, they will chase morsels such as these copepods. The fish shown here is *Chaetodon lunula*.

Specialist Feeders

There are a number of groups of fishes that need special attention. Some chaetodons are difficult to feed. They are usually those that eat live coral or similar live foods in nature. You can start them on a live coral head if you feel affluent. Then switch to live tubificid worms stuffed into a piece of coral. See if that fools them. If this is successful, stuff other meaty foods into the coral. With any luck the chaetodons soon will be eating normally. Some angels can be treated in the same manner. Alternatively, they can be persuaded to eat newly opened live shellfish. Some otherwise difficult angels will eat brown bread. It must look like their favorite diet of sponges. After getting them started on something, many difficult fishes eventually eat most of the regular foods. They soon learn to compete with the others at feeding time.

Predators, such as
lionfishes and groupers,
may have to be taught to eat
dead food. This can be
managed by switching from
live *Gambusia* or guppies to
a dead one impaled on the
end of a stick or needle. Be
sure that the tip is hidden
and will not harm the
predator. Keep the
predator hungry. Then
wave the dead prey in front

Rock lobsters are active
predators, seeking out their
victims at night; the blue starfish
is a sedentary detritus feeder.
This shows the diversity of feeding
methods you may have to contend
with in your reef tank.

of him gently. Move away rather than toward him. He will quite likely seize it. If he does not, try again later. It is quite unusual for more than a week or so to pass before he will be taking any meaty chunks that way. Soon he will be looking for it as you approach the aquarium. He may eventually snap up any chunk thrown into the water. Anglerfishes can be trained in the same way. With both anglerfishes and lionfishes, it is a pity to lose the display they normally exhibit when stalking or attracting their prey. Also, an occasional feed of whole, preferably live, fishes is of benefit to predators since they get all the parts and their contents that are missing from mere chunks of flesh.

Slow feeders can be a problem when housed with a community of other fishes that gobble everything up before the finicky ones get it. There are several solutions. The finicky fishes can be kept together only with others of the same disposition— seahorses, pipefishes, and dragonets, for example. They can be kept with relatively small predators that will not harass them or be interested in their tiny

food. A few can be kept in a reef aquarium. They will find enough to eat as they forage around all day long. Too many will deplete supplies.

Cleaners do not get much of their normal food in an aquarium. As in nature, they can feed on other than parasites if they must. Although useful in keeping down parasitism, they can be a pest if kept with only a few other fishes. They chase the fishes around and annoy them. Not many fishes retain a cleaning habit into adulthood. The cleaner wrasse, *Labroides dimidiatus*, and the neon gobies, *Gobiosoma* spp., are exceptions. However, many young fishes are cleaners as well as being plankton feeders. As they grow up they switch over to whatever is the adult diet. In the aquarium, this is whatever is offered. This is so unless they happen to belong to a species that does not readily make the change to anything but its natural food. Many young chaetodons must be given frequent feeds of newly hatched brine shrimp or plankton-like foods. They are hard to persuade to change to other foods. This is one reason why it is unwise to buy these fish.

A common kitchen turkey baster is handy when feeding such things as anemones or featherduster worms. It causes little disturbance in the tank (the worms will not contract) and places the food exactly where it is intended.

DISEASES AND PARASITES

No fish is free of disease, any more than you or I. Each of us, including our pets, has a number of infections and parasites that cause little or no trouble. If something abnormal happens we get ill. We can also catch something new, to which we have no immunity, and then we again get ill. The extent to which we suffer depends on how rapidly we can deal with the enemy. Usually we build up an immunity to it. This in turn will often depend on how healthy we are initially. In aquarium terms, this means how well the aquarium is maintained and the fishes properly fed and cared for. Fishes have an immune system less well developed than mammals. It still works and they can develop resistance to some diseases. An aquarium may carry an infection that behaves as above, not causing any trouble although the fishes have it. Then something goes wrong. The fishes' resistance is lowered, perhaps by a wave of ammonia, and down they go. Or a new fish with no immunity to that particular disease is introduced. The fish catches it and releases so many bacteria or whatever that even the old

inhabitants succumb.

This sort of thing happens in aquaria because they are closed systems. There are probably many more fishes per unit of volume than occurs in the wild. Thus there is a much greater chance of passing on diseases or parasites. In fact, even normally harmless inhabitants of a fish's skin or gut can become a cause of disease in captivity. Protozoa that have never been known to cause disease in wild fishes do so in the aquarium. Luckily, the contrary also happens. Parasites, like many flukes that infest fishes in the wild, need an intermediate host, such as a bird. Therefore, the cycle cannot occur in the aquarium.

Stress

Opportunities for upset—stress, in other words—are many in the marine tank. Failure of equipment or carelessness on our part can cause unwanted changes in pH, temperature, salinity, concentrations of toxins, aggression, and so on. Stress operates in two recognized ways. Sudden severe stress causes hormonal changes that can

result in gut ulceration, muscular degeneration, and even death. Chasing a fish around an aquarium to catch it stresses it and often does no good at all, even though the fish usually survives. Prolonged milder stress gradually exhausts an animal. It may eventually cause a breakdown, either of the hormone balance, leading to death, or of resistance to disease, in part a reflection of the same thing. White blood corpuscles that fight disease in various ways are under hormonal control.

Feeding small fishes to anemones can also be accomplished with any grasping instruments as seen here. The idea is to get the food item past the fast moving fishes to the slow anemone.

In stress, the primary disturbance is in the pituitary gland at the base of the brain and in the adrenal glands, both under nervous control. The pituitary gland regulates growth and reproduction and the activities of various other glands that produce their own hormones. Acute stress causes the sudden release of adrenal gland hormones. Prolonged stress exhausts these and other glands. This results in emaciation, muscular wasting, loss of bone structure, poor wound healing, and eventual death with or without a recognized disease occurring. A very common occurrence in a marine aquarium is bullying, as marine fishes are often aggressive. This causes stress. The unfortunate victim shows the symptoms outlined above if he does not manage to find his place in the community. He can do this by hiding away until left alone, or by asserting himself and giving as good as he gets. Some fishes, notably anemonefishes, have attitudes of submission, like a dog exposing his throat, that inhibit further attack.

The Selection of Cures

A physician's first consideration when treating a patient is to do no harm. This is a precept that aquarists would do well to bear in mind. When we decide that a tank of fishes must be treated for disease, we should always consider whether any proposed treatment is likely to cause further trouble. However effective a cure may be, we do not want deeply colored water, stained equipment or decorations, or a filthy precipitate that damages other organisms and is hard to remove. For these reasons, you will not find agents like methylene blue or potassium permanganate recommended here. Antibiotics and some other drugs affect biological filters. They must be avoided unless administered by harmless routes not involving high concentrations in the water. Some cures, such as copper, are dangerous to fishes in doses not much higher than are needed to kill parasites. Their concentration in the aquarium should be monitored closely. Nearly all common cures are harmful or lethal to various invertebrates. Thus they present a problem when present in the tank.

Treatments that must be given outside the aquarium in temporary baths can also be a problem. It is often difficult to catch the fishes affected by a particular disease. Also, we do not want to wreck a beautiful aquarium. Such procedures will be recommended only when felt to be absolutely necessary or are alternative to other treatments.

Commercial cures are often shotgun treatments aimed at a number of common diseases. They contain a mixture of drugs or chemicals. If you are reasonably sure of the nature of a disease, it is best to give a specific treatment, a single drug or recommended combination. If you mix drugs yourself, try them out in a jar of aquarium water. If no color change, cloudiness, or precipitate occurs, the mixture may be judged safe to use. Do your best to identify a disease or parasite before starting treatment. Use a single recommended procedure. If possible, get an experienced marine aquarist to inspect your fishes before treating them. Do not expect him to diagnose trouble over the telephone.

Using a Cure

Carbon filters, resins, and similar absorbents take up most cures very efficiently. They must be turned off or removed while treatment is in progress. Afterward, they can be very useful for mopping up the now unwanted chemicals. Biological filters, except for the "dry" part of a wet/dry filter, cannot be turned off for long. A few hours at a time is the most they can stand. A "dry" or trickle filter can stand two days and in some setups it can be isolated and kept going indefinitely. Copper depresses the action of a biological filter but in the recommended dosage it does no permanent harm. Antibiotics have a similar but much more potent action. The filter must be turned on only intermittently—a few hours off, an hour or two on, and so on. Disinfectants, like formalin, kill the bacteria and must not be placed in the aquarium.

Symptoms

Nearly all aquarists depend on symptoms visible in the aquarium for the diagnosis of disease in fishes. With some diseases, diagnosis is easy and pretty certain. White spot, velvet, and flukes are examples. Other diseases present problems. For example, ulcers may be bacterial or fungal in origin, or even due to toxins in the water. Odd behavior without other

Table 3
Visible Physical Symptoms

Symptoms	Probable Cause
Small white spots on fins or skin	White spot (*Cryptocaryon*)
Fine whitish peppery coating on fins or skin	Velvet (*Amyloodinium*)
White patches or spots on skin	Bacterial or Microsporiasis
Spawn-like gray-white patches on fins or skin	*Lymphocystis*
Cauliflower-like patches on fins or skin	*Lymphocystes* (later stage)
Larger "tapioca" patches on fins or skin	Myxosporiasis (*Henneguya*)
Blue-white or pinkish patches on skin	Ciliates
Red streaks on body or fins	Bacterial septicemia
Tattered fins or tail (fin or tail-rot)	Bacterial septicemia (later stage)
Red-ringed craters on skin	*Argulus* bites
Irregular bulges (cysts) under skin	Microsporiasis
Swellings on body or base of fins	Tumors or Sporozoans
Yellow-to-black cysts up to ⅛ in. bursting through body	*Ichthyophonus*
White to dark nodules under skin	Larval tapeworms (Cestoda)
Black or reddish nodules under skin or in eyes	Metacercaria
Ulcerated patches on skin	Bacterial septicemia or *Ichthyophonus*
Cloudy skin and gills, blood flecks	pH too extreme
Blisters beneath skin	Gas embolism
Cloudy eyes or blindness	Severe velvet or white spot, metacercaria, toxins
Sunken eyes	Chlorine excess, dropsy
Exophthalmos (pop-eye)	Gas embolism, toxins, copper, some diseases
Lateral line erosion	Lateral line disease (viral?)
Emaciation, paleness, perhaps skin ulcers	Tuberculosis, *Nocardia*, or Tripanoplasmiasis
Emaciation, paleness, light-colored or bloody feces	Coccidiosis
Emaciation, swollen belly, odd movements	Cestoda (tapeworms)
Red and swollen gills with gray or white spots	Bacterial gill disease
Brown gills	Nitrite poisoning
Eroded or tattered gills	Chlorine poisoning
Gill covers protruding	Gill parasites, goiter, velvet
Swollen body, often scales protruding	Dropsy, intestinal blockage
Scales protruding, body normal	Bacterial infection of scales
Flukes about ⅛", fins perhaps torn	*Benedenia*
Flukes ¹⁄₂₅" or less on gills or body	*Gyrodactylus*, etc.
Large "lice" up to ⅛" on skin	*Argulus*
Wood-louse-like crustacea in mouth or on skin	Isopods
Threadworms hanging from anus	Nematodes
Egg-sacs hanging from gills or body	Copepods
Hemorrhages, spinal curvature, pop-eye, cloudy eyes, convulsions (2 or more of these together)	Vitamin deficiency

Behavioral Symptoms	Probable Cause
Flashing, fins clamped	Velvet, white spot, toxins
Listlessness, loss of balance	Sleeping sickness, metallic poisoning, chill
Swimming on the spot (Shimmies)	Velvet, chill
Severe loss of balance, belly up	Swim bladder disease
Gasping at surface	Oxygen lack, toxins, gill diseases, tank overheated
Sudden dashes, jumping out of tank	pH wrong, toxins
Unusual colors, especially dark	Toxins, *Ichthyophonus*
Failure to eat	Stress, wrong food, various diseases

Table 4

Smears of Skin or Gills	
Small flukes	*Benedenia* larvae, *Gyrodactylus*, etc.
Pear-shaped flagellate	Velvet *(Amyloodinium)*
Ciliates:	
Ovoid, no hooks	White spot *(Cryptocaryon)*
Ring of hooks	*Trichodina*
Heart-shaped or oval, no hooks	*Brooklynella* or *Chilodonella*
Single stalked ciliate	*Trichophrya, Glossatella*, etc.
Branching stalks	*Epistylis*

physical signs may present a real problem resolvable, if at all, only by laboratory examinations and tests. If fishes are behaving oddly, it always pays to think back over possible recent causes. Consider insect and other sprays, changes in the water supply, or a dead fish or invertebrate that may be hidden away.

Table 3 gives a list of the symptoms of many common diseases and parasites. You can then make a good guess at what ails your fishes and turn to the relevant section of this chapter to see if it confirms your idea. A fish removed from the aquarium, alive or dead, will enable you to make skin or gill smears. These can be of further help. Skin smears are taken by rubbing a blunt instrument, like the back of a knife, gently along the skin from the head backwards. Gill smears are taken by removing a little mucus from the gills with a needle or forceps. A smear is spread on a clear glass slide. It is then preferably covered with a cover slip (a thin square of glass about 2 cm²). Slides and cover slips are available from microscopists or their suppliers. The smear can be examined with a hand lens or the low power of a microscope. **Table 4** lists some possible findings.

Viral Diseases

A virus cannot reproduce on its own. It is virtually a packet of genes in a protein envelope. It has to inject itself, or rather the genes, into a living cell. There it will multiply sooner or later to produce many copies of itself. These copies will infect other cells if possible. A virus is minute, around 0.0003 mm. There are few recognized viral diseases of fishes. This only reflects ignorance. Most of the viral diseases affect freshwater fishes or salmon and trout.

Lymphocystis This virus causes the connective tissue cells in the skin to swell up to enormous proportions. It often affects the mouth or fins, but it can occur anywhere. The groups of swollen cells look like spawn sitting on the surface, white and scattered. These may enlarge further to cauliflower-like tumors, particularly around the mouth. Sometimes it looks very much like a fungus and is often mistaken for it. The condition is rarely fatal. It usually disappears without leaving scars, but may take months to do so. It is highly infectious and an individual fish showing it is best removed. A tank of infected fishes is best left to recover. If the fishes are discarded, the tank must be thoroughly disinfected before more fishes are introduced. No specific cure is known. A treatment may appear to work because of the natural course of the disease.

Lateral Line Disease
This condition is suspected to be viral in origin. It begins at the head end of the lateral line system and may spread down along the body. The lines appear to have been gouged out. They may be infected with what are probably secondary invaders. There is no known treatment. Improvements in tank hygiene are said to be followed by the disappearance of the disease in some cases. This disease develops slowly. It may affect only a limited number of fishes in a tank. However, it is best to remove an infected individual.

Zebrasoma desjardinii with an advanced case of lateral line disease. The diseases can be seen here to have followed the course of the lateral line.

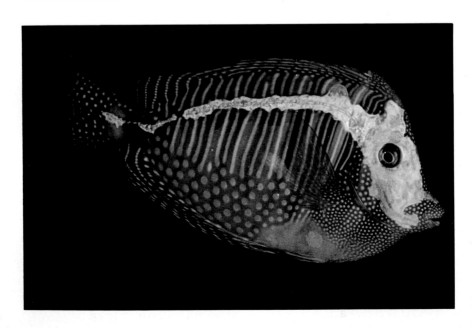

Bacterial Diseases

Bacteria are primitive but fully developed cells, although they do not have the cell nucleus of higher organisms. They typically reproduce rapidly in a suitable food supply. Some species find this in another living organism and may cause disease. Many live on or in multicellular creatures without causing harm, while some are even essential to the host. A particular bacterial disease can sometimes be recognized by the symptoms it causes. Very often, though, only laboratory tests can tell which species is responsible. Tests are rarely available or sought by the aquarist. Such testing would often be too late anyway for an effective cure. I shall deal with bacterial diseases under general headings and shall indicate, where possible, which types of bacteria may be responsible.

Bacterial Septicemia
When bacteria get into the blood stream and are not successfully dealt with by its defenses, bacterial septicemia results. Red streaks show up on the fins and body which may proceed to ulceration.

Parts of the fins and tail may drop off. These conditions are known as red pest and fin and tail rot respectively. Internally, any organ may be attacked. Death may follow kidney, liver, or heart failure.

Right at the beginning of an attack, with only a fish or two mildly affected, general disinfection combined with a good clean-up and light feeding for a short period may effect a cure. The fishes' own resistance may be able to overcome the disease. The antibacterials acriflavine (trypaflavine) or monacrin (monoaminoacridine) are suitable. They both color the water mildly, the first yellow, the second a very attractive blue. Both colors fade in a few days and are good indicators that a further dose may be needed. In both cases, a stock solution at 0.2% in fresh water is used (2 gm per liter) and added to the tank at up to 1 standard (5 ml) teaspoon per gallon. With monacrin, even 2 teaspoons per gallon is safe. No further water changes than usual need be given as the drugs biodegrade. An exception is in the possible instance where the fishes may be used for breeding. At least acriflavine is suspected of causing sterility if used for a prolonged period.

If the treatment above does not work within a few days, switch to the following. If the condition is other than very mild, omit disinfection and start with the following straight off. Bacterial septicemia usually progresses slowly, therefore, there is a chance to try more than one cure if necessary. This method uses antibiotics in the food and depends on the fishes continuing to eat. The best wide-spectrum antibiotic for fishes is chloromycetin (chloramphenicol sodium succinate). It is odorless and colorless and is taken readily by fishes. If it is not available, use terramycin, kanamycin, erythromycin, gentamycin, furanace (a nitrofuran), or ampicillin. Do not use aureomycin. This froths up and turns red in sea water. Mix up to 1% of the antibiotic in a dry food or frozen food, such as brine shrimp. A 250 mg capsule should be thoroughly mixed with 25 gm (just under 1 oz) of food. Feed at least twice daily and keep up the treatment for 2 weeks even if the fishes seem cured earlier.

It is not advisable to dissolve antibiotics in the water. Although a little will get in by the feeding method, it will not be enough to matter. However, if the fishes are not eating, it will have to be done. In that case, 50-100 mg per gallon is needed. Repeat every 2-3 days for 2 weeks. Carbon filters and any other absorbent filters must be turned off. Biological filters must be protected as far as possible by periodic turnoffs. In some reef aquaria, the biological filter can be isolated and kept going with ammonium salts. Bacterial septicemia may be caused by organisms in the genera *Vibrio, Aeromonas, Pseudomonas,* and others. This is why a wide-spectrum antibiotic must be used. This is particularly a disease of tropical fishes.

Bacterial Gill Disease This usually follows ammonia poisoning. The gills become red and swollen and the fishes gasp at the surface and are obviously in distress. The bacteria—any of those above or many others—then attack. They are visible as whitish or gray spots (tiny colonies) on the gills. Two courses of action must follow. First, the cause of

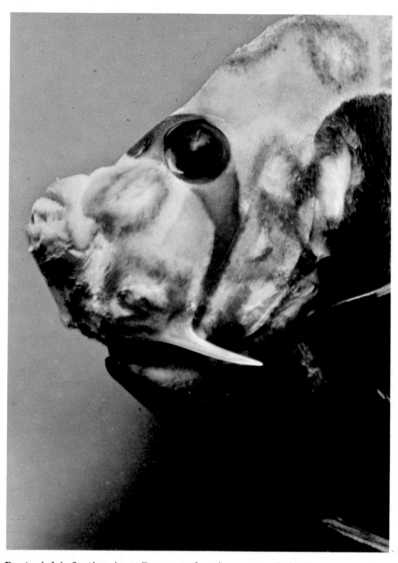

Bacterial infection in a *Pomacanthus imperator*. Infections as severe as this one are probably fatal. However, they can be cured if caught and treated early.

ammonia or possibly some other toxin must be sought. Water changes and a general clean-up must be instituted. Second, the bacteria must be treated. Early on, acriflavine or monacrin treatment as described above may be sufficient. This is usually when no bacterial colonies are seen yet. Later, use the antibiotic treatments just as for septicemia. If bacterial gill disease happens in a newly set-up tank, you have a case of the new tank syndrome. The biological filter has not been adequately conditioned. It will be necessary to start again unless you are prepared to make frequent and considerable water changes. You must do this until the biological filter has caught up. Gradually reduce the changes as it does so.

Tuberculosis

This is caused by a specific bacterium, *Mycobacterium marinum*. It is closely related to other mycobacteria that cause tuberculosis in freshwater fishes and other vertebrates, including ourselves. Symptoms are varied. They range from listlessness, emaciation, and fading of color, to ulcers on the skin. Spinal curvature, pop-eye, and tattered fins and tail may also occur. On opening up a dead fish, you may see gray to dark nodules here and there in the body tissues. Infected fishes are best destroyed as a cure is difficult and takes time. The disease is highly infectious. It is a sign of poor husbandry.

If it is decided to try for a cure, the necessary drug is isoniazid. It is either fed at 1% in the food or at 50 mg per gallon in the water. Repeat every third day with a 25% water change. In addition, rifampin may be fed at 0.01% in the food, together with isoniazid either in the food or the water. Rifampin may be stored in made-up foods, like Gordon's formula (beef liver and Pablum made into a puree), in the refrigerator as it is very stable. A cure may take up to 2 months. It should not be undertaken lightly as it is possible to become infected with *M. marinum* which will appear as a local sore on the hand or arm after servicing the tank.

Another organism, *Nocardia asteroides*, recently has been found to cause very similar

This diseased kyphosid was photographed at Steinhart Aquarium. It probably has tuberculosis.

symptoms to tuberculosis. It can affect humans, also. So far it has been reported in freshwater fishes but it seems likely that it or a near relative may turn up in marines. No cure is known.

Ascites

Dropsy (ascites) is a swelling of the body, especially the abdomen. It is usually caused by kidney or heart failure and a leaking of fluid into the body cavity. The scales may protrude because of the internal pressure, but they are not infected. Already weakened fishes are typical victims. They often become infected with the genus *Corynebacterium*, susceptible to penicillin, as well as, perhaps, *Pseudomonas*, which is not affected by penicillin. It is best to use a wide-spectrum antibiotic from the list given under bacterial septicemia. Preferably give the treatment orally.

Ascites does not often occur as an epidemic and the odd sufferer is best killed or removed for treatment. A large fish can

The raised scales and pop-eyes are symptoms of abdominal dropsy. Scales may protrude in the absence of dropsy if they themselves are infected.

be given immediate relief by aspirating the fluid with a hypodermic syringe. Pierce the belly just in front of the anus in a forward direction. Do not go in too far!

Scale Protrusion
The scales may protrude in the absence of dropsy because the scales themselves are infected. They will show reddening possibly accompanied by septicemia. Various bacteria may be responsible. Exactly the same medications as recommended for bacterial septicemia are appropriate—disinfectants for very mild cases and antibiotics for anything else. White or gray spots on the skin may precede scale protrusion. The spots are colonies of the bacteria in question. *Flexibacter columnaris*, the cause of "mouth fungus" in freshwater fishes, does not take the same form in marines but causes spots on the skin or gills. Immediate treatment on seeing them is indicated. Do not confuse them with white spot (*Cryptocaryon irritans*).

Protozoal Diseases

Protozoa are single-celled animals that cause many of the known diseases of aquarium fishes. They are generally larger than bacteria, but they are still tiny. Even the biggest are just visible to the naked eye. In nature, they are pretty harmless and it is only in the aquarium that they build up to disease proportions. This reflects the crowding that aquarium fishes undergo compared with the wild state, and the much greater chance of a parasitic protozoan finding a host. Protozoa are a great advance organizationally on bacteria. They have a cell nucleus, complex cell structure, and the ability to ingest vegetable or animal foods. Most of them lack the ability to use carbon dioxide and water to build simple organic compounds.

As in previous publications, I shall use a simplified classification of the protozoa for convenience. I will not use the seven separate phyla into which they are now grouped. The Flagellata have one or more whip-like flagella with which they swim around and latch onto a host. The Sporozoa are all parasitic. They produce spores, resting stages usually tough and resistant to destruction that eventually produce new sporozoans. The Ciliata have small hair-like swimming organs, the cilia, which are usually found in large numbers.

Flagellata

Many of the flagellates found on marine fishes have never been described as causing disease whether in nature or the aquarium. This is odd because their fresh water relatives (*Ichtyobodo (Costia), Hexamita, Cryptobia*) do so. Salmon are believed to transmit *Hexamita salmonis* to their eggs and hence the young, but only while the fish are in fresh water.

Velvet

The protozoa that cause this very common disease are so peculiar that zoologists classify them as such, while botanists are apt to claim them as algae. They are dinoflagellates. They possess chlorophyll and are capable of living like a plant. Another dinoflagellate, *Gymnodinium microadriaticum*, is the "alga" that lives in corals. It enables the coral to thrive in sunlight. Velvet is caused by *Amyloodinium ocellatum*. This is related to

Chelmon rostratus infected with both white spot (*Cryptocaryon irritans*) and velvet (*Amblyoodinium ocellatum*).

Ichthyophtherius sp. can most easily be seen on the clear fins of some species.

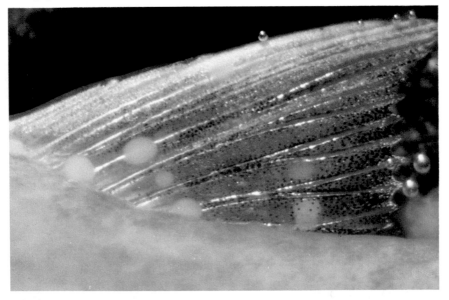

the fresh water *Oodinium* spp. that cause diseases with the same name. Marine velvet is white. Therefore, it must have little chlorophyll in contrast to the yellowish *Oodinium*.

In the free-swimming infective stage, the dinospore of *A. ocellatum* has two flagellae, a long and a short one. It is about 0.01 mm in size. It has to find a host within a day or two or it dies. Most of those that do find a host are filtered off in the gills, with some coming to rest on the fins and body. By the time many are seen externally, you can be sure that many more are in the gills. The parasite adheres by the long flagellum. The dinospore grows larger and pear-shaped and is attached by the narrow end by finger-like processes, the pseudopodia, that penetrate the surface. It feeds on the cells of the host. At this stage the invader is called a trophont. The trophont enlarges further to about 0.1 mm in diameter, becomes more spherical, and forms a cyst. The cyst may fall off or remain stuck in the mucus coat of the fish. It eventually releases about 200 dinospores. The

cycle takes about 10 days at tropical temperatures but takes longer at lower temperatures and even fails to complete in really cold water.

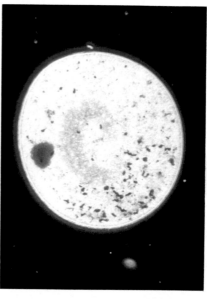

A dark-field photograph of the disease causing organism *Ichthyophthirius multifiliis.*

Velvet gives a white, powdery appearance to the surface of the fins or body. It is hard to see, unless viewed in oblique light. Look at a suspected fish as it swims toward you or shine a flashlight onto it in a dim light. Early in an attack, when the gills are primary targets, behavioral changes are likely to alert you to the possibility of velvet disease. The fishes rub themselves on objects in the aquarium, a movement known as flashing. Fishes often gasp at the surface and show rapid respiration. Later, fins may become clamped, ragged, and inflamed. The eyes may be cloudy. Death commonly follows. As soon as any symptoms occur, even if you cannot yet see the whitish coating, assume that velvet is present and treat accordingly.

The best and simplest treatment for velvet is copper given as copper sulfate, copper citrate, or a mixture of copper sulfate and citric acid. The presence of citrate keeps the copper in solution better, but copper sulfate alone is satisfactory. More complicated (chelated) preparations keep the copper in solution longer but they are less effective and difficult to monitor. This is because the copper kits do not accurately measure their concentration in the water. Citrate is a kind of chelator, but it does not suffer from this drawback. Also, highly chelated preparations are hard to remove when a cure has been affected. They make an aquarium unfit for invertebrates for months afterward. Copper kills algae and invertebrates and should never be used in their presence.

Fishes are susceptible to copper poisoning. Although a minimum concentration of metallic copper ions must be maintained to kill the dinospores, it is best to monitor the level during treatment to make sure that it does not go too high or fall too low. Copper is readily taken up by coral sand or skeletons (it kills live coral!) and so additions have to be made and monitored during treatment. It kills the dinospores and causes the fishes to secrete a lot of mucus from the skin. The mucus makes it difficult for any survivors to obtain a hold and appears to hasten the shedding of cysts. A concentration of 0.15 ppm (0.15 mg per liter) is needed to effect a cure. Most fishes can stand up to 0.4 ppm, so there is a fair safety margin. This is the concentration of copper itself, not of the salt used. The salt is generally composed of copper sulfate, $CuSO_4 \cdot 5H_2O$, the blue crystals which contain 20% copper.

A butterflyfish *(Chaetodon bennetti)* infected with velvet. This diagram shows how it attacks the body and fins.

For treatment, a 1% solution of the blue crystals is made up in distilled or demineralized water (tap water often forms a precipitate). Dissolve 10 gm per liter, 38 gm per U.S. gallon, or 45 gm per U.K. gallon. The stock 1% solution is added to the tank at 0.25 ml per U.S. gallon, or one standard 5 ml teaspoon per 20 gallons. Measure the concentration of copper in the aquarium shortly afterward. It should be between 0.15 and 0.20 ppm. As it falls, take measurements at least every other day and add more of the stock solution to keep up the concentration to at least 0.15 ppm. Maintain treatment for at least 10 days at 75-80°F, longer if the temperature is below 75°F. During treatment, turn off all filters with carbon, resins, etc., in them but leave the biological filter going. Should you have to treat with copper without monitoring the concentration, add an extra half dose on days three, six, and nine of the treatment period and hope for the best. Signs of copper poisoning in fishes are gasping, loss of equilibrium, and pop-eye (exophthalmos). If these are seen, turn on the absorbent filters or add carbon or a suitable resin to any filter other than the biological one.

Treatment in the

presence of invertebrates is problematic. Additions of quinine hydrochloride or sulfate at 2 g per 100 liters of tank water (77 ppm), or quinacrin at 4-6 mg per liter (4-6 ppm) have been suggested. However, these cannot be monitored. Purely guessing, I would add a further half dose with a part water change every three days. The doses recommended are quite low and harmless to fishes. I do not know about the reactions of invertebrates.

Sleeping Sickness

Individual fishes caught in the wild may be suffering from sleeping sickness. This is caused by *Tripanoplasma* sp. or *Trepanosoma* sp. They probably cannot pass it on to others as an intermediate host is needed. Although the intermediate host has not been identified, it is not likely to be in the aquarium (I hope!). In fresh water it is leeches. Symptoms are listlessness, emaciation, and odd swimming movements. There is no known cure and the disease is likely to be confused with tuberculosis. For safety, remove any fishes showing the symptoms listed as it may be tubercular.

Sporozoa

Here comes a dismal tale. All sporozoans are parasitic and there are plenty of them. In no case has an effective cure been described. They may be grouped under a number of general headings, causing typical syndromes by which they may be identified, but that is all. Except for the sexual "sperm" cells, sporozoans have no cilia or flagellae.

Coccidiosis

Many species of protozoa live in the gut of fishes. They reproduce in the cells of its lining, eventually freeing forms that pass out with the feces and infect other fishes via the food. These spores can sit around for months waiting to be picked up. Symptoms of coccidiosis are the rather common listlessness and emaciation coupled with light colored or bloody feces. Internally, the walls of the gut are blistered and the gut is full of fluid. The usual marine form is caused by one of forty species of *Eimeria*. They can also attack the liver, testes, swim bladder, and other organs. Suspected cases should be destroyed as they can infect others. The aquarium should be given a good clean-up.

Microsporiasis

Protozoa in the phylum Microspora live in the cells of different tissues. They cause nodules containing masses of spores. The muscles are commonly infected, but other organs, including the reproductive system, may be implicated. The spores infect other fishes by being eaten with the flesh either by predation or after death. Herbivores and plankton eaters are less liable to infection, but they could ingest spores freed by decaying dead fishes. In addition, rupture of surface nodules can free spores into the water. On entering a host a spore becomes like an amoeba. It penetrates the wall of the gut and travels in the blood stream to a suitable site where it penetrates a cell. It causes the cell to enlarge and possibly rupture freeing spores that can continue the process.

In the advanced stage affected fishes may show microsporans beneath the skin with color changes or large irregular bulges due to cysts. The disease causes emaciation as well. Well known marine genera are *Pleistophora*, *Glugea*, *Thelohania*, and *Nosema*, each with various species. The first three often disfigure victims severely while *Nosema* species typically produce small white tumors that cause little trouble. Once more, an affected fish is best destroyed in the hope of limiting the spread of the disease. A thorough siphoning of the bottom of the tank to remove any spores that may have been shed is advisable.

Myxosporiasis

Hundreds of species of the phylum Myxozoa have been found in fishes. Unlike the microsporans, myxozoans live between cells or in various body cavities. Many cause few external symptoms and only a few have been identified in marine aquaria. Of them, *Henneguya* causes large cysts almost anywhere in fishes that, when showing externally, resemble *Lymphocystis*. They have been called tapioca disease because they are white and about the right size. Spores are taken into the mouth and develop into amoeba-like trophozoites that penetrate the gut and are carried in the blood to their destinations. Destroy infected fishes and clean up as usual.

Ciliata

The ciliates that swim around with the aid of a coating or patches of cilia (hair-like organs) can cause trouble in the aquarium although they rarely do so in nature. Genera like *Epistylis*, *Trichophrya*, and *Glossatella* are attached by stalks to any substrate, including fishes. They are usually harmless unless aquarium conditions favor them and not the fishes, at which time they can proliferate to plague proportions. If toxins or poor nutrition cause shedding of the fishes' surface cells, for instance, the parasites feed on them. The ciliates may become visible as bluish or pinkish patches on the skin or gills and may cause irritation, respiratory difficulties, and open the way for other parasites. In such circumstances, treatment with acriflavine or monacrin as described for bacterial septicemia is indicated. There should be a rise in temperature to 82-86°F, which will cause some species of ciliates to die off.

White Spot

Marine white spot, which resembles the same disease in freshwater fishes, is caused by a different organism, *Cryptocaryon irritans*. Rarely causing trouble in nature, it is common in the aquarium and joins velvet disease as the two most frequent pests in the marine tank. It also is a gill and skin infection and causes symptoms similar to velvet—flashing, gasping respiration, ulceration, and ragged fins in advanced cases. It is easier to see than velvet, causing fewer and larger white spots up to pin-head size. The infective stage is the free-swimming tomite. This is a cell about 0.05 mm long that digs into the gills or skin as a trophont. Then a cyst, 1 mm in diameter called a tomont, is formed. Meanwhile it feeds on the tissues of the fish and causes the irritation that earned it its systematic name. Eventually, the cyst drops off or may get stuck in the mucous coating. It produces up to 200 new tomites after about eight days in the tropical tank. These are said to die after 24 hours if they do not find a host.

Marine white spot responds to copper treatment the same as velvet disease. This is in contrast to freshwater white spot which does not respond. The same procedures and precautions as described under velvet should be followed. The treatment should be kept up for two weeks, as marine white spot is rather more difficult to eradicate than velvet. The spots will not disappear immediately, but no new ones should form during treatment. If copper cannot be used, as when invertebrates are present, quinine or quinacrin may be tried, again as for velvet. If you are sure that the disease is white spot, sulfathiazole or its sodium salt may be employed at one level teaspoon (5 ml) per 5 gallons. First dissolve in a glass of fresh water, then mix this well into the aquarium. Repeat the treatment in a few days.

Trichodina etc. This ciliate has many species. Most have a saucer-shaped body about 0.05 mm across and circles of cilia around the edge inside of which is a circle of hooks. There is also a ring-shaped sucker for attachment to the host. All this forms an effective abrasive instrument since the protozoan rotates on the skin surface while the hooks wear it away. Severe attacks have been described only in freshwater aquaria, although there are eight known marine species. The same story is true for some other ciliates, *Brooklynella* and *Chilodonella* for example. These live on the surface of fishes but seem to cause little harm.

Fungal Diseases
Fungi used to be called plants without chlorophyll but there is now a tendency to place them in a separate kingdom. They may be uni- or multi-cellular, from yeasts to mushrooms and are the cause of a number of vertebrate diseases. Like animals, they need to feed on organic compounds already manufactured for them by other life forms, whether dead or alive. A typical multicellular fungus produces filaments called hyphae that form networks and mats called mycelia in or on their food supply. These send up fruiting bodies—the mushroom is an example—that produce spores that are released in quantity into the air or water. The spores are tough, being resistant to drying, heat, and even disinfectants.
Fungi affecting fishes have not received the

attention they perhaps deserve. Only one, *Ichthyophonus*, is a serious known disease-causing organism in marine fishes. A group of fungi that attack freshwater fishes is often referred to as affecting marines as well, but it is doubtful whether they do. This is the *Saprolegnia, Achlya*, etc., group, generally referred to as "fungus." They resemble tufts of cotton-wool-like hyphae on and in the surface of the fish. This group is in fact killed by immersion in salt water so it must be some others that attack marine fishes if any do. None has been identified, however. I have never seen fungus on a marine fish, nor have my colleagues. *Lymphocystis* can look very much like it at certain stages and I wonder if that is the explanation of reports of its occurrence.

Ichthyophonus hoferi
The organism causing this fungal disease is often referred to as *Ichthyosporidium hoferi*. It attacks fishes in general—cold or warm water, marine or fresh. It also attacks some amphibians and copepods that can pass it on to fishes. The original infection is via the gut and can happen when a fish swallows a cyst lying dormant on the substrate or eats another fish or its tissues carrying cysts. Once a cyst is in the gut, it frees a mass of small bodies called amoeboblasts. These penetrate the walls of the gut and enter the blood stream which carries them around the body. The liver usually gets infected first, but other organs and tissues get their share sooner or later. External signs usually appear later, too late for any hope of a complete cure. The amoeboblasts form thick-walled cysts wherever they settle down. The cysts are up to ⅕″ in diameter and yellow to black in color. These original cysts may put out hyphae that give rise to daughter cysts or daughter cysts form inside the original ones and break out of them. Some will be voided in the feces; most remain in the tissues.

Symptoms of diseases caused by *Ichthyophonus* are varied, depending on internal damage and whatever is visible externally (if anything). Internal damage may result in deformation, loss of balance, emaciation, and a host of other symptoms.

The skin may become rough from underlying cysts and some cysts may burst through as brown nodules or sores. The disease may never be discovered without dissection of some of its victims. A whole stock of fishes may be infected and gain a degree of immunity, the cysts being encapsulated within the tissues and prevented from spreading to other individuals. The odd fish may die because of damage to a vital organ, yet the disease remains undiagnosed. This enables the disease to be widespread yet undetected.

When *Ichthyophonus* is actually recognized, there are two suggested treatments. Neither treatment is likely to be a sure cure, but they may help. Phenoxethol (2-phenoxyethanol), an oily liquid that is sparingly soluble in water, is used with a stock solution at 1% in distilled water. It is then added to the aquarium at 40 ml per gallon, stirring it carefully into the water. Repeat the treatment only once a few days later. Additionally, or instead, the 1% solution can be used to soak the food or 1% by weight of the oil is added to the food. Some fishes will not eat such prepared food. The related drug, parachlorophenoxethol, is too toxic, but has been recommended. Chloromycetin may be added to the food as for bacterial septicemia, but I do not think that it would do much good as it is not a fungicide. Any treatment should be accompanied by a good clean-up including siphoning the bottom of the tank to get rid of cysts.

Trematodes

These worms, or flukes, are all parasitic. Some have complicated life histories requiring several hosts and so are unlikely to be propagated in an aquarium. Others, the monogenetic trematodes, require only one host. In our case, this is a fish. Such organisms are found in the mouth and gills, on the skin, and sometimes in body cavities. Each fluke possesses both male and female reproductive organs and most species lay eggs that may remain on the fish or fall off it. Some produce live young that may remain to infect the same host or leave it. The flukes feed on mucus and shed epithelia. They can also cause irritation and even blood loss. This is caused by the hooked organ, the haptor, by which they adhere to the host. There may also be suckers, one on the rear haptor and another at the front end.

Both juvenile and adult flukes may leave a host to seek another one. Some species are more host specific than others and it seems that Pacific Ocean flukes often have only one species or genus as host. In addition, they sometimes leave the hosts entirely for a period each day in the aquarium, reappearing perhaps each morning and disappearing later. Adults may be quite tiny and easily mistaken for white spot. Their movements on or from the fishes and the usual presence of dark eye spots give them away.

Benedenia melleni
This fluke has come to prominence in recent years as a frequent aquarium pest affecting all species of fishes. It breeds rapidly to plague proportions if not checked. It grows to around ⅙ inch long, causes intense irritations, and juveniles are commonly mistaken for white spot. Eggs are laid on the fish and hatch in about a week

as ciliated larvae. They die in a few hours if they fail to find a host, as the eggs will usually have fallen to the bottom of the aquarium. Symptoms of a *Benedenia* attack are flashing, paleness, drooping or torn fins, rapid respiration if gills are involved, and sometimes blood spots.

With a light infestation, particularly if in a reef tank or with invertebrates in any tank, biological control with cleaner wrasses, *Labroides dimidiatus*, or neon gobies, *Gobiosoma oceanops*, may be successful. With a heavy infestation, this would be asking too much. Problems arise unless the fishes are the only inhabitants or you are willing to kill off any unremovable crustaceans. If so, an organophosphate insecticide is the solution. Most are made from DTHP, dimethyltrichlorohydroxy-ethyl phosphonate—i.e. Dibrom, Dipterex, Dylose, Trichlorfon, Neguvon, and Masoten. Make sure of the concentration of DTHP in any brand used. Give 0.25-0.4 ppm in the water (0.25-0.4 mg per liter or 1.0-1.6 mg per gallon). Another drug, Lindane, has been recommended at 0.01 ppm (0.01 mg per liter or 1.0 mg per 25 gallons). Stir the

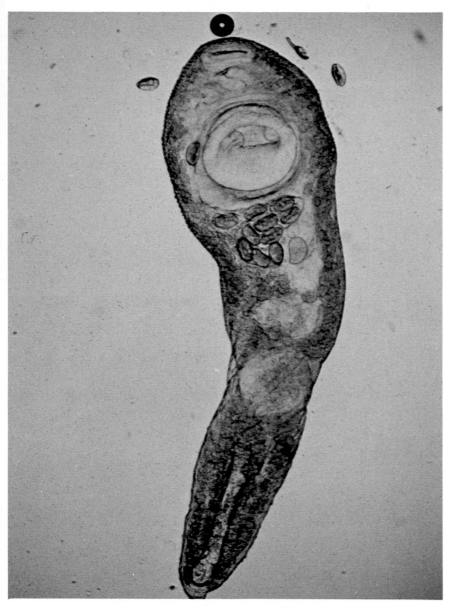

A digenetic fluke from the gut of *Amphiprion ephippium*. Digenetic flukes are not uncommonly found within newly collected fishes.

total amount needed into a gallon of tank water. Add it to the aquarium over a period of several minutes, with good stirring if necessary. Turn off carbon, etc., filters but leave the biological filters on.

It is best to treat the whole tank to destroy eggs, juveniles, and unattached flukes. If you prefer to remove the fishes for an external bath, use 1 standard teaspoon of formalin (40% formaldehyde) per 5 gallons for one hour. This may have to be repeated if the fishes get reinfected on being returned to the aquarium.

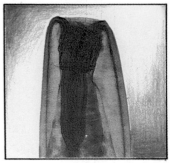

Acanthocephalan

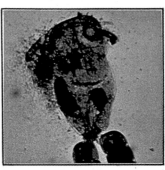

Parasitic copepod

Gyrodactylus

Two large families of tiny flukes, each with dozens of species, are fairly common in aquaria. They are the Gyrodactylidae and Dactylogyridae. They are mostly under ¹⁄₂₅ inch in size (1 mm) and mistaken for white spot. *Gyrodactylus* species bear live young; *Dactylogyrus* species lay eggs. There are also about 30 other families of monogenetic flukes, many marine and tiny. This is why a diagnosis is impossible without expert advice. Reproduction resembles that of *Renedenia*, and so does treatment. The important thing is to distinguish them from white spot so that an appropriate treatment is given.

Digenetic Trematodes The adults of digenetic trematodes are found in the gut, gall bladder, or urinary bladder of the host. Sometimes they are found

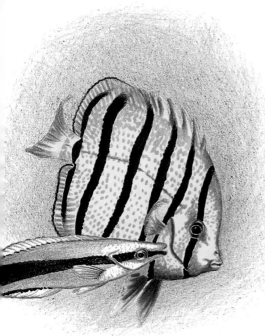

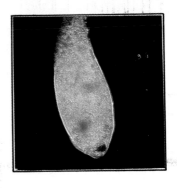

Marine Flukes

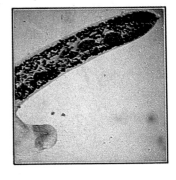

Parasitic copepod
(*Ergasilus*)

in other places. Neither they nor the young are external parasites but most are hermaphroditic (possessing both sexes) and release eggs into the excreta of the host.

The primary host houses the adults; the intermediate hosts the larval forms. The first intermediate host is a mollusk (often a gastropod)

A butterflyfish (*Chaetodon octofasciatus*) being cleaned by a cleaning wrasse (*Labroides dimidiatus*). Cleaner wrasses will help fishes rid themselves of pests, examples of which are shown in the satellite photos around the illustration.

or an annelid worm. The larva enters the mollusk or worm via the skin and is called a miracidium. In the host, the miracidium becomes a redia, or sporocyst. This multiplies asexually to produce hundreds of cercariae, free-swimming larvae that infect the second intermediate host, often a fish. Alternatively, the fish may eat the mollusk and get infected that way. The cercariae penetrate the skin, gills, or gut and find the part of the body where they characteristically develop into metacercariae. In the simplest of cycles they wait for the primary host to eat the second intermediate host, but there may be more to it than that. There may be as many as five intermediate hosts, for example. A common example is a bird as the primary host, a mollusk infected by the bird's droppings as the first intermediate host, and a fish, to be eaten by the bird, as the second intermediate host.

Metacercaria Disease
Here the fish is the second or later intermediate host. It bears masses of metacercariae that are recognized only if they show up under the skin or in the eyes as black or red nodules. If the fish is a primary host, you will probably never know it. There is no treatment for this disease, but fortunately it is not transmissible in the aquarium except in the very rare circumstance that all three hosts are present. If an affected fish is not uncomfortable or excessively ugly, there is no need to destroy it. It will not get any worse.

Tapeworms
Almost every known vertebrate has its particular tapeworms that in the adult stage lives in the intestines of the vertebrate. The cestodes hang on to the wall of the gut by the head end by means of a scolex, an organ with hooks or suckers or the like. From the head hangs a series of segments, the proglottids. Each bear reproductive organs that get larger and larger as one travels down the tapeworm. Finally, they break off and release eggs to be eaten by the next host. This may be another vertebrate or an invertebrate or, in the case

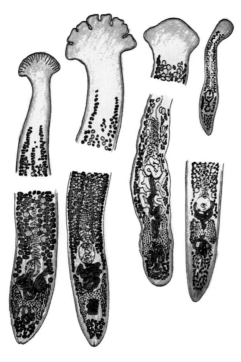

Fish parasites. Above is the flatworm *Caryophyllaeus*, below are various hookworms.

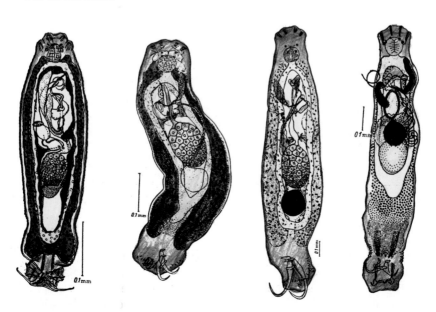

of fishes, another fish or a worm or crustacean, etc.

The eggs release ciliated larvae when in water. Each is identifiable by possessing six hooks. As with digenetic flukes, there are many different life cycles and intermediate hosts. The larvae may be eaten, for example, by a shrimp, the first intermediate host, the shrimp by a fish, the second intermediate host, the fish by a bird, the primary host. In the fish, the larvae may settle almost anywhere. If they form large cysts, this can be a cause of trouble. Some tapeworms are species specific; others are not. When a fish is the final host, the tapeworms in its gut will be recognized only upon dissection and will usually be a nuisance only if present in large numbers. Either as an intermediate or a final host, a fish may show various symptoms: emaciation, swollen belly in tapeworm infestations, irregular swimming, etc. A fish will not normally be a source of infection to others. The most likely, but still improbable scenario, would be a cyst-carrying fish eaten by a predator as a final host.

There is no treatment for encysted larvae. If adult tapeworms are suspected or seen upon dissection of a specimen taken and killed, they can be eliminated. Mixed in the diet at 0.5%, di-N-butyl tin oxide or dilaurate fed for three days can get rid of them. Yomesan at 0.1% or phenothiazine at 2% work as well. The effects of these compounds on invertebrates seem not to have been reported.

Nematodes

These are the threadworms or roundworms. They are well known to us in the form of whiteworms and microworms but there are thousands of parasitic species as well. They may infest the gut or various organs but they are rarely detected unless seen hanging from the anus. They may pass directly via eggs from fish to fish or have intermediate hosts, frequently copepods, that pass the nematodes on in the aquarium. Treatment with phenothiazine at 2% in the diet is sometimes effective. As suggested by Goldstein, a cat or dog worm food containing thiabendazole might be used if the fishes will eat it.

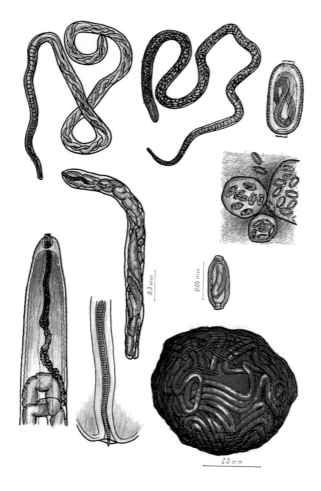

Hepaticola petruschewskii, showing male, female, egg with embryo, and cyst with eggs in fish liver; and *Cystoopsis acipenseris*, showing front end of male, caudal end of male, general view of male, general view of female, and egg.

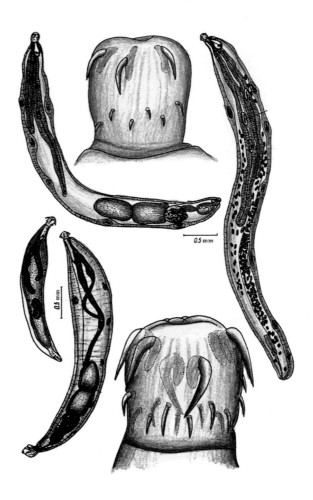

Thorny-headed Worms

These worms have rings of hooks on the head by which they attach themselves to the wall of the intestine and hang along the gut like a tapeworm. They have intermediate hosts with the larvae forming cysts. Fishes may be either primary or intermediate hosts. As with cestodes, you will not be

Thorny-headed worms, *Neoechinorhynchus* sp. A, showing young female, male, and proboscis; and *N.* sp. B, showing male, female, and proboscis.

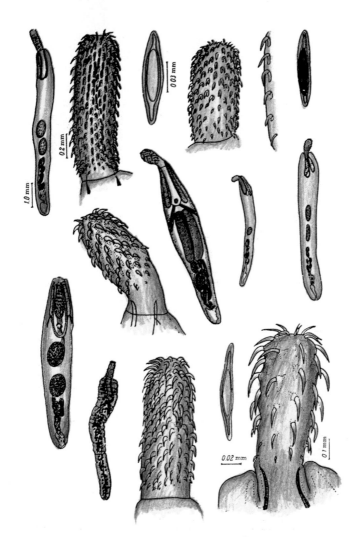

Several species of thorny-headed worms of the genera *Metechinorhynchus,* *Echinorhynchoides,* and *Acanthocephalus,* showing the males, females, and proboscises.

aware of an infestation unless large numbers are in the gut or a cyst forms in a vital organ. Even then only dissection would tell you what is wrong. When a fish shows signs of distress or damage and worms are suspected, the same treatments as for nematodes or tapeworms should be tried.

Crustaceans

There are over a thousand species of parasitic crustaceans in fishes. Many involve intermediate hosts, so their danger in the aquarium is limited. However, crustaceans are also a danger in that they carry the larval or adult stages of other fish parasites. Copepods in particular act both ways, as parasites themselves and carriers of them. Crustaceans usually have a series of larval stages; the first is a nauplius, unsegmented and part of the plankton, that is followed by a metanauplius showing the first signs of segmentation. Other forms looking more and more like the adult follow. Finally they settle down as a crab, shrimp, etc. Some larvae are parasitic and leave the plankton for a host—intermediate usually—while most species are parasitic as adults.

Copepods

These are common fish parasites. They are also a nuisance in the aquarium from time to time. Even though the species concerned is not parasitic, they can form unwanted swarms that annoy both the

A selection of parasitic copepods. It is amazing how changed from "normal" type copepods these parasitic forms appear.

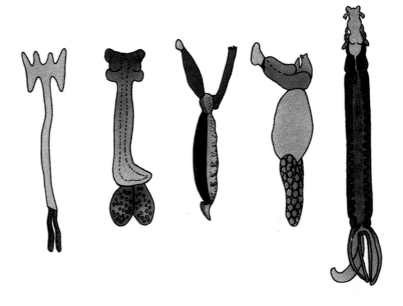

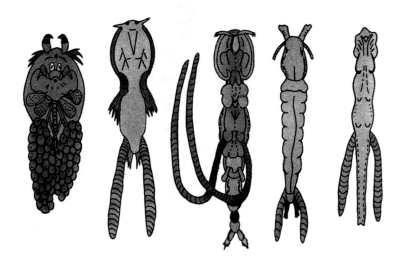

More parasitic copepods. The parasitic copepods become a nuisance in the aquarium from time to time but can be dealt with relatively easily.

fishes and their owner. Females are easily recognized by their paired egg sacs and are normally the parasitic sex. Males are rarely so, except that the larvae of both sexes may be parasites.

Anchor Worms
These copepods, family Lernaeidae, have both fresh and salt water representatives. *Lernaeocera branchialis* is a gill parasite that loses its crustacean appearance in the adult female. She becomes worm-like with chitinous anchors on the head and long coiled egg sacs. She is quite large, up to 1½ inches long and does not attack small fishes, but may be found on cods, groupers, etc. The anchors are long and may penetrate the heart, causing considerable blood loss. The presence of the parasite is shown by the long, coiled egg sacs protruding from the gills. *Lernaeenicus encrasicholi* and *L. sprattae* are found in smaller fishes—*L. encrasicholi* is found in the muscles and *L. sprattae* on the body and eyes. Both penetrate as does *L. branchialis* and have similar, but smaller, protruding egg sacs. Other small copepods are found

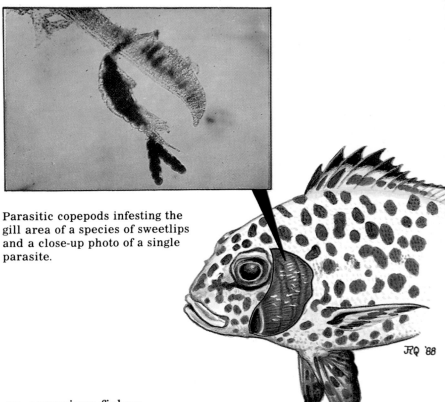

Parasitic copepods infesting the gill area of a species of sweetlips and a close-up photo of a single parasite.

on aquarium fishes, revealed by the usual egg sacs. The sacs may be long and thin, coiled or not, short and fat, or even spherical.

Treatments are the same as for monogenetic flukes. The treatments need to continue for a month since some of the life cycles take that long. There is always the possibility that eggs have been released. Remember that other invertebrates, especially crustaceans, may be affected. The dose of trichlorfon or whatever insecticide is used should be repeated for each of four successive weeks. A fish removed from the aquarium can be given a 2-3 minute bath of DFD (difluorodiphenyltrichloro methylmethane) at 1 ml in 2.5 gallons of sea water. Do not use DFD in the aquarium.

Fish Lice

The Argulidae or fish lice are sometimes classed as copepods and sometimes not. The best known to aquarists are in the genus *Argulus*. This genus has many species, most with adults about ⅕ inch long. The parasitic adult has an almost circular, flat body with two hooks at the base of the first pair of antennae, and suckers and a stiletto beneath a pair of eyes. The stiletto injects a poison that can kill small fishes and sucks blood from the fish. *Argulus* often swims from fish to fish, leaving nasty circular wounds behind. The female lays batches of eggs on rocks, etc. which should be removed promptly. The larvae, which may hatch in an advanced stage looking like young adults, take 6-14 weeks to mature and mate.

Adults on fishes are easy to see and can be removed with forceps. They occur particularly on seahorses but are harder to spot than on an ordinary fish. If they escape notice and treatment is needed, the same drugs recommended for flukes and copepods should be used.

A bright field photo of the carp louse, *Argulus foliaceus.*

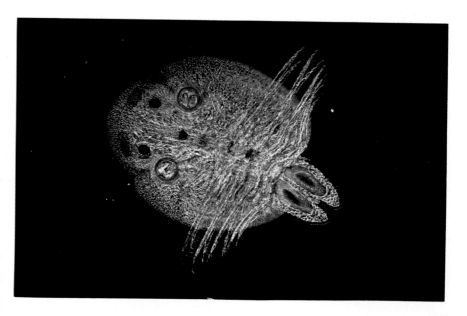

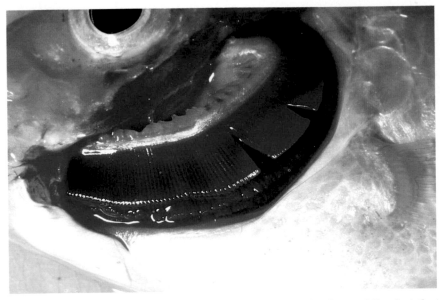

Healthy gills are recognized because of their bright red color and the fact that the gill filaments are not stuck together.

Diseased gills lack color and stick together (necrosis). This photo shows a gill infected by columnaris disease.

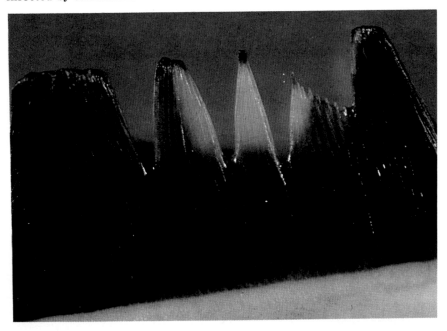

NON-INFECTIOUS DISEASES

Genetic deficiencies, poor nutrition, or poor aquarium conditions can all cause disease. So can stress. It may be quite difficult to decide whether a sick fish is suffering from such causes rather than from an infectious disease or parasites. Loss of appetite, emaciation, color changes, and odd behavior can be symptoms of water changes, from pH or poor gas exchange, to a toxic build-up of some kind. Gradual deterioration of the fishes would suggest that type of cause while sudden onset would suggest a recent catastrophic event, such as absorption of a household spray or a hidden dead fish or other creature. Always act when any change occurs. Do your best to identify the cause since a marine aquarium can go bad very fast if a serious cause of trouble is neglected.

Nutrition

We have seen that fishes require a diet high in protein and low in carbohydrates and fat. As long as their food contains the essential proteins and is not loaded with other constituents, they will do well, even on a restricted diet. However, it must contain the needed vitamins and trace elements as well. Too much fat (saturated or unsaturated) will probably cause liver damage. Too much saturated fat can cause intestinal obstruction as well. Old batches of food can accumulate saturated fat since oxidation gradually transforms unsaturated fats to saturated ones, despite the addition of antioxidants. Too much carbohydrate causes both liver and kidney damage arising from the storage of excessive amounts of glycogen. Therefore, check the composition of any prepared foods you purchase and do not buy old stock or too much at a time. Fish foods do not carry "use by" dates so buy from a busy pet shop that is not likely to have a slow turnover in your chosen purchases.

Vitamin deficiency can be avoided by the routine addition of supplements to the diet—not to the water. Lack of vitamin A can cause eye abnormalities, including exophthalmos (pop-eye). Insufficient doses of the B-complex causes poor growth, muscle degeneration, skin lesions, convulsions, and cloudy eyes. Lack of vitamin C in those fishes that cannot manufacture their own can lead to blood spots, poor bone and cartilage development, including spinal curvature, poor appetite, and exophthalmos. Vitamin D deficiency allegedly causes muscle spasms as well as poor growth.

Minerals are needed in the diet rather than in the water. They may have toxic effects in the water not seen by dietary administration. Sea water is rich in calcium and magnesium and may supply all of the minerals needed. However, the concentrations of many elements is very low. Thus, iron should be around 100 ppm in the diet, zinc should be around 200 ppm—concentration that in the water would kill the fishes. The following trace elements are also needed in the diet for normal functioning—iodine, copper, cobalt, manganese, aluminum, and molybdenum. All should be at 1 ppm or less. A human multivitamin plus minerals capsule added at the rate of one capsule per kg of made-up diet can supply much that is needed. Do not use a mega-dose! Water

conditioners often have vitamins and minerals of one kind or another in their make-up. These are harmless amounts in the water and do little or nothing for the fishes, although they may help some invertebrates.

Exophthalmos

Eye protrusion—pop-eye—is a symptom of some diseases and nutritional defects, but it may also arise from gas embolism, toxins, or as a genetic condition. It is exaggerated in some fancy goldfishes. One or both eyes stick out and, if severe, the condition may lead to blindness or loss of the eye. Water under high pressure can have an excess of dissolved gases that can invade the blood and lead to a deposition of bubbles, usually of nitrogen, in the tissues. In the eye, they can lead to exophthalmos. Elsewhere they may be causing trouble since the fish, in effect, has a case of the "bends." Even a sudden rise in temperature that causes gases to have a lower solubility could release enough nitrogen to start the bends or pop-eye. So if bubbles are seen in an eye, look for them under the skin in other fishes as well. Look also for tell-tale behavior that suggests the bends. Treatment is to lower the temperature as far as feasible, cut down aeration, and, in a big fish, remove large bubbles with a hypodermic syringe and fine needle. This is best done by a veterinarian.

Another cause of exophthalmos is excess copper, usually accompanied by some loss of balance. So if copper has been used or may be contaminating the water, do a copper test or renew carbon or similar filters just in case. A further possibility is excess thyrotrophic hormone of the pituitary gland. There is no feasible cure in that particular case. Often the condition rights itself even when left untreated.

Swim Bladder Disease

The swim bladder normally keeps the overall specific gravity of the fish the same as the water surrounding it so that little effort is needed for the fish to remain afloat. Inflammation or pressure from other organs can cause the swim bladder to malfunction, either by blocking the duct or preventing it from working

Exophthalmos in the freshwater cichlid *Pelvicachromis pulcher.*

properly. The fish may then have difficulty in swimming normally or even float belly up at the surface. If the condition is caused by a disease, it may be relieved on treatment. Otherwise there is no specific treatment except that it may be possible to remove excess gases in a large fish with a hypodermic syringe.

Tumors

There are many causes of tumors in fishes: diseases, parasites, irradiation, chemicals, hormonal imbalance, and heredity. However, fish cancers do not behave like human cancers, even when malignant, and do not usually spread around the body in the blood stream. Many tumors are harmless and can be ignored unless they are unsightly or interfere with some bodily function. There is usually no successful treatment and a sick or ugly fish is best destroyed. An exception is a thyroid tumor. This can cause a swelling that interferes with eating. It can be cured with iodine-containing Lugol's solution. Add it at weekly intervals at the rate of 2 ml per gallon and expect a cure to take several weeks.

The cause of many tumors is uncertain. The causes can be a fish's genes, a virus, or a chemical, or the interaction of two or all three of these. Not that it matters in practice, as there are no treatments except surgery. Toxins can get into prepared foods. An example is aflatoxins, produced by a fungus, *Aspergillus flaxus*, that grows on stored cereals if they are not kept cool and dry. Incorporated into a food, the toxins cause liver tumors. These are visible as body swellings if they are large enough. They are likely to cause trouble even if they are not visible.

Deformities

Most fishes with obvious deformities, such as spinal curvature, or missing or misshapen fins, will have been rejected before you are offered them. Occasionally, however a late development can occur that is not due to a disease. We have yet to see, however, if tropical marine fishes in general breed true as do most freshwater tropicals. The evidence so far is that they do and they do not behave like the goldfish. The goldfish, for reasons unknown to me, throws runts and abnormals that need constant culling even after centuries of breeding in captivity. This fact about marines is good news for aquarists. We may expect that as we learn to breed more and more marines, we shall not have the goldfish problem. Until feeding the fry is on a firmer basis, we may well lose many of

Spinal curvature in a freshwater guppy (*Poecilia reticulata*) with tuberculosis.

them. There may be severe size differences at an early stage, but that is a different problem.

pH

Although most of us like to keep the pH of the aquarium water around 8.2, marine fishes do not seem to suffer between 7.5-8.6. Some invertebrates, though, must be kept over 8.0. If the carbonate hardness is kept up, so is the pH. The fishes are happy around 10-15, although some coelenterates in particular are not. As long as regular water changes are made, it pays to add about 1 level teaspoon of sodium bicarbonate per 25 gallons per week. A low pH damages the skin and gills causing cloudiness and even capillary hemorrhage. There may also be a greater susceptibility to any toxic substances that are present. With ammonia, however, a low pH protects because the release of free ammonia gas is lessened. A high pH causes similar symptoms and also increases the damage ammonia causes. It can be brought about by plenty of algae in bright light with a low carbonate hardness. With both pH extremes, the fishes may gasp at the surface or dash around the tank. A test will tell which extreme is happening if pH is, in fact, to blame.

Temperature

Fishes are comfortable between 75°F and 82°F or a little higher. They can withstand up to 88°F in most cases if they are not crowded or short of oxygen. If the temperature does go so high, the fishes should be cooled down slowly when possible, say not more than 2°F per day. If they are severely chilled due to heater failure or chilling in transport, they should be raised rapidly to a normal temperature within an hour or two. Excessive heat makes fishes dash around the tank, gasp at the surface, and possibly try to jump out. Chilling slowly causes them to crowd around a working heater or to sit quietly at the bottom of the tank. They may show a peculiar swimming motion known as the shimmies. They swim slowly with a sinuous motion but do not get anywhere. The fishes are more likely to do this when suddenly or rapidly chilled. Anemonefishes often swim in this fashion when alarmed or even just when they feel like it.

Oxygen and Carbon Dioxide

The symptoms of lack of oxygen or the rare occasion of carbon dioxide excess are similar: increased respiration and gasping at the surface. Carbon dioxide does not cause pH changes in a well-buffered tank as it would in a freshwater aquarium because of the capacity of sea water to neutralize its effects. If the carbonate hardness is low, it can lower the pH. It is now possible to measure the amount of oxygen in the water with a simple kit. It should be used if any suspicion of oxygen lack arises. The solution is a good clean-up, increased aeration, and partial water changes. A low dose of ozone can be delivered to the tank, although potentially a rather hazardous procedure. These are first-aid measures. They are to be followed by a serious inquiry into the cause of the trouble—overcrowding, filter problems, overfeeding, insufficient light, temperature too high, etc. Excess of oxygen is hardly possible. However, oversaturation could occur in a static tank with algae in bright light. "Bends" caused by oxygen bubbles will soon disappear when the situation is corrected.

Chlorine

This gas is often present in tap water. It is detectable by its smell if present in quantity but it is still poisonous even if the water is odorless. If only chlorine is there, it will soon evaporate with aeration. It can be neutralized by the addition of 0.5 grains (30 mg) per gallon of sodium thiosulphate (photographer's "hypo"), or by a preparation from your pet shop. Do not add more than 10% of newly made salt mix to the aquarium without treating it for chlorine. Better still, always treat for chlorine no matter how little water you add.

The main effect of chlorine is on the gills. They become pale and break down, either as a result of a short period of high dosage or a long period of a lower dose. Death results in a few days from as little as 0.1 ppm. Recovery from all but the early stages is impossible. First, there are signs of irritation and rapid respiration. Then there is listlessness and death.

Chloramines

These are compounds of ammonia and chlorine. They are frequently added to tap water because they are more persistent and kill bacteria and neutralize some toxins better than chlorine alone. They are also more difficult to remove. They may be broken down with twice the dose of sodium thiosulfate recommended above, or by preparations from your pet shop. Inquire from local aquarists or the authorities whether they are in your tap water and act accordingly.

Ammonia

We have seen how ammonia acts on the gills to cause bacterial gill disease to set in. The gills become red and swollen and may push out from under the gill covers. Later they are attacked in all likelihood by various bacteria. Ammonia is a particular danger in marine aquaria because the relatively high pH causes a significant release of free ammonia gas, the particularly dangerous component of the system. Less than 0.1 ppm of ammonia causes distress. A safe concentration is less than 0.01 ppm. In testing with a kit, take action if any measurable quantity is present. Make several successive water changes of up to 30% and renew carbon filters or Polyfilters. Check the biological filter in particular, it may need improvement. If these measures are not followed by a permanent improvement, the biomass in the aquarium needs reducing. Switching to a trickle filter (a dry or wet/dry filter) will probably solve the problem for good.

Hydrogen sulfide

This is "rotten egg gas," named from the smell of even small amounts. It is more poisonous than chlorine or ammonia and is formed, in the absence of oxygen, in deep sand not subject to a flow of water such as when an underground filter malfunctions. Sometimes it happens in stagnant corners of a reef aquarium, too. No smell is actually detectable at the very low concentrations that are nevertheless lethal—down to 0.002 ppm. Any gray or black areas detected should be eliminated promptly. Prevention is the real solution: no deep areas of unfiltered sand or gravel, adequate maintenance, and no overfeeding. Symptoms of hydrogen sulfide (H_2S) poisoning are just death of fishes.

Ozone

Another highly lethal gas is ozone (O_3). It is about as poisonous as H_2S. It should not be metered into the aquarium directly, only via a protein skimmer and a carbon filter to remove the excess. A concentration of 0.005 ppm causes gill enlargement and changes in blood chemistry. Ozone has an "electrical" smell. If you can detect it, you are using too much.

Toxic Metals

Very low concentrations of various metals are lethal in the aquarium. They are particularly likely to affect marine fishes because of the liability of sea water or mixes to attack and dissolve any metal with which they come in contact. Tap water used for making up a mix, although fit for human consumption, may carry toxic levels of several metals, particularly copper, lead, and zinc. It should always be tested for copper. Unfortunately, we do not have lead and zinc kits so it may be necessary to ask local authorities for a water analysis if any suspicion arises. Always run the cold tap for several minutes to clear static water that may have dissolved metal from the pipes. Never use water from the hot water system. There are various brands of water conditioners that help to neutralize the effects of moderate levels of heavy metals by chelation or similar chemical bonding, but they do not remove them from the aquarium unless they are taken up by carbon or resins, etc.

Poisoning by metal is difficult to diagnose when sublethal amounts are present. Only general malaise is apparent. Acute poisoning causes blood patches on skin or gills and fins, cloudy eyes, and listlessness. Pop-eye and loss of balance also occur with copper. Always suspect it when no other cause is obvious. Avoid any contact of the water with metals. Guard against drip-back from overhead fittings which is liable to occur in reef aquaria with open tops and often fluorescents close to the surface. If you feel a tingling or mild shock on touching the water, look for the cause immediately. It indicates not only danger to you but

Cyanide toxicity in the angelfish *Holacanthus ciliaris.* Note the corneal opacity.

to the inhabitants as well because there is a metallic contact somewhere.

Copper

New copper piping is a common source of poisoning in the aquarium. So is old copper piping that is suffering from breakdown due to electrolytic action. Brass joints are a potential cause of this. If a fish-only tank has been treated with copper several times, it is possible for a flood of the metal to be released from coral sand or other materials within the tank following a pH fall. Allegedly, copper taken up in high concentrations by brine shrimp has caused the death of fishes that have eaten them. Apart from such effects, fishes usually can stand a concentration of up to 0.5 ppm for some time. A safe permanent concentration is thought to be less than 0.02 ppm.

Zinc

The danger from zinc is that humans can drink it at 40 ppm without obvious effects while this is a concentration that kills fishes. Galvanized pipes, tanks, buckets, and other implements are sources of trouble. Luckily the danger is decreasing as such materials are going out of use. People in country areas should be more careful than most of us as galvanized roofs and tanks still persist and water is still collected from them for domestic use. Concentrations over 1 ppm are dangerous to some fishes and a permanent level of less than 0.05 ppm is recommended.

Brass is an alloy of copper and zinc. It dissolves in water more readily than copper alone and is more toxic than either metal separately. It should be avoided completely.

Iron

This metal can be a danger in fresh water at a low pH but it is relatively safe in sea water. In acid water, it affects the gills. A concentration of about 0.1 ppm is maintained by some aquarists to help the growth of algae. It can be monitored by a suitable kit. Anything over 0.5 ppm is a potential danger.

Lead

In soft, acid water, lead can dissolve in dangerous amounts. In harder water it forms an insoluble coating inside the pipes and, therefore, lead piping is safe for domestic supplies. Lead sinkers must never be used in marine tanks as the water attacks them. Concentrations over 0.03 ppm are reported as dangerous.

Aluminum etc.

As it is more soluble the higher the pH, aluminum is to be avoided for any tank fittings. This is the same for any metal except high grade stainless steel. Contact of all metals used as coatings on domestic articles, except perhaps gold, should be avoided as well. Silver, nickel, cadmium, and chromium are all toxic in sea water. They dissolve in it and attain levels far above the minute amounts already there.

Nitrites and Nitrates
Although less toxic than ammonia, nitrites combine with the hemoglobin in the blood to form methemoglobin, which cannot carry oxygen. Some marine fishes appear to be remarkably resistant to nitrites. For safety, though, the concentration should not exceed 0.5 ppm, preferably 0.2 ppm, of nitrite-nitrogen. Susceptible fishes show signs of oxygen deprivation and may have brown gills as a sign of what is happening. The next step in the nitrification process is nitrates, which are relatively harmless to fishes up to 40 or 50 ppm. This is not harmful to all invertebrates. Aquaria easily go over 100 ppm with poor husbandry, a concentration that may prove dangerous even to some fishes.

Phenols
Decaying algae and foods, like tubifex, can release phenols into the water, one of the reasons for removing them from the aquarium. Household sources are numerous: carbolic acid (phenol), cresols, resorcinol, and polychlorinated biphenols are examples. They will only get into the aquarium through carelessness (they occur in some sprays. They kill fishes at around 100 ppm or less and damage gills and internal organs at lower concentrations.

Household Sprays and Paints
These have become a real danger to aquaria. Nearly all are toxic to fishes if taken up by the air pump or allowed to settle onto the water surface. Present-day open tanks with semi-exposed external trickle filters present a particular problem. They should be protected as far as possible. Regard all sprays as dangerous: hair, deodorant, mold-killing, carpet-cleaning, air-sweetening, insect-killing, etc. If a room containing an aquarium must be sprayed or painted and the tank cannot be removed, turn off the aeration, if possible, and cover the tank tightly.

Marine aquaria come in a variety of designs. This one is quite modern and fits in well with the clean, simple lines of the room.

MARINE ALGAE

Algae in the aquarium can enhance it, or it can become a considerable nuisance. If conditions are not right, the desirable algal types, in particular the fronded macroalgae, will not flourish. Instead, encrusting and hair algae may take over. They will blanket everything unless constantly removed. Even algae-eating fishes or mollusks can fail to control them. Probably the most common complaint of reef aquarium owners is that they find it hard to maintain a desirable balance in the algal population. In other aquaria, where the algae present are usually encrusting types on rocks, dead coral, or the sides of the tank, periodic partial clean-ups should leave some, but not too much algae.

Control of algae depends on lighting and water chemistry. Of course, it also depends on the extent to which the algae are grazed by vegetarians. The natural history of an aquarium is also a factor. The types of algae that tend to predominate vary at different stages of its development. At first diatoms, golden-brown algae, are likely to flourish, much to the annoyance of the aquarist. They are soon displaced by green algae, usually unicellular or filamentous species. If you are unlucky, blue-green algae that do not always look blue-green, may start to take over as slimy masses. These are difficult to control.

Lighting

All algae possess chlorophyll-a. This is a light sensitive pigment that enables plants to convert water and carbon dioxide into glucose, which the plants require. Green algae have chlorophyll-b as well. Both chlorophylls absorb red and blue light and reflect most of the rest of the spectrum so that the plants look green. Brown algae have fucoxanthin; so do diatoms and dinoflagellates (the zooxanthellae of corals, etc.). This and other pigments present in most of them absorb in the middle of the spectrum and pass the resultant energy to the chlorophyll-a. Red algae have phycoerythrin and phycocyanin that absorb in the green, yellow, and orange end, thus taking advantage of the greater part of the spectrum. Another pigment, Bcarotene, is present in all algae. It absorbs blue and green light. These additional pigments explain why red and brown algae can grow at greater depths in the ocean than green algae—they can make better use of the dimmer light available. This also explains why a dimly lit aquarium suits the reds and browns. It does not explain why they tend to disappear in brighter light. This is particularly so with the browns.

Light saturation studies are a guide to the requirements of various types of algae. They measure the maximum amounts of light needed by algae and above which they do not grow any faster. It may be assumed that any particular type of alga will flourish at well below its saturation value; certainly half the value should be adequate in the aquarium. Saturation values depend in part on the carbon dioxide concentration, the more CO_2, the higher the value. So the tests must be made at a constant CO_2 level. The results are probably higher than would be found in the sea with its low CO_2 levels.

Using cool white fluorescents, studies of macroalgae show that tropical green types need 13,000 to 16,000 lux. Lux is a measure of light intensity. Tropical browns need 7,500 to 11,000 lux, and reds 2,000 to 8,500 lux. Observations in the ocean agree when several methods of measuring photosynthesis (the building of organic substances from water and carbon dioxide) are used. They show that too much light depresses photosynthesis. Algae grow best in tropical seas a few fathoms below the surface and not where they receive maximum illumination. Note that this refers to free-living algae, not the zooxanthellae of coelenterates, etc. They require more light as they are within the tissues of their hosts. Little is known about how much light penetrates to them. Note also that studies of the effects of the ultraviolet component of sunlight or other types of illumination seem to be unavailable. There is a suspicion that it plays a part in inhibiting growth and photosynthesis at high intensities.

The conclusions from the information above are that the macroalgae we wish to cultivate in the aquarium can be expected to grow satisfactorily at relatively low levels of illumination. Our observations so confirm. The greens should flourish at around 8,000 lux. The reds and browns need even less. These levels are easily obtained,even way down in the tank in modern aquaria using either fluorescents or metal halides. The levels are far exceeded at the surface in reef aquaria. The average fluorescent tube gives around 5,000 lux at the surface but varies from less than 1,000 to 8,000 lux according to brand. An average metal halide would give around 40,000 lux at 175 watts, but this depends on how far it is placed from the water's surface. Even two fluorescent tubes running the length of the aquarium should be sufficient for algal growth as long as we choose suitable types—whites or daylights, or one of these together with an actinic-03. This does not look very bright but it gives out masses of blue rays needed by chlorophylls.

Algal Nutrition

With the aid of the chlorophylls, algae take carbon dioxide from the air and water. They combine it with the water to manufacture simple sugars. Hence the name carbohydrates to denote that they are made of carbon, hydrogen, and oxygen. These are then combined with nitrogen and phosphorus to build up more complete organic compounds. In addition, all species of algae studied need traces of iron, copper, magnesium, zinc, manganese, and molybdenum. Various individual species of algae need other elements as well. Cobalt, in vitamin B_{12}, is needed by some, silicon by diatoms, and boron by *Ulva* species. Some of the vitamins needed by fishes are manufactured by algae, which themselves need a few to grow or reproduce. Many need B_{12}; some greens need thiamine (B_1) as well; and some of the single-celled algae need biotin. No other vitamins have been found to be necessary to any algae.

Iron is present in sea water at a concentration of 0.002 ppm. It is often present in artificial mixes at 10 times this amount, as it is hard to keep in solution and it is needed for algae. Some aquarists add even more, up to 0.1 ppm. They find that the algae green up nicely under such treatment. However, some of the trace elements needed are toxic, both to algae and to various fishes and invertebrates if given in excess. Copper and zinc are examples. Hence, it is dangerous to use copper to cure velvet or white spot in the presence of a lush growth of algae. Crustaceans need copper in minute quantities to form their blood pigments but an excess accumulates in their system and kills them. An excess of phosphorus in the water is loved by hair algae that tend to swamp the aquarium if it is present at over 2 ppm. This is easily achieved if decaying material is not removed from the tank. Phosphates as pollutants at sea cause algal "blooms." These are vast numbers of red or green unicellular algae floating in the water.

CLASSIFICATION

Botanists do not use the term phylum. Their equivalent to it is division. Algae fall into quite a number of different divisions, perhaps surprisingly in that even the single-celled types

manage to have evolved along distinct lines. There is also confusion with primitive animals, fungi, slime molds, and other lowly creatures. They tend to be more and more placed into distinct kingdoms rather than divisions. At that level, we cannot really distinguish animals from plants and the rest. A good example is the "algae" of corals and other coelenterates, the zooxanthellae. They are classed by zoologists as animals, the dinoflagellates, despite their having chlorophyll.

In the scheme we shall employ, there are seven algal divisions. Cyanophyta, or blue-green algae, are ancient unicellular types. Possibly they should be separated from the others because they have no cell nuclei. Euglenophyta are flagellated unicellular algae. Chrysophyta, the yellow-green and golden-brown algae, include the diatoms. They have siliceous shells and are all unicellular. Pyrrophyta, mostly with two flagellae, include the dinoflagellates and are unicellular also. Chlorophyta are the green algae. They are both uni- and multicellular. Phaeophyta are the brown algae, all multicellular. Rhodophyta are the red algae, also multicellular.

Cyanophyta

The early earth, four or so billion years ago, had an atmosphere of mixed gases, but no oxygen. The first living creatures were all anaerobes that derived their energy and substance from gases like methane and ammonia. Then along came blue-green algae. They produced oxygen. Together with later divisions, they changed the atmosphere into what it is today—mainly nitrogen and oxygen with traces of other gases, like carbon dioxide and helium. Oxygen was a poison and most living creatures had to adapt to living with it, evolving into our present oxygen-dependent majority. Some, mostly bacteria and some other unicellular forms, can still do without oxygen, some are still poisoned by it.

The Cyanophyta had originated chlorophyll that is just floating around in the cell. They have none of the organization of more typical plant or animal cells: no chloroplasts, nucleus, or mitochondria. That is why many classify them as distinct from the true algae. They reproduce by fission and have no sexual modes of reproduction as far as has been discovered. They do not always look blue-green. Red colored species that possess additional pigments are common. Also, they may form filamentous masses, connected by slimy material that is rather characteristic of them. The genus *Trichodesmium* can be responsible for maroon-colored, slimy collections of filaments that rapidly coat everything in the marine aquarium if left to do so. By some miracle of organization, the tips of the filaments can move and will push up through sand if covered by it. The best treatment of this and other more typical blue-greens seems to be no water changes until they die off. Presumably this happens because of the exhaustion of essential elements.

Euglenophyta

Although the euglenophytes are not known to reproduce sexually, they are ahead of the blue-greens in having chromatophores containing chlorophyll, the equivalent of chloroplasts, a nucleus, and cell organization like the higher algae. They even have "eye spots" which are sensitive to light. Some class them as protozoa, flagellated and possessing chlorophyll. Most are found in fresh water.

Chrysophyta

Except for the diatoms, nearly all chrysophytes are in fresh water. Diatoms, though, are an important part of the marine ecology. They form the usual brown algal coating that occurs in a newly set-up aquarium. Their chlorophyll is masked by other yellow or brown pigments. These enable the algae to absorb light not taken up by the chlorophylls and to increase the total energy uptake of the cells. They reproduce by fission and by sexual processes as well, although not all Chrysophyta do so.

Diatoms are found in the plankton, of which they are usually a major part. They also are found, alive or dead, on various surfaces in the ocean. The shell has two valves, like a pill box, but it may be more complex in shape. The valves separate when fission occurs. A new valve fits into the inside of each old one to form a new pair of individuals. This means that the new diatom possessing the old outside valve is as big as the original, but the other one is smaller. When it in turn undergoes fission, one of its offspring will be smaller still, and so on. However, the process is halted before long by sexual reproduction in which two cells each produce a large gamete. These fuse to form a new, larger individual once more.

Despite their siliceous shells, diatoms are eaten by many invertebrates and small fishes. Their contents is digested and the shells voided to float down to the bottom of the ocean. The same thing happens when they die, eaten or not. Deposits of "diatomaceous earth" are found up to 3,000 feet thick, the product of millions of years of sedimentation.

Pyrrophyta

This division, mostly marine, contributes to aquarium science in several ways. Its members usually have two unequal flagellae. Among them are the dinoflagellates, *Oodinium* and *Amyloodinium* species, the causes of velvet disease in fresh water and salt water

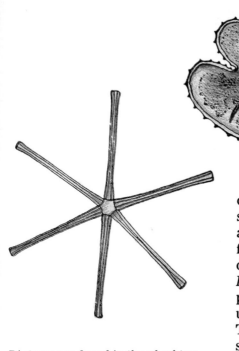

Diatoms are found in the plankton, of which they are a major part. The range of shapes can be seen in these two examples.

respectively. Other dinoflagellates constitute the zooxanthellae that can live independently or in the tissues of corals and other marine organisms. The majority are planktonic, neither parasitic nor commensal, and carry chromatophores with yellow to brown pigments. Those in animal tissues are frequently brown, hence the predominant colors of the corals and anemones carrying them. Some strains are brightly colored and so are their hosts. I saw for the first time the other day a brilliantly blue *Radianthus* anemone, presumably colored by an unusual strain of the alga. These algae perform several functions in corals, not only mopping up their waste products and donating organic chemicals to them, but also contributing to their ability to lay down their calcareous skeletons.

Chlorophyta

Green algae are the first division to form large, multicellular plant bodies, the thalli, that often resemble the leaves of land plants. Some are unicellular, however. All possess various chlorophylls and are found in both fresh and salt water. The macroalgae, multicellular types, do not have roots but are attached to the substrate by holdfasts. These often look very much like a root when they are thin and penetrate sand. They characteristically show alternation of generations, a plant version of the types of reproduction seen in many invertebrates, such as jellyfishes and flukes.

The seaweeds with which we are familiar are usually the sporophyte generation, formed by sexual reproduction. Therefore, they are diploid (having two sets of chromosomes as we do). They produce motile spores that swim by flagellae. They settle down somewhere suitable, if lucky, and produce a gametophyte generation. This often is quite different in appearance from the sporophyte. The gametophytes produce haploid gametes (they have a single set of chromosomes) that may be equal or unequal in size that fuse to produce a new sporophyte generation. There are many green algae. Here are some of those suitable for the aquarium.

Caulerpaceae

This family is very popular in marine aquaria. The genus from which it takes its name, *Caulerpa*, is unusual in that it will grow from cuttings and has a variety of species. It is losing out gradually to other, less obliging macroalgae as they become better known. This is not only because of the novelty, but because *Caulerpa* tends to blanket the tank almost as badly as hair algae. It has periods of collapse when it degenerates rapidly, loses color, and dies. Some species are worse than others in this respect. In all species, fronds arise from a rapidly growing stalk, the stolon, that also grows root-like rhizomes that attach themselves to various substrates. In bright light, growth can be very rapid. The algae will survive in fairly dim light, but they tend to be pale and spindly. When growing rapidly, *Caulerpa* is a great remover of nitrates and phosphates and helps to purify the water better than most algae.

The genus offers a nice selection of species, mostly from subtropical to tropical Atlantic and Pacific Ocean coasts. *C. prolifera* has simple, flat, green fronds (thalli), and grows rather slowly compared with most. Divided, fern-like fronds are produced by a number

Two types of the very popular algal genus *Caulerpa*, the flat-leafed *C. prolifera* and the feather-leafed *C. mexicana*.

of species. *C. mexicana* is a common example while *C. crassifolia, C. lanuginosa, C. sertularoides,* and *C. ashmedii* are others. The last named has very finely divided, feathery fronds and is very attractive. *C. paspaloides* goes further, having secondary pinnules that give it a very delicate appearance. *C. verticillata* produces fine tufts like a shaving brush, but it is not 'Neptune's shaving brush.' That is in a different genus, *Penicillus. C. racemosa* and *C. peltata* produce branching fronds with many spherical or cup-like knobs. They look like bunches of green berries.

These are two of the species particularly liable to collapse in the aquarium. A final mention is of *C. cupressoides,* which resembles a long-stemmed cactus plant.

Caulerpa is a soft alga. Other members of the family include varying amounts of calcium in their tissues and may be rigid and hard to the touch. They do not grow from fragments and must be obtained as whole plants, attached by their holdfasts to a piece of rock or other substrate. In some cases they can be gotten as monocultures, in vials, that attach and grow in the aquarium. They are generally slower growers than *Caulerpa* and are also more likely to survive the herbivores because of their calcareous fronds. They also seem to grow more readily in established aquaria than in one recently set up. The calcareous genera are placed in a separate family, Codiaceae, by some botanists.

The best known of these calcareous algae is the genus *Penicillus,* various species of which are sold under the names of Neptune's or merman's shaving brush. *P. capitatus* and *P. dumetosus* both grow to about 6 inches high, the former having a smaller brush than the latter, whose brush can spread out to around 6 inches wide. The genus *Rhipocephalus,* with two species, *R. phoenix* and *R. oblongus* looks like a *Penicillus* but the 'hairs' of the shaving brush are fused into plates.

The sea fans, genera *Udotea* and *Avrainvillea,* have flat, calcified thalli. They vary from 3 inches to around 6 inches in height. The smallest is *U. spinulosa* and has a nice fan-like appearance. The larger *U. cyanthiformis* has a curved thallus that gives it the look of an inverted cone.

Avrainvillea species differ from *Udotea* species in having a longer stalk bearing the fan-like thalli.

The genus *Halimeda* is strongly calcified, having a rigid hold-fast bearing branching thalli consisting of hard plates joined by flexible connections. Baby bows, *H. discoidea,* has rows of nearly circular plates and is a most attractive alga. *H. copiosa* has pendulous square-shaped plates and should be placed where they can hang down as in nature. *H. opuntia,* from shallow water, forms dense masses of small plates entangled together. Other members of the genus are deep water forms, despite being green.

Ulvaceae

This family, the sea lettuces, is found all over the world. The genera *Ulva* and *Monostroma* are alike, except that *Ulva* has a double layer of cells in the thallus while *Monostroma* has only one layer. Both look like loosely folded sheets of semi-transparent green material. They grow best in bright light. The green hair or ribbon algae that grow unwanted in the aquarium belong to this family. They also are widespread geographically. *Enteromorpha* and

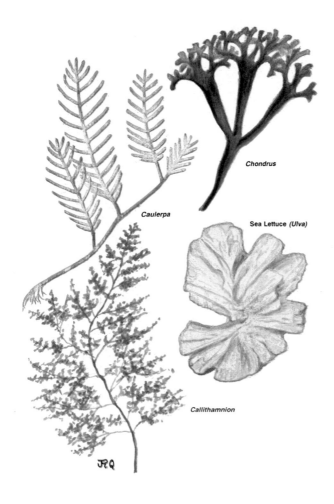

Various types of common marine algae: *Chondrus, Caulerpa, Ulva* (sea lettuce), and *Callithamnion.*

Blindingia, like *Ulva*, have two cell layers as well as alternation of generations, although the two generations of *Ulva* look alike. Species of *Monostroma* do not alternate in generations and so they all look alike. They also look very much like the *Ulva* species, from which they cannot be differentiated by appearance alone. Another worldwide genus of hair algae with dozens of species is *Cladophora*, family Cladophoraceae. Its members form mats of filaments composed of single cells laid end to end.

Dasycladaceae

This is a large family containing members with an upright central stalk with whorls of branches covering its entire length. In some species the whorls are confined to the top. The genus *Cymopolia* has species looking like strings of beads while *C. barbata* has branching strings with tufted beads, an attractive alga.

Phaeophyta

Brown algae are the dominant marine algae, though mostly not welcome in the aquarium. There are no unicellular types known, and nearly all have alternation of generations with the sporophyte as the usual 'seaweed.' Their carotenoid and other pigments allow them to grow at greater depths than the greens. Some are immense plants. The common rock species mostly belong to the Fucales where the gametophyte generation is retained in the thallus of the sporophyte generation. 'Egg' cells are housed in organs called conceptacles whereas the 'sperm' cells are produced elsewhere on the thallus. This parallels the course of sexual development seen in higher plants and animals. In other orders of Phaeophyta, separate sporophyte and gametophyte generations are seen, the sporophyte usually large and the gametophyte tiny, but sometimes they are alike.

In the aquarium, brown algae often degenerate to a soggy mess. It is usually a mistake to introduce types other than the small, calcified ones. Those growing up on living rock are normally harmless and quite attractive. They are rather rare, though, except for the brown hair algae of the genus *Ectocarpus*.

Caulerpa is a great remover of nitrates and phosphates and helps to purify the water better than most other algae.

Dictyotaceae

These small algae are safe in the aquarium, forming brown fans or branches, calcified or otherwise. *Dictyota* and *Dilophus* are brown counterparts of *Ulva* or *Monostroma*. They have flat thalli, sometimes greenish-brown, branching into a cluster of folds.

Padina is a fan-like genus with small fronds usually split into several segments. *P. vickersiae* grows to about 3 inches and is lightly calcified. *P. sancticrucis* is similar in appearance, but is heavily calcified. *Stypopodium* is bigger and has blades that branch. Some species have green-brown stripes. *Pocockiella* species have variously shaped plates. They tend to lie flat in the case of *P. variegata*.

Sargassaceae

The alga of the Sargasso Sea is *Sargassum natans*. It floats in isolated masses separated by clear areas and is believed to be the most abundant plant in the world. Reproduction is solely by fragmentation. Any formation of a gametophyte generation, even in conceptacles, would be useless. *Sargassum* species that do not float reproduce by alternation of generations. *S. natans* could be a useful aquarium plant, although it does not seem to have been tried. It has fronds and stems like *Caulerpa* and presumably could be anchored if desired so as not to float. In the sea, the floating masses wander under the surface for some distance and are the home of many organisms, including the anglerfish, *Histrio histrio*. The greenish brown genus *Turbinaria* has members, such as *T. tricostata* and *T. turbinata* that have stalks bearing groups of thalli that are pyramidal in shape. They have air-filled vesicles and do well in the aquarium.

Rhodophyta

The red algae are by no means always red in appearance. They may

Besides the berry-type *Caulerpa racemosa*, one can see a group of *Acetabularia* on the right. It is obvious why this latter alga has become known as the Mermaid's Wineglass.

carry a blue pigment, phycocyanin, which, combined with the masking red of phycoerythrin, can cause them to look purple or brown. These auxiliary pigments enable them to survive in relatively poor light as they pass energy over to the chlorophylls. Hence, these algae are found in quite deep waters. They are mostly small plants and are safe in the aquarium. Many of them are attractive as well.

Reproduction in most genera is complicated. There are three generations: the sporophyte, gametophyte, and carposporophyte. The latter is interpolated between the more usual two generations.

The sporophyte generation produces non-motile spores. They develop into separate male and female plants of the gametophyte generation and usually look the same as the sporophytes. In turn, these produce non-motile sex cells. The male cells are set adrift like pollen and stick to the sex organs of the female plants. They fertilize the ovules or 'egg' cells which grow into the carposporophyte generation. They remain on the female gametophyte plants and produce carpospores that develop into sporophytes after being set free. So the plant you see may be a sporophyte or a gametophyte, as with *Ulva*. If there is a difference, though, it may be the gametophyte in the red algae that is the large, typical alga and the sporophyte that is small.

All the red algae are placed in a single class, Rhodophyceae. Further classification is based on rather erudite details of cellular structure and behavior. Our interest in them is of a more practical nature, so that division into the soft and coralline types is adapted here. The corallines are heavily impregnated with calcium salts. They are decorative in the aquarium and are not eaten by most fishes although crabs may destroy them. They are now recognized as being more important than corals in the formation of 'coral' reefs, being able to form reefs on their own. They pack down between corals, etc., to form limey concretions that give the reef solidity and a base for further growth.

A tank without some algae looks quite barren. Actually, a well-lighted tank will soon have a growth of filamentous and other algae covering the decorations as well as the glass.

Soft Red Algae

Most red algae are found below tidal levels, but some of the filamentous types occur in rock pools and between tides. Two such genera are *Bangia* and *Polysiphonia*. Both have many species that form slippery masses on rocks. They are purplish or brown in color and are composed of patches of entangled filaments. *Bangia* is solid throughout, but *Polysiphonia* can usually be recognized by its typical

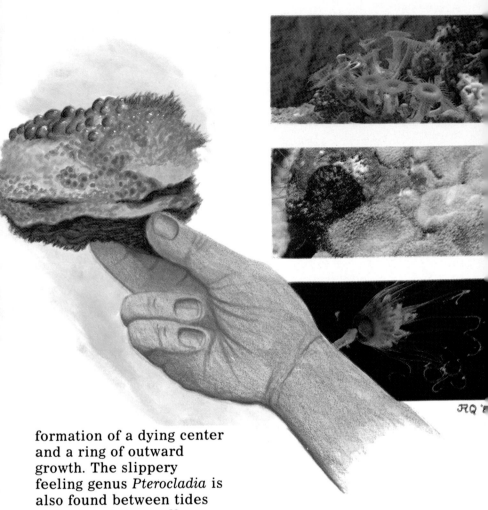

formation of a dying center
and a ring of outward
growth. The slippery
feeling genus *Pterocladia* is
also found between tides
and grows into small
'bushes' about 4 inches
high. It does well in the
aquarium.

At lower levels,
Laurencia is a fern-like,
small, and attractive alga
that has many widely
distributed species.
Porphyra is larger and has
leathery blades up to a foot
long. Many other genera

Living rock may have many
different types of animals and
plants growing on or in it. Only
time and the proper conditions
will let the aquarist know what
has been introduced into his tank.
The photos at the right show some
of the organisms that may appear.

exist that have not been reported as suitable for aquaria or otherwise. Many of them certainly are, though.

Coralline Red Algae Some of the heavily calcified red algae are minute and encrusting causing the attractive pink and other colored patches on rocks or dead corals. The genera *Melobesia* and *Lithothamnion*, divided into further genera by some authors, are typical of the encrusting types. Small, bushy species also cover rocks, with jointed stems branching continuously to form interlocking carpets or isolated masses. The genera *Amphiroa* and *Arthrocardia* have stems that are circular and oval in cross-section respectively. Both are decorative in the aquarium, slow-growing, and tough. They are pink or mauve in color and 3 to 4 inches in height. When dried out, there is sufficient lime to remain as a white, bleached skeleton with ¼ inch joints.

A more delicate genus is *Corallina*, which has tiny individual segments about ¼₀th of an inch long and a feathery appearance with pastels colors. *Jania* is similar. Quite different in

structure, the leathery *Peyssonnelia* has species looking like a tree fungus or Chinese mushrooms with flat, rounded plates an inch or two across overlapping each other. They are dark red or brown in color. This genus is less calcified and more readily eaten or destroyed in the aquarium.

Higher Plants

Some of the flowering plants have adapted to salt water. They form the 'eel grasses' or 'turtle grasses' seen on sale in pet shops. They are not grasses, but they look very much like them. *Zostera* species have rhizomes like an Amazon sword plant that creep along and put up thin leaves that are short and bright green in color. *Posidonia* has longer, wider leaves. It is dull in color and less attractive in the aquarium. *Thalassia* (turtle grass) has quite broad leaves. *Halophila* has normal looking blades and is found near coral reefs. All, however, are found in sandy or muddy substrates and do not grow on reefs proper. They are important in nature as breeding grounds for fishes or their foods.

BREEDING MARINE FISHES

An exciting period in marine fishkeeping is opening up right now—that of success in breeding various species. The basic problems are several. First, we know much less about how many species breed in nature than we do about freshwater fishes. Second, we are dealing in many instances with fry that cannot eat the kinds of first foods we are accustomed to providing. Third, we are dealing with fry that are part of the zooplankton in nature. Thus, they live in a boundless volume of water that is not readily imitated in the aquarium. There are lots of difficulties to solve that may not be the same from one species to another. We are making good progress, though, and it looks as though it will not be too long before tank-bred marine fishes will be available on a significant scale. This is important since we must not go on depleting the seas of coral and fishes that may become endangered at the present rate of collection.

If you cannot keep a tank like this one healthy you are probably not experienced enough to tackle the breeding and raising of marine fishes.

We have not been greatly helped by the quite early and successful culture of food fishes by the marine scientists of the late nineteenth and early present century. They obtained eggs and milt (sperm) by stripping fishes and fed the fry on newly hatched brine shrimp. Neither technique, though, is feasible with the great majority of aquarium species. Present-day methods are similar and we are seeing an uprising of fish culture that has followed a period of quiescence. Work on cod began in 1878. By 1917, the output of only three American hatcheries was over 3 billion fry per annum. Many other countries contributed like numbers for release into the sea. However, the release of such numbers of fry, covering cod, haddock, flounder, plaice, and others, could not be shown to result in useful increases in catches. By the 1950s the practice was abandoned. Present-day methods retain the fry and raise them in enclosures.

Physiology of Reproduction
Sex determination in fishes is in a more primitive state than in most

land vertebrates that have a defined sex once and for all. Many fishes, particularly marine ones, can and do change sex. Sometimes they change from male to female, sometimes the other way around, and sometimes they change from juveniles to whichever sex is needed at the time. In rare cases, a fish can be hermaphroditic (possessing organs of both sexes) and may fertilize its own eggs, as in the sea perch, *Serranus subligarius.* However, at any one time, a fish usually has either ovaries or testes, producing either eggs or sperm. The reproductive system is under the control of the pituitary gland. This is a small organ at the base of the brain, controlled in turn by the brain itself.

The pituitary gland puts out hormones among which are gonadotrophins. They stimulate the growth and functioning of the gonads, the ovaries or testes. The gonads produce other hormones, androgens and estrogens, that stimulate the development of male or female sex organs and behavior. Then, in some fishes, the pituitary gland puts out a hormone, prolactin, that stimulates parental behavior. This may extend only to guarding the eggs, and sometimes the early fry. The name 'prolactin' comes from the fact that the same hormone in mammals stimulates milk production.

The stimulus to the pituitary gland varies in different species, often depending on where the fishes live. Changes in day length may be effective in temperate latitudes, where most fishes spawn in spring or early summer. Changes in water quality can operate also, such as dilution brought about by monsoonal rains. Changes in temperature are often effective in freshwater fishes, probably less often in marines—at any rate in tropical marines. Some fishes, like the grunion, *Leuresthes tenius*, spawn following a full or new moon. It is not certain whether moonlight or tidal changes are responsible.

Amphiprion frenatus laying (upper photo) and guarding (lower photo) eggs. The eggs are placed within the defensive perimeter of the anemone's tentacles to keep potential predators at bay.

What triggers the gland in deep sea fishes remains a mystery. Some, such as the deep sea anglers that never rise above 3,000 feet depth, spawn at definite seasons. Apart from bioluminescence, they live in the dark at a temperature of 35-40°F.

The usual fate of the eggs of marine fishes is to become part of the zooplankton and be eaten. Just the one in a million ever comes to anything. Shallow water species quite often guard their eggs for a period, but when the eggs hatch they follow the same route and become planktonic. Losses have been reduced by the early parental care. Very few species take any care of the newly hatched fry; most are much more likely to eat them instead. Nest builders and/or egg guarders are found among the blennies, gobies, damsels, basslets, and sticklebacks. The anemonefishes clean a patch near their anemone, lay a few hundred eggs onto it, and guard them until hatching, after which they take no further interest. Quite a number of species mate readily in the aquarium—it is raising the young that presents the problems.

Breeding Pairs

Several factors make it difficult to obtain true pairs of most marine fishes. Their intolerance of members of the same species, in the case of angelfishes and many others, is one factor. Difficulties of sexing in some cases is another. Males are often slimmer, smaller, and more colorful than females, and also exhibit male behavior when the time comes. Seahorse and pipefish males have a brood pouch in which they carry the eggs until hatching. Thus they are easy to sex, even when the pouch is empty. The most obliging are the anemonefishes and some other damsels. They are happy in groups and change sex on demand. A typical group of a single species of anemonefishes presents the appearance of a mother and father with their young, there being

Chromis viridis spawning. Damselfishes are prime candidates for a first spawning. Like the anemonefishes, they will guard their eggs but abandon their fry once they hatch.

two adult fishes and a group of small ones. Not so! They are all of about the same age. Two have developed into a female, the largest, and a male. The rest have been halted in development by chemicals (pheromones) produced by one or both of the pair. If the female dies, the male becomes a female and one of the 'young' takes his place. If the male dies, the same thing happens. So any batch of say, half a dozen small fishes, may be expected to produce a mated pair.

In other cases, as long as a group of young fishes can be kept together, about six should be selected to give a good chance of finding a true pair. If the sexes are equal in numbers (not always the case), six give a 31:1 chance of at least one pair, five give a 15:1 chance, and four a 7:1 chance if all live. Whether they decide to pair up and mate is another question, but we hope for the best.

Some fishes spawn in groups. In such cases we may be in real trouble. This is typical of ocean fishes, wrasses, and many damsels other than *Amphiprion*. So, to chance one's ability at breeding, it is best to choose seahorses or anemonefishes. Seahorses will mate in the aquarium performing a rather pretty courtship dance when the female inserts eggs into the pouch of the male. The pouch swells up during the course of development and the male eventually expels the tiny replicas of himself with quite violent contortions. The young can be fed initially on newly hatched brine shrimp. Some species will go on eating them to maturity, but they will remain small in comparison with their parents. An acquaintance of mine raised three generations of such dwarfed seahorses (*Hippocampus whitei*) solely on newly hatched brine shrimp. Instead of the usual batch of several hundred young, the dwarfs produced only about 30 at a time.

A spawning sequence involving *Dascyllus aruanus*. The site selected was at the end of a coral branch but on its lower side. The spawners must therefore turn upside-down to lay and fertilize the eggs.

Spawning Arrangements

Marine fishes are usually kept in a special breeding tank for spawning. It is the eggs or fry that are removed and not the parents, for if they are left undisturbed, the parents often will go on spawning at frequent intervals. If removed to another tank, they will likely stop. So tanks for mated pairs should be set up and maintained on   more or less permanent bases. They should be large enough to accommodate the mating actions of the particular species. For demersal, bottom-spawning fishes that lay their eggs on a rock, a shell, or in a nest, a minimum of 20 gallons is needed. For others, a much larger tank may be necessary since they must have space to swim around and release the eggs and sperm. Depending on the size of the fishes, tanks of 50-100 gallons may be required.

Tropical marine fishes rarely need a change of conditions to induce spawning as long as a normal temperature of 78-82°F and a 12-12 to 14-10 lighting schedule are maintained. There are accounts of a lowered salinity stimulating some demersal fishes to spawn, from 1.022 to 1.016 or 1.026 to 1.019. This may be imposed within as little as an overnight period. These have not gone on to say how the fry were raised—at the lower or some other salinity? Such fishes produce a few hundred eggs as a rule, and so do small pelagic fishes, but they produce that number per day totaling typically in the thousands. Larger pelagic fishes may release tens or hundreds of thousands per day with totals in the millions. The demersal eggs take 10 or more days to hatch whereas eggs of pelagic fishes hatch rapidly, usually within a day. If they did not, there would not be sufficient left to give a viable hatch of fry. Some marine fishes are mouthbrooders, producing even fewer eggs than demersals, for obvious reasons. These also have a prolonged hatching period. Eggs and fry often are spat out at intervals and sucked back again. Presumably this is to keep them clean and well aerated.

Courtship and Spawning

Quite a lot has been written about the mating behavior of various marine fishes. I shall now present a

A male *Chromis atripectoralis* in courtship coloration.

A female *C. atripectoralis* in the inverted position as she lays her eggs. The male remains close by ready to fertilize the eggs.

few examples most likely to be relevant to the home aquarist. Seahorses already have been described so we shall pass on to the anemonefishes. These tend to form permanent pairs. There is little in the way of courtship apart from preparing the site for laying the eggs. All *Amphiprion* species and *Premnas biaculeatus* behave similarly as far as I know. They differ only in the degree of belligerence shown. This is manifest even without spawning, as they are territorial. The belligerence increases as spawning and care of the eggs approaches.

The *Amphiprion* pair chooses a site under or near their anemone. They guard it against all comers, especially other anemonefishes. In nature, an exception is made for the immatures that may be with them, while in the aquarium they may tolerate well-known tank mates as well. In a spawning tank, they will, of course, be on their own. The female grows quite fat as she fills up with eggs. At this stage both the male and female start to clean the site they have chosen, biting at it for several days. While this is going on, there is a little

display from time to time— both shimmy at one another in a pause during cleaning. Spawning starts late in the afternoon and takes one or two hours. During this time, row after row of usually orange-colored eggs is laid down. The female is followed by the male as he fertilizes the eggs. Finally, the cleaned area and often some of the surroundings are covered by a dense sheet of adhesive eggs. These are not only guarded, but they are mouthed and cleaned by both parents. Any infertile eggs are dislodged. After eight or ten days the eggs will have developed eye spots and will wave around in water currents. This is a hint that they will hatch soon. This happens shortly after dark. Therefore, strategies must be developed for collecting the fry. More about this later.

Other damselfishes may lay eggs onto patches of rock or coral, much as the anemonefishes do. Alternatively, they may make nests or burrows in which the eggs are deposited. The blue devil, *Chrysiptera cyanea*, has been described by DeBernardo as spawning in pairs and depositing only 15-45 eggs. Moe and Young,

A group spawning of the wrasse *Thalassoma lucasanum*. The group in the center is ascending in a spawning burst. A previous sperm cloud can be seen above them.

A pair of *Labroides dimidiatus* spawning. At this point they are spiraling toward the surface in preparation for spawning.

on the other hand, described the yellow-tailed damsel-fish, *Microspathodon chrysurus*, as producing batches of eggs covering up to 80 square inches of coral and numbering up to 92,000. They did not state how many males or females were concerned in such a feat but one guesses that at least one male and several females must have been involved. The translucent, pinkish eggs hatched in about three days at 80°F. The demoiselle, *Pomacentrus coelestis*, has spawned in my own tanks. As the male digs out a nest in the coral sand and entices females into it, a useful account of what goes on is limited. The male, brilliantly colored, 'shimmied' in front of each of the three females. They then followed him down into the hidden nest. He guarded this fiercely for three days and then deserted it. Presumably this was the hatching period. Since all this took part in a community tank, no young survived.

A number of other fishes pair off and care for the eggs in a similar fashion to the demoiselles. Neon gobies, *Gobiosoma oceanops*, behave rather like *P. coelestis*, except that only one permanent female is concerned. Some of the other gobies and blennies also behave similarly, a few of which have been raised to at least a juvenile stage in aquaria. Rather surprisingly, some of the egg scatterers that take no interest in their eggs which join the plankton, also pair up for at least a period. Among them is the harlequin bass, *Serranus tigrinus*, which pairs off with another of the same species, although both are hermaphrodites. They share a territory when spawning, rushing up through the water after a display of body contortions and flicking of the fins. Mandarins, at least in the aquarium, form loose pairs that tend to wander around together. Both *Pterosynchiropus splendidus* and *Synchiropus picturatus*, despite their demersal habits, are pelagic spawners. The male chases the female around the aquarium as he displays and circles over her with erect fins. Eventually they both rise up into the water several times releasing eggs and milt side by side. Only a few hundred eggs are laid despite their planktonic nature.

Some of the angelfishes

A male *Chaetodon rainfordi* is courting the female by "dancing" in front of her.

The male is intent on moving into spawning position with the female. All this is happening over the reef. The eggs are released in open water to become part of the plankton.

This spawning group of *Chromis atripectoralis* find the algae growing at the base of the coral to their liking for depositing eggs. Two pairs can be seen in the algae while the others are busy courting.

and chaetodons that also produce planktonic eggs nevertheless pair off. In the wild, this apparently happens for long periods. So does the jackknife, *Equetus lanceolatus*, although it also forms schools. Pairing in pelagic spawners thus is not all that rare in coral fishes. Also, there would seem to be cases of different behavior in related species. The male of the Atlantic anglerfish, *Antennarius striatus*, is said to guard the female when she is full of spawn. My own observations of the Pacific *A. striatus* show no such behavior in several cases of spawning. All I have seen is indifference to each other until a male chases a female around the tank quite briefly. This is followed by the release of a roll of jelly containing thousands of eggs. The developing embryos hatch from the disintegrating roll after four or five days. These observations were made in the early 1950s. Attempts at raising the young back then were unsuccessful.

Mouthbrooders have their problems. Young's account of the behavior of the yellowhead jawfish, *Opistognathus aurifrons*, shows that the male who broods the eggs still eats. The eggs are kept in his mouth nearly all of the time, but are spat out and cleaned or oxygenated at intervals. When he eats, for a short period of time they are placed in the tunnel where he lives. They are not separate, the 150-300 eggs being attached by fibers to a stem and look like a bunch of grapes. They are yellowish in color and a little over 1/25 inch in diameter. Young did not describe mating behavior or give other details.

The Rearing Tank

A tank in which fry are to be reared should be bare, fairly deep, and gently aerated. Its size should be limited to the number of larvae to be raised as it will require frequent partial water changes. Otherwise it should be as large as possible. Tailor it to 10-20 gallons per 100 eggs or fry to be kept. Provide good top lighting and the temperature and pH should be about the same as the tank from which the larvae came. One airstone per twenty gallons, which gives as fine a spray of bubbles as possible, is best. As constant a temperature and

A pair of Butter Hamlet, *Hypoplectrus unicolor*, in a spawning
embrace. These fish are hermaphroditic and either one can be acting
as the female.

This is a male *Assessor macneilli* brooding eggs in his mouth.
Mouthbrooders are not common in groupers or related families.

pH as possible is an advantage as the fry do not respond well to changes. The biomass of the young will be so small that little ammonia will be produced. The frequent water changes will take care of it as long as there is no overfeeding.

Eggs or larvae may be placed in the tank according to species and convenience. With the demersal fishes that lay visible eggs onto a rock or other surface, it may be convenient to remove it before the eggs hatch and place it in the caring tank. This avoids losing the fry when they hatch. However, it may provide too many fry and you may need to net some out and discard them. Alternatively, use a small vessel or siphon, not a net, and place them in another tank. Signs of readiness to hatch are the developing eye spots and in many cases the larvae waving about on stalks as they are fanned by the parents. The eggs of *Dascyllus* and other damsels hatch after two to three days. *Amphiprion* and *Premnas* hatch after 8 days, or a little longer if below 78°F. Hatching occurs an hour or two after dark and sometimes takes place over two nights.

In order to disturb a breeding pair as little as possible, it is better to wait for the eggs to hatch in the breeding tank. Then, a few hours after dark and a probable hatching, shine a flashlight along the top of the tank, preferably in a corner. The fry are normally phototrophic and will swim up to the light to be caught—not with a net, but with a vessel or siphon. The larvae of pelagic spawners can be caught in a similar manner or their eggs, which are fat-laden and float, can be removed. With such fishes, it is probably best to collect the eggs, if possible.

The fry will be in the rearing tank for several weeks. Cleaning and water changes are easy as the light keeps them up in the water. Siphon about 10% per day from the bottom, preferably every day. Do not wait longer than every three days, and then change 30%. The fry are sensitive and do not appreciate too much new water. The water must be prepared and conditioned carefully. It must be as similar as possible to that already in the tank. Use the same mix or batch of sea water at the same temperature and pH and gently introduce it into the tank, best siphoned in at the bottom.

Fry Foods

Newly hatched fry need no feeding for the first two days as they are still absorbing nourishment from the yolk-sac. They will need food on the third day and will die if not fed by the end of that day. Most species of tropical marine aquarium fishes cannot take newly hatched brine shrimp as a first food so that some alternative must be available. Marine fry are not attracted to the types of substitutes used with many freshwater species; egg yolk suspensions, finely ground foods of various types, yeast, and so on are all useless. Live foods, necessarily minute, must be used. An obvious substance is marine plankton if you can get it. Tropical plankton lives long enough to be eaten in a tropical tank, but that from colder water does not. Plankton also contains predators and parasites that may cause a lot of trouble so it is best to forget it, even if available. Culture a suitable food.

Young fishes do not eat planktonic algae, it is zooplankton that they need.

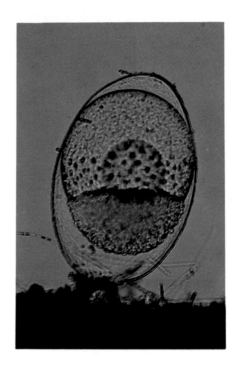

A developing egg of the damselfish *Chromis atripectoralis.* The pomacentrid eggs are usually attached to the substrate by short filaments as shown here.

However, zooplankton cannot be cultured—at least not the tiny stuff that is needed—without phytoplankton or a suitable substitute on which to feed. The normal routine that has been developed is a two stage job. First an algal culture is developed. Next an animal culture that feeds on it and in turn can be fed to the fry is developed. The fry can eat

919

only while they can see the zooplankton, therefore an extended lighting period is indicated. Freshwater fry are sometimes illuminated constantly. That seems to do no harm. It also causes them to grow rapidly because they eat continuously. No such practice has been reported with marines. It would be worth trying, however.

Algal Culture

It is not as easy to start a marine algal culture containing the desirable one-celled varieties as it is to start a freshwater one. Boiling some plant material in sea water and leaving it to develop a culture does not work, except to produce a bacterial culture. It is best to inoculate a sterile mix, fortified with nutrients, with a specific organism purchased commercially. The mix needs the various algal nutrients (already discussed earlier), to be supplied as directed on containers, or in the form of a fertilizer used for house plants. Use a liquid fertilizer made as directed for watering the plants and add one fluid ounce (28 ml) per gallon of sea water. This can be added per day to a flourishing culture that will be mopping it up—the culture will have up to 100 million cells per 5 ml teaspoon. Keep such cultures constantly illuminated and use reasonably large containers—around 2-5 gallons is best. Employ several of them to ensure a good supply of algae. If anything goes wrong, an individual culture can collapse quite suddenly and you are left without a supply.

Suitable algae that can be purchased are *Chlamydomonas, Anacystis, Dunaliella, Isochrysis,* and *Chlorella.* The latter is in vogue right now and easily obtained in pure culture. An alternative is *Torula* yeast. It will grow in sea water and can be bought at health stores. It can be used on its own, or to supplement the algae. It is even possible to feed baker's or brewer's yeast suspensions that need no complicated culturing. The results with yeasts, though, are not as good as with algae. The amounts of any culture to be fed to the zooplankton must be judged from what happens in the zooplankton culture tanks. These should be kept nicely green, or milky, as the case may be. They should not be thickly so, just mildly colored. Be careful not to transmit any back to the algae or yeast cultures when topping up the animal cultures or you will finish up with only zooplankton.

Zooplankton Culture Two types of zooplankton are in favor at the present time, although many others would be perfectly feasible as well. The two are copepod larvae and rotifers. Rotifers have been used more extensively today, but there is evidence, that the older use of copepod larvae produces better stocks of artificially raised fishes. Copepods of the genus *Acartia* were the original favorites. They are brackish water calanoid copepods that swim in the water all the time and can take full strength sea water if subjected to it. The adults are too large for the fry; it is the larvae that have to be collected and fed to them. The females have paired egg sacs hanging from the rear. From these the ripe eggs drop into the water. They are so prolific that it has been calculated that one female, should all of her progeny and their progeny survive, etc., could give rise to over four billion young per annum.

Copepods can be cultured by placing the females, about 100 at a time, in a small vessel. The container should be about ½ gallon with a base consisting of a fine screen. The eggs can drop through this but not the adult females. The eggs are received into a surrounding vessel of about one gallon and are siphoned off regularly and transferred to 5-gallon jars. Here the nauplii will hatch. The females and the nauplii are both fed a rich algal culture. Their jars are kept well lit to keep the algae flourishing. So far, only calanoids appear to have been used. Others, particularly harpacticoids, which live at the surface and have free-swimming nauplii, seem equally suitable. As with algae, it is essential to keep several cultures going to insure against loss.

Rotifers are easier to cultivate than copepods. This is because the adults are small enough in most species for the larval fishes to eat. Many different species could be used. The present choice, readily available, is *Brachionus plicatilis*. This is a marine and brackish water species about half the size of a newly hatched brine shrimp. It has a nice, meaty, rounded body. A rotifer gets its name from the feeding mechanism; two circlets of cilia around the gullet that wave rapidly and give the appearance under the microscope of two rotating wheels. They vary in size from ⅛ to ⅟₅₀₀ inch in length; *B.plicatilis* is less than ⅟₁₀₀ of an inch. It reproduces mostly by parthenogenesis, the production of unfertilized eggs that hatch over as all females. They are larger than males and are just what is wanted. Growth is rapid. The generations follow each other at two-day intervals.

The optimum salinity for breeding is that of sea water. The best temperature is 80-90°F. Considerable variations in both can be tolerated as long as there is plenty of food. Container sizes can be from ½-5 gallons. The larger is better as far as stability and life of the cultures are concerned. Breeding females are added to a culture fluid with plenty of algae. Keep illuminated. More algae will need to be added as the rotifers increase in numbers. Mild aeration is necessary in any ordinary-sized tank. Start with a good-sized batch of thousands of females. This helps to swamp out any chance of some other infusorians taking over and spoiling the brew.

All of this will be set in motion well ahead of need. Then, if your fishes spawn, you will be ready to feed the larvae. The problem, therefore, arises with keeping either colonies of copepods or rotifers in good shape but not overproducing. With rotifers, you can lower the temperature to about 60°F and feed about a quarter of the rate needed for full production. With copepods, maybe lower feeding a little, but your main control is via collecting and hatching the eggs. Rotifer colonies may need up to two weeks to build up to full production again, so you have to be good at anticipating the need for them. Remember that rotifers can be fed yeast if necessary. I am not sure about copepods. This is something to try out. I guess that it would work with *Torula* yeast at least.

Feeding Larval Fishes

Fish larvae do not so much seek out their food as just snap it up when they come across it. This means that one larva in a body of water needs just as many planktonic organisms per unit volume as do 100. It has to come across one sufficiently often to feed at a satisfactory rate or it will starve. So, whatever the larval number happens to be, we must aim at so many organisms per gallon. The desirable number is 5-10 per ml, 20-40,000 per gallon. This is to be kept up by frequent additions as the larvae eat them. The more larvae, the more frequent the additions. Larvae feed at a rate of around one organism per five seconds, or 72,000 per hour per 100 larvae. You can see the need to keep up the additions.

About 20 rotifers or other organisms per drop is a good culture. This is about 2,000 per standard teaspoon of 5 ml. It is best to filter them through a 40-micron mesh rather than to pour cultures into the larvae tank. You risk contaminating with too many algae. A culture with 20 organisms per drop will need to be supplied to the larvae tank at about 3 oz (90 ml) per gallon, strained or otherwise. If a 20-gallon tank holds 100 larvae, it will start off with 800,000 organisms. It will need topping up at least every four or five hours when illuminated as the larvae eat 72,000 organisms per hour!

I have talked a lot about so many organisms per drop or per gallon. How do you determine this? One drop is $\frac{1}{20}$ ml—at least one 'standard' drop, the sort given by an eye dropper or a pH test kit dropper. It is easy to check any of them; just drip 20 drops into an appropriate measure and check. If your drops differ in volume from 20 per ml, make an adjustment in the calculations. Place one drop of a culture on a microscope slide or any clean piece of glass. Warm it over a flame to evaporate the water and count the corpses. You can stain with methylene blue after drying. Use a hand lens or the lowest power of a microscope—you do not need much magnification.

The count in an aquarium should be 25-50 per teaspoon (5 ml). To determine this, take a sample from the middle of the water with a dip tube. Fill a standard teaspoon of equivalent measure. Dry it in a watch glass or any suitable small glass container and count as

The courtship pattern of a male *Amblyglyphidodon leucogaster.*

before. In both cases, repeat your count several times unless the second count is about the same as the first. Average the counts. A count of 41 followed by, say, 36 would be okay; a count of 41 followed by 25 would need a few repeats to get a good average figure. After some experience you can estimate densities roughly by eye. Do not depend on this forever, as your guess will probably drift and needs occasional checks.

The larvae should be observed at all stages. See that they are getting enough food and are eating it. Their little bellies should be round, full, and milky-colored. After a week or so, try some newly hatched brine shrimp and see if they are eaten. It is easy to tell; some stomachs will look pink if the shrimps are eaten. Not all the larvae will be likely to eat them on the first try. Keep a mixture going until all bellies are pink; then you can switch over completely. Brine shrimp are an adequate diet for a few weeks but add other tiny foods as soon as they are taken. Fry at two to four weeks often can eat newly hatched mosquito wrigglers, fine dry food, ground prawn, clam, and such. They benefit from the varied diet. You can float mosquito rafts on sea water. They will hatch whether they come from fresh water or tide pools.

Transferring the Fry

You will want to move the fry to a normal aquarium, on their own of course, as soon as this is feasible. Then they are cared for by a biological filter. The onerous duties of frequent cleaning up and water changes can be relaxed somewhat. The time to do it is when they start to show color and act like juvenile fishes. They seek food and even joust a bit now and then. The actual time in weeks varies—anywhere from five or six weeks onwards. They could be ¼ inch long or more. Take great care in moving them—do not use nets. Bale them out gently with a suitable vessel or siphon. You will want plenty of their present water to go with them, so it is quite in order to siphon them into their part-filled new home. Do not overcrowd them. The new tank must be several times the volume of the old one, otherwise only a fraction of their number should go in. Put the rest into other tanks or give them away.

As they grow, you will have a batch of practically disease-free fishes that will have little or no immunity to disease. Thus they are very tender and likely to fall to the common diseases to which other fishes are mostly immune. Be wary when transferring them to communities. Observe a kind of reverse quarantine by dosing the community with copper before the transfer. This will knock off any white spot or velvet parasites at least—the diseases most likely to hit them first. Just a single dose of 0.15 ppm copper a day before the transfer is harmless to all. It is enough to eradicate the infective stage of two aquarium pests. Do not forget that you cannot use copper if invertebrates are present. The copper also can hit the algae, although a single dose, as advised, would not be too dangerous to most algal species.

Index

This index covers only the Marine Aquarium Setup and Maintenance section (beginning on page 673). Indices for the other sections of this book begin on page 929.

INDICES

Scientific Name Index

This index immediately following is an index to the scientific names of the fishes shown in the Pictorial Identification Section, showing the page numbers on which illustrations of those fishes can be found.

An index that combines both scientific and common names of marine fishes follows this index.

A

Common Name Index

For your convenience, the following index contains the common names of all the fishes included in DR. BURGESS'S ATLAS OF MARINE AQUARIUM FISHES, along with the scientific names to which they apply. If you know only the accepted common name of a fish it can be found in this index so that the scientific name can be obtained. The scientific name can then be looked up in the Scientific Name Index elsewhere in this book to find the page number on which a photograph of the fish appears. If the scientific name is not listed in this latter index, the fish is not included in the MINI-ATLAS (though it is included in the larger atlas).

Ambon Wrasse, *Halichoeres amboinensis*

Ambonor Goby, *Vanderhorstia ambonoro*

Analspot Demoiselle, *Pomacentrus stigma*

Anchovy, unidentified, *Anchoviella* sp.

Anemone Demoiselle, *Amphiprion ocellaris*

Angel Blenny, *Coralliozetus angelica*

Angle-mouthed Lanternfish, *Gonostoma elongatum*

Anomalus Tripodfish, *Triacanthodes anomalus*

Antenna Featherfin Scorpionfish, *Hipposcorpaena* sp. 2

Arabian Angelfish, *Arusetta asfur*

Arabian Blenny, *Ecsenius pulcher*

Arc-eyed Hawkfish, *Paracirrhites arcatus*

Argus Comet, *Calloplesiops argus*

Argus Grouper, *Cephalopholis argus*

Argus Wrasse, *Halichoeres argus*

Armored Pipefish, *Solenostomus armatus*

Arrow Blenny, *Lucayablennius zingaro*

Arrow Cardinal, *Rhabdamia gracilis*

Arrow Stargazer, *Gillelus greyae*

Arrowhead Soapfish, *Belonoperca chabanaudi*

Arrowhead Wrasse, *Wetmorella nigropinnata*

Arrowtooth Cardinalfish, *Cheilodipterus lachneri*

Ascencion Cherubfish, *Centropyge resplendens*

Ash-colored Rudderfish, *Kyphosus fuscus*

Ashen Conger-eel, *Conger cinereus*

Ashen Drummer, *Kyphosus cinerascens*

Asian Trumpetfish, *Aulostomus chinensis*

Assarius Butterflyfish, *Chaetodon assarius*

Astrolabe Wrasse, *Halichoeres margaritaceus*

Atlantic Bearded Goby, *Barbulifer ceuthoecus*

Atlantic Bumper, *Chloroscombrus chrysurus*

Atlantic Clown Wrasse, *Halichoeres maculipinna*

Atlantic Cod, *Gadus morhua*

Atlantic Fanged Blenny, *Ophioblennius atlanticus*

Atlantic Hawkfish, *Cirrhitus atlanticus*

Atlantic Moonfish, *Selene setapinnis*

Atlantic Needlefish, *Strongylura timucu*

Atlantic Porkfish, *Anisotremus virginicus*

Atlantic Rainbow Wrasse, *Halichoeres pictus*

Atlantic Spadefish, *Chaetodipterus faber*

Atlantic Spotted Scorpionfish, *Scorpaena plumieri plumieri*

Atlantic Tarpon, *Megalops atlanticus*

Atlantic Thread Herring, *Opisthonema oglinum*

Atlantic Torpedo Ray, *Torpedo nobiliana*

Australian Banded Pipefish, *Corythoichthys intestinalis*

Australian Bandfish, *Cepola australis*

Australian Black Porgy, *Acanthopagrus australis*

Australian Blenny, *Ecsenius australianus*

Australian Blowfish, *Torquigener pleurogramma*

Australian Brown Grouper, *Acanthistius pardalotus*

Australian Damsel, *Pomacentrus australis*

Australian Globefish, *Diodon nicthemerus*
Australian Gregory, *Stegastes apicalis*
Australian Guitarfish, *Rhinobatos vincentiana*
Australian Numbfish, *Hypnos subnigrum*
Australian Salmon, *Arripes trutta*
Australian Slender Wrasse, *Eupetrichthys angustipes*
Australian Spotted Pipefish, *Stigmatopora argus*
Australian Stripey, *Atypichthys strigatus*
Axelrod's Blenny, *Ecsenius axelrodi*
Azure Damsel, *Chrysiptera parasema*

B

Baldchin Tuskfish, *Choerodon rubescens*
Bali Threadfin Bream, *Nemipterus balinensis*
Ballan Wrasse, *Labrus bergylta*
Ballieu's Scorpionfish, *Scorpaena ballieui*
Balloonfish, *Diodon holacanthus*
Ballyhoo, *Hemiramphus brasiliensis*
Balteata, *Redigobius balteata*
Banda Blenny, *Ecsenius bandanus*
Banded Basslet, *Liopropoma fasciatum*
Banded Blanquillo, *Malacanthus brevirostris*
Banded Blenny, *Paraclinus fasciatus*
Banded Boarhead, *Evistias acutirostris*
Banded Butterflyfish, *Chaetodon striatus*
Banded Cardinalfish, *Cheilodipterus zonatus*
Banded Catshark, *Chiloscyllium plagiosum*
Banded Cleaner Goby, *Gobiosoma digueti*

Banded Coral Goby, *Eviota fasciola*
Banded Damsel, *Dischistodus fasciatus*
Banded Fairy Basslet, *Pseudanthias fasciatus*
Banded Goby, *Amblygobius phalaena*
Banded Guitarfish, *Zapteryx exasperata*
Banded Hawkfish, *Cirrhitops fasciatus*
Banded Jawfish, *Opistognathus macrognathus*
Banded Knifejaw, *Oplegnathus fasciatus*
Banded Longfin, *Belonepterygion fasciolatum*
Banded Longnose Pipefish, *Corythoichthys amplexus*
Banded Moray, *Gymnothorax rueppelliae*
Banded Parma, *Parma polylepis*
Banded Pipefish, *Doryrhamphus dactyliophorus*
Banded Pony Fish, *Leiognathus fasciata*
Banded Prawn Goby, *Amblyeleotris* sp. 2
Banded Rudderfish, *Seriola zonata*
Banded Sweep, *Scorpis georgianus*
Banded Threadfin Bream, *Pentapodus caninus*
Banded Triplefin, *Tripterygion bucknilli*
Banded Wobbegong, *Orectolobus ornatus*
Banded-head Goby, *Priolepis boreus*
Banded-tail Coral-cod, *Cephalopholis urodelus*
Bandtail Frogfish, *Antennarius strigatus*
Bandtail Puffer, *Sphoeroides spengleri*
Bandtail Searobin, *Prionotus ophryas*
Band-tailed Cardinalfish, *Apogon aureus*

Band-tailed Goatfish, *Upeneus vittatus*
Bandtooth Conger, *Ariosoma impressa*
Bank Butterflyfish, *Chaetodon aya*
Bannerfish, *Heniochus acuminatus*
Bantayan Butterflyfish, *Chaetodon adiergastos*
Bar Jack, *Carangoides ruber*
Barbed Hunchback Poacher, *Agonomalus proboscidalis*
Barber Perch, *Caesioperca rasor*
Barber's Lionfish, *Dendochirus barberi*
Barbier, *Anthias anthias*
Barbu, *Polydactylus virginicus*
Bar-cheeked Cardinalfish, *Apogon sangiensis*
Bar-cheeked Toadfish, *Amblyrhynchotes hypselogenion*
Barchin Scorpionfish, *Sebastapistes strongia*
Bar-faced Weever, *Parapercis nebulosa*
Barfin Sculpin, *Gymnocanthus detrisus*
Bargibant's Seahorse, *Hippocampus bargibanti*
Bar-head Damsel, *Paraglyphidodon thoracotaeniatus*
Barnacle Bill Blenny, *Hypsoblennius brevipinnis*
Barnacle Blenny, *Acanthemblemaria macrospilus*
Baroness Butterflyfish, *Chaetodon baronessa*
Barramundi, *Lates calcarifer*
Barred Blenny, *Hypleurochilus bermudensis*
Barred Cardinalfish, *Apogon binotatus*
Barred Flagtail, *Kuhlia mugil*
Barred Garfish, *Hemiramphus far*
Barred Grunt, *Conodon nobilis*
Barred Hamlet, *Hypoplectrus puella*

Barred Morwong, *Cheilodactylus quadricornis*
Barred Pargo, *Hoplopagrus guentheri*
Barred Sandperch, *Parapercis multifasciata*
Barred Serrano, *Serranus fasciatus*
Barred Snake Eel, *Myrichthys colubrinus*
Barred Surfperch, *Amphistichus argenteus*
Barred Thicklip, *Hemigymnus fasciatus*
Barrier Reef Anemonefish, *Amphiprion akindynos*
Barrier Reef Chromis, *Chromis nitida*
Barrier Reef Prawn Goby, *Amblyeleotris* sp. cf *japonicus* 1
Barspot Cardinalfish, *Apogon retrosella*
Bartail Flathead, *Platycephalus indicus*
Bartail Fusilier, *Pterocaesio tile*
Bartail Grunter, *Amniataba caudavittatus*
Bar-tail Moray, *Gymnothorax zonipectus*
Bar-tailed Goatfish, *Upeneus tragula*
Bartlett's Fairy Basslet, *Mirolabrichthys bartletti*
Basking Shark, *Cetorhinus maximus*
Batavian Batfish, *Platax batavianus*
Batavian Parrotfish, *Scarus bataviensis*
Batesian Mimic, *Petroscirtes fallax*
Bath's Blenny, *Ecsenius bathi*
Bay Blenny, *Hypsoblennius gentilis*
Bayer's Moray, *Enchelycore bayeri*
Beach Conger, *Conger japonicus*
Beady Pipefish, *Hippichthys penicillus*
Bearded Brotula, *Brotula barbata*
Bearded Codfish, *?Lotella* sp.
Bearded Goby, *Sagamia geneionema*

Bearded Rock Cod, *Pseudophycis breviusculus*

Bearded Scorpionfish, *Scorpaenopsis barbatus*

Beardie, *Lotella fuliginosa*

Beau Brummel, *Stegastes flavilatus*

Beau Gregory, *Stegastes leucostictus*

Beautiful Angelfish, *Holacanthus venustus*

Beautiful Cheek Prawn Goby, *Amblyeleotris callopareia*

Beautiful Fusilier, *Caesio teres*

Bellus Lyretail Angelfish, *Genicanthus bellus*

Bellyband Damselfish, *Plectroglyphidodon leucozona*

Bellystriped Puffer, *Arothron inconditus*

Belted Cardinalfish, *Apogon townsendi*

Belted Morwong, *Cheilodactylus zonatus*

Belted Sandfish, *Serranus subligarius*

Belted-tail Cardinalfish, *Apogon taeniopterus*

Bengal Sergeant, *Abudefduf bengalensis*

Bengal Snapper, *Lutjanus bengalensis*

Beniteguri, *Repomucenus beniteguri*

Bennett's Butterflyfish, *Chaetodon bennetti*

Bennett's Toby, *Canthigaster bennetti*

Bermuda Chub, *Kyphosus sectatrix*

Bermuda Porgy, *Diplodus bermudensis*

Bicolor Basslet, *Lipogramma klayi*

Bicolor Chromis, *Chromis bicolor*

Bicolor Damselfish, *Stegastes partitus*

Bicolor Fairy Basslet, *Mirolabrichthys bicolor*

Bicolor Goatfish, *Parupeneus barberinoides*

Bicolor Parrotfish, *Cetoscarus bicolor*

Bicolor Scorpionfish, *Scorpaenodes* sp. 1

Bicolored Butterflyfish, *Chaetodon dichrous*

Bicolored Cleaner Wrasse, *Labroides bicolor*

Bicolored Coral Goby, *Gobiodon quinquestrigatus*

Bicolored Parma, *Parma bicolor*

Bicolored Toby, *Canthigaster smithae*

Big Lip Damsel, *Cheiloprion labiatus*

Big Red Cardinalfish, *Apogon crassiceps*

Big-belly Seahorse, *Hippocampus abdominalis*

Bigeye, *Priacanthus arenatus*

Bigeye Kingfish, *Caranx sexfasciatus*

Bigeye Monotaxis, *Monotaxis grandoculis*

Bigeye Scad, *Selar crumenophthalmus*

Bigeye Soldierfish, *Myripristis murdjan*

Bigeye Stumpnose, *Rhabdosargus thorpei*

Bigeye Thresher, *Alopias superciliosus*

Bigeye Tuna, *Thunnus obesus*

Bighand Thornyhead, *Sebastolobus macrochir*

Bigscale Goatfish, *Pseudupeneus grandisquammis*

Bigscale Scorpionfish, *Scorpaena scrofa*

Bigscale Soldierfish, *Myripristis berndti*

Big-scaled Parma, *Parma oligolepis*

Bigtooth Cardinalfish, *Apogon affinis*

Bird Wrasse, *Gomphosus varius*

Black Anemonefish, *Amphiprion melanopus*

Black and White Striped Blenny, *Meiacanthus kamoharai*

Black and Yellow Butterflyfish, *Chaetodon mitratus*

Black Back Butterflyfish, *Chaetodon melannotus*

Black Beauty, *Macolor niger*

Black Belt Flagfin Goby, *Pterogobius elapoides*

Black Bream, *Acanthopagrus berda*

Black Cardinalfish, *Apogon niger*

Black Damsel, *Paraglyphidodon melas*

Black Demoiselle, *Chrysiptera niger*

Black Dorsal Grubfish, *Parapercis bivittata*

Black Dragonfish, *Melanostomias* sp.

Black Grouper, *Mycteroperca bonaci*

Black Hamlet, *Hypoplectrus nigricans*

Black Jack, *Caranx lugubris*

Black Long-nosed Butterflyfish, *Forcipiger longirostris*

Black Margate, *Anisotremus surinamensis*

Black Marlin, *Makaira indica*

Black Patch Triggerfish, *Rhinecanthus verrucosus*

Black Pomfret, *Parastromateus niger*

Black Rockfish, *Sebastes melanops*

Black Scraper, *Thamnaconus modestus*

Black Sea Goby, *Gobius elanthematicus*

Black Snapper, *Apsilus dentatus*

Black Snook, *Centropomus nigrescens*

Black Spot Sergeant, *Abudefduf sordidus*

Black Surfperch, *Embiotoca jacksoni*

Black Sweetlips, *Plectorhinchus nigrus*

Black Triggerfish, *Melichthys niger*

Black Velvet Angelfish, *Chaetodontoplus melanosoma*

Black Widow, *Stygnobrotula latibricola*

Black-and-White Chromis, *Chromis iomelas*

Black-and-White Striped Bream, *Pentapodus* sp. 2

Black-and-Yellow Rockfish, *Sebastes chrysomelas*

Black-axil Chromis, *Chromis atripectoralis*

Black-axil Damsel, *Pomacentrus nigromanus*

Black-backed Damsel, *Dischistodus melanotus*

Blackband Threadfin Bream, *Pentapodus* sp. 3

Blackbanded Amberjack, *Seriolina nigrofasciata*

Black-banded Cardinalfish, *Apogon endekataenia*

Black-banded Demoiselle, *Amblypomacentrus breviceps*

Black-banded Seaperch, *Hypoplectrodes nigrorubrum*

Black-bar Chromis, *Chromis retrofasciata*

Blackbar Soldierfish, *Myripristis jacobus*

Blackbelly Eelpout, *Lycodes pacifcus*

Blackbottom Goby, *Eviota nigriventris*

Blackburn's Butterflyfish, *Chaetodon blackburnii*

Blackcap Basslet, *Gramma melacara*

Blackcheek Moray, *Gymnothorax breedeni*

Blackear Wrasse, *Halichoeres poeyi*

Black-eared Moray, *Muraena clepsydra*

Black-eared Surgeonfish, *Acanthurus maculiceps*

Blackedge Fairy Wrasse, *Cirrhilabrus melanomarginatus*

Blackedge Triplefin, *Enneanectes atrorus*

Black-edged Puffer, *Arothron immaculatus*

Black-edged Sculpin, *Gymnocanthus herzensteini*

Blackeye Goby, *Coryphopterus nicholsi*

Black-eyed Spinefoot, *Siganus puelloides*

Blackfin Barracuda, *Sphyraena qenie*
Blackfin Blenny, *Paraclinus nigripinnis*
Blackfin Gurnard, *Lepidotrigla microptera*
Blackfin Hogfish, *Bodianus macrourus*
Blackfin Snapper, *Lutjanus buccanella*
Blackfin Squirrelfish, *Neoniphon opercularis*
Blackfin Sweeper, *Pempheris japonicus*
Black-finned Anemonefish, *Amphiprion nigripes*
Black-finned Butterflyfish, *Chaetodon decussatus*
Black-finned Grunt, *Hapalogenys nigripinnis*
Black-finned Melon Butterflyfish, *Chaetodon melapterus*
Black-finned Morwong, *Cheilodactylus nigripes*
Blackfish, *Girella tricuspidatus*
Blackfoot Cardinalfish, *Apogon nigripes*
Blackfoot Lionfish, *Parapterois heterurus*
Black-headed Blenny, *Lipophrys nigriceps*
Black-lined Blenny, *Meiacanthus nigrolineatus*
Blackloin Butterflyfish, *Chaetodon nippon*
Black-margined Damsel, *Pomacentrus nigromarginatus*
Black-mouth Damsel, *Paraglyphidodon nigroris*
Blackmouth Goosefish, *Lophiomus setigerus*
Black-nosed Butterflyfish, *Pseudochaetodon nigrirostris*
Black-nosed Cardinalfish, *Rhabdamia cypselurus*
Blackrag, *Psenes pellucidus*
Blacksaddle Goatfish, *Parupeneus rubescens*
Black-saddled Leopard Grouper, *Plectropomus laevis*
Black-saddled Toby, *Canthigaster valentini*
Black-shoulderspot Cardinalfish, *Archamia melasma*
Black-shoulderspot Threadfin Bream, *Nemipterus upeneoides*
Blacksmith, *Chromis punctipinnis*
Blackspot Butterfish, *Psenopsis anomala*
Blackspot Cardinalfish, *Apogon notatus*
Blackspot Demoiselle, *Chrysiptera melanomaculata*
Black-spot Emperor, *Lethrinus harak*
Blackspot Goatfish, *Parupeneus pleurostigma*
Blackspot Lyretail Angelfish, *Genicanthus melanospilus*
Blackspot Moray, *Gymnothorax melatremus*
Black-spot Snapper, *Lutjanus ehrenbergii*
Black-spot Surgeon, *Acanthurus bariene*
Blackspot Tuskfish, *Choerodon schoenleinii*
Blackspot Wrasse, *Decodon melasma*
Black-spotted Crevalle, *Carangoides orthogrammus*
Black-spotted Goby, *Istigobius nigroocellatus*
Black-spotted Puffer, *Arothron nigropunctatus*
Black-spotted Ray, *Taeniura melanospila*
Black-spotted Rubberlip, *Plectorhinchus gaterinus*
Black-spotted Tamarin, *Anampses geographicus*
Blackstripe Parrotfish, *Scarus tricolor*

Black-stripe Sweeper, *Pempheris schwenki*

Black-striped Cardinalfish, *Apogon nigrofasciatus*

Black-striped Worm Eel, *Gunnelichthys pleurotaenia*

Blacktail Humbug, *Dascyllus melanurus*

Black-tail Sergeant, *Abudefduf lorenzi*

Blacktail Sweetlips Snapper, *Lutjanus* sp. *(lemniscatus?)*

Black-tailed Chromis, *Chromis nigrura*

Black-tailed Wrasse, *Symphodus melanocercus*

Blackthorn Sawtail, *Prionurus scalprus*

Blackthroat Seaperch, *Doederleinia berycoides*

Black-throated Threefin, *Helcogramma decurrens*

Blacktip Bullseye, *Pempheris analis*

Blacktip Grouper, *Epinephelus fasciatus*

Blacktip Reef Shark, *Carcharhinus melanopterus*

Blacktip Soldierfish, *Myripristis adustus*

Blacktongue Unicornfish, *Naso hexacanthus*

Blackwing Searobin, *Prionotus rubio*

Bleeker's Coralline Velvetfish, *Cocotropus dermacanthus*

Bleeker's Damsel, *Stegastes simsiang*

Bleeker's Hawkfish, *Cirrhitichthys bleekeri*

Bleeker's Monocle Bream, *Scolopsis bleekeri*

Bleeker's Rainbowfish, *Pseudocoris bleekeri*

Bleeker's Sea Bream, *Argyrops bleekeri*

Blenny 1, *Cirripectes* sp.

Blenny 2, *Salarias* sp. 2

Blenny 3, *Istiblennius* sp.

Blind Goby, *Brachyamblyopus coesus*

Bloch's Tripodfish, *Tripodichthys blochii*

Bloody Frogfish, *Antennarius sanguineus*

Blotched African Goby, *Gnatholepis* sp. 1

Blotchy Grouper, *Epinephelus fuscoguttatus*

Blotchy Mudskipper, *Periophthalmus koelreuteri*

Blotchy Sillago, *Sillago maculata*

Blue Angelfish, *Holacanthus isabelita*

Blue and Gold Snapper, *Lutjanus viridis*

Blue and Yellow Grouper, *Epinephelus flavocoeruleus*

Blue Bird Wrasse, *Gomphosus caeruleus*

Blue Blanquillo, *Malacanthus latovittatus*

Blue Bumphead Parrotfish, *Scarus cyanescens*

Blue Cheekline Triggerfish, *Xanthichthys mento*

Blue Chromis, *Chromis cyaneus*

Blue Croaker, *Bairdiella batabana*

Blue Damsel, *Pomacentrus pavo*

Blue Damselfish, *Pomacentrus coelestis*

Blue Devil, *Chrysiptera cyanea*

Blue Groper, *Achoerodus gouldii*

Blue Gudgeon, *Ptereleotris microlepis*

Blue Hamlet, *Hypoplectrus gemma*

Blue Hana Goby, *Ptereleotris hanae*

Blue King Angelfish, *Pomacanthus annularis*

Blue Maomao, *Scorpis aequipinnis*

Blue Marlin, *Makaira mazura*

Blue Moki, *Latridopsis ciliaris*

Blue Moon Angelfish, *Pomacanthus maculosus*

Blue Moon Parrotfish, *Scarus atrilunula*

Blue Parrotfish, *Scarus coeruleus*
Blue Rockfish, *Sebastes mystinus*
Blue Runner, *Caranx crysos*
Blue Sleeper, *Ioglossus calliurus*
Blue Smalltooth Jobfish, *Aphareus furcatus*
Blue Spangled Angelfish, *Chaetodontoplus cyanopunctatus*
Blue Sprat, *Spratelloides robustus*
Blue Surgeonfish, *Paracanthurus hepatus*
Blue Tang, *Acanthurus coeruleus*
Blue-and-Gold Blenny, *Ecsenius lividanalis*
Blue-and-Gold Fin Goby, *Cryptocentrus* sp. 1
Blue-and-Gold Fusilier, *Caesio caerulaurea*
Blue-and-Gold Striped Snapper, *Symphorichthys spilurus*
Blue-and-Gold Triggerfish, *Pseudobalistes fuscus*
Blueback Pygmy Angelfish, *Centropyge colini*
Bluebacked Puffer, *Takifugu stictonotus*
Blue-backed Silver-Biddy, *Gerres abbreviatus*
Blue-barred Orange Parrotfish, *Scarus ghobban*
Blue-blotched Butterflyfish, *Chaetodon plebeius*
Blue-bronze Sea Chub, *Kyphosus analogus*
Blue-dorsal Mudskipper, *Periophthalmus papilio?*
Bluedotted Damselfish, *Pristotis cyanostigma*
Blue-eared Wrasse, *Halichoeres pardaleocephalus*
Bluefin Kingfish, *Carangoides caeruleopinnatus*
Bluefin Searobin, *Chelidonichthys spinosus*

Bluefin Trevally, *Caranx melampygus*
Bluefin Tuna, *Thunnus thynnus*
Blue-finned Shrimp Goby, *Cryptocentrus fasciatus*
Bluefish Tailor, *Pomatomus saltatrix*
Blue-girdled Angelfish, *Euxiphipops navarchus*
Blue-green Reeffish, *Chromis viridis*
Bluehead Wrasse, *Thalassoma bifasciatum*
Bluejean Squirrelfish, *Neoniphon laeve*
Blue-line Demoiselle, *Chrysiptera caeruleolineata*
Blue-lined Angelfish, *Chaetodontoplus septentrionalis*
Blue-lined Leatherjacket, *Meuschenia galii*
Blue-lined Parrotfish, *Scarus cyanotaenia*
Blue-lined Snapper, *Symphorus nematophorus*
Bluelined Wrasse, *Stethojulis albovittata*
Bluemask Angelfish, *Chaetodontoplus personifer*
Blue-nosed Monocle Bream, *Scolopsis personatus*
Blue-sided Fairy Wrasse, *Cirrhilabrus cyanopleura*
Blue-sided Wrasse, *Leptojulis cyanopleura*
Blue-spot Blenny, *Salarias guttatus*
Bluespot Cardinalfish, *Apogon nitidus*
Blue-spot Damsel, *Pomacentrus grammorhynchus*
Bluespotted Boxfish, *Ostracion immaculatus*
Blue-spotted Coral Goby, *Gobiodon histrio*
Bluespotted Cornetfish, *Fistularia tabacaria*

Blue-spotted Dascyllus, *Dascyllus carneus*

Blue-spotted Goatfish, *Upeneichthys lineatus*

Bluespotted Jawfish, *Opistognathus* sp. 3

Bluespotted Mudskipper, *Boleophthalmus polyophthalmus*

Blue-spotted Puffer, *Omegophora cyanopunctata*

Blue-spotted Ray, *Taeniura lymma*

Blue-spotted Sharp-nosed Puffer, *Canthigaster epilamprus*

Blue-spotted Shrimp Goby, *Cryptocentrus caeruleomaculatus*

Blue-spotted Snapper, *Lutjanus rivulatus*

Blue-spotted Spinefoot, *Siganus corallinus*

Bluespotted Stargazer, *Gnathagnus elongatus*

Blue-spotted Stingray, *Amphotistius kuhlii*

Bluespotted Tamarin, *Anampses caeruleopunctatus*

Blue-spotted Trevally, *Caranx bucculentus*

Bluespotted Wrasse, *Macropharyngodon cyanoguttatus*

Bluestreak Cleaner Wrasse, *Labroides dimidiatus*

Blue-streak Devil, *Paraglyphidodon oxyodon*

Bluestreak Mudskipper, *Periophthalmus* sp.

Bluestreak Prawn Goby, *Cryptocentrus cyanotaenia*

Bluestreak Weever, *Parapercis pulchella*

Bluestripe Pipefish, *Doryrhamphus excisus excisus*

Blue-striped Butterflyfish, *Chaetodon fremblii*

Blue-striped Dottyback, *Pseudochromis cyanotaenia*

Blue-striped Grouper, *Gracila albomarginata*

Bluestriped Grunt, *Haemulon sciurus*

Blue-striped Kasmir Snapper, *Lutjanus kasmira*

Blue-striped Orange Tamarin, *Anampses femininus*

Blue-striped Snapper, *Lutjanus notatus*

Bluetail Trunkfish, *Ostracion cyanurus*

Bluethroat Pikeblenny, *Chaenopsis ocellata*

Blue-throat Triggerfish, *Xanthichthys auromarginatus*

Blue-tip Longfin, *Paraplesiops poweri*

Bluetrim Monocle Bream, *Scolopsis taeniopterus*

Blunthead Triggerfish, *Pseudobalistes naufragium*

Blunt-headed Grouper, *Epinephelus amblycephalus*

Blunt-snout Gregory, *Stegastes lividus*

Blushing Hogfish, *Bodianus* sp.

Boarfish, *Antigonia capros*

Bocaccio, *Sebastes paucispinis*

Boenacki Grouper, *Cephalopholis boenack*

Boga, *Inermia vittata*

Bonefish, *Albula vulpes*

Boomerang Triggerfish, *Sufflamen bursa*

Borbon Fairy Basslet, *Holanthias borbonius*

Bottome's Blenny, *Emblemaria bottomei*

Bougainville's Frogfish, *Histophryne bougainvillei*

Bow-tailed Dottyback, *Pseudochromis melanotaenia*

Boyer's Smelt, *Atherina boyeri*

Brackish-water Frogfish, *Antennarius biocellatus*

Braun's Hulafish, *Trachinops brauni*

Breaksea Grouper, *Epinephelides armatus*

Bridle Cardinalfish, *Apogon aurolineatus*

Bridle Triggerfish, *Sufflamen fraenatus*

Bridled Burrfish, *Chilomycterus antennatus*

Bridled Goby, *Coryphopterus glaucofrenum*

Bridled Leatherjacket, *Acanthaluterus spilomelanura*

Bridled Weed Whiting, *Siphonognathus radiatus*

Bright Eye, *Plectroglyphidodon imparipennis*

Bright-saddled Goatfish, *Parupeneus pleurotaenia*

Brill, *Scopthalmus rhombus*

Brilliant Red Hawkfish, *Neocirrhitus armatus*

Brindlebass, *Epinephelus lanceolatus*

Bristle-feathered Anchovy, *Setipinna breviceps*

Broad Alfonsino, *Beryx decadactylus*

Broad-barred Butterflyfish, *Amphichaetodon howensis*

Broad-barred Spanish Mackerel, *Scomberomorus semifasciatus*

Broadnose Sevengill Shark, *Notorynchus cepedianus*

Broadsaddle Cardinalfish, *Apogon pillionatus*

Broad-striped Cardinalfish, *Apogon angustatus*

Bronzeback Damsel, *Pomacentrus albimaculus*

Bronze-streaked Cardinalfish, *Archamia lineolata*

Broom Filefish, *Amanses scopas*

Broom Grouper, *Epinephelus cometae*

Broomtail Wrasse, *Cheilinus lunulatus*

Brown and Yellow Blenny, *Enchelyurus flavipes*

Brown Blenny, *Ernogrammus hexagrammus*

Brown Carpet Shark, *Chiloscyllium griseum*

Brown Comber, *Serranus hepatus*

Brown Demoiselle, *Neopomacentrus filamentosus*

Brown Dottyback, *Pseudochromis fuscus*

Brown Electric Ray, *Narcine brunneus*

Brown Hakeling, *Physiculus maximowiczi*

Brown Moray, *Gymnothorax prasinus*

Brown Puller, *Chromis hypselepis*

Brown Rockfish, *Sebastes auriculatus*

Brown Sailfin Tang, *Zebrasoma scopas*

Brown Smoothhound, *Mustelus henlei*

Brown Triggerfish, *Sufflamen* sp.

Brown-and-White Butterflyfish, *Hemitaurichthys zoster*

Brown-backed Puffer, *Lagocephalus gloveri*

Brown-banded Dottyback, *Pseudochromis perspicillatus*

Brown-banded Razorfish, *Xyrichthys* sp.

Brown-banded Rockcod, *Cephalopholis pachycentron*

Brown-barred Rosy Hawkfish, *Amblycirrhitus oxyrhynchus*

Browncheek Blenny, *Acanthemblemaria crockeri*

Brown-eared Surgeonfish, *Acanthurus nigricauda*

Brownfield's Wrasse, *Halichoeres brownfieldi*

Brown-gold Blenny, *Ecsenius frontalis*

Brown-spot Blenny, *Ecsenius paroculus*

Brown-spotted Cat Shark, *Chiloscyllium punctatum*

Brown-spotted Spinefoot, *Siganus stellatus*

Brown-spotted Wrasse, *Pseudolabrus parilus*

Brownstreak Prawn Goby, *Amblyeleotris gymnocephala*

Brownstripe Wrasse, *Suezichthys gracilis*

Bucchichi's Goby, *Gobius bucchichii*

Bucktooth Parrotfish, *Sparisoma radians*

Buffalo Sculpin, *Enophrys bison*

Bullethead Rockskipper, *Istiblennius periophthalmus*

Bullis' Squirrelfish, *Sargocentron bullisi*

Bullseye, *Pempheris multiradiata*

Bullseye Jawfish, *Opistognathus scops*

Bullseye Puffer, *Sphoeroides annulatus*

Bullseye Stingray, *Urolophus concentricus*

Bullseye Tripodfish, *Johnsonina eriomma*

Bumphead Coris, *Coris bulbifrons*

Bumphead Damselfish, *Microspathodon bairdi*

Bumphead Parrotfish, *Scarus gibbus*

Bumpytail Ragged-tooth, *Odontaspis ferox*

Bundoon, *Meiacanthus bundoon*

Burger's Spotted Cat Shark, *Halaelurus buergeri*

Burgess's Butterflyfish, *Chaetodon burgessi*

Butler's Frogfish, *Tathicarpus butleri*

Butterfly Bream, *Pentapodus setosus*

Butterfly Dragonet, *Synchiropus papilio*

Butterfly Goby, *Amblygobius albimacula*

Butterfly Gurnard, *Paratrigla vanessa*

Butterfly Ray, *Gymnura micrura*

Butterfly Threadfin Bream, *Nemipterus zysron*

Butterflyfin Goby, *Fusigobius* sp. 2

C

Cabezon, *Scorpaenichthys marmoratus*

Cadenat's Chromis, *Chromis cadenati*

Caesiura Goby, *Trimma caesiura*

Caffer Goby, *Caffrogobius caffer*

Caledonian Devilfish, *Inimicus caledonicus*

Calico Rockfish, *Sebastes dallii*

California Corbina, *Menticirrhus undulatus*

California Kelp Bass, *Paralabrax clathratus*

California Moray, *Gymnothorax mordax*

California Sand Bass, *Paralabrax nebulifer*

California Scorpionfish, *Scorpaena guttata*

California Sheephead, *Semicossyphus pulcher*

Camouflage Grouper, *Epinephelus microdon*

Camurum Spotted Boxfish, *Ostracion meleagris camurum*

Canary Demoiselle, *Chrysiptera galba*

Canary Rockfish, *Sebastes pinniger*

Candy Basslet, *Liopropoma carmabi*

Candystripe Hogfish, *Bodianus opercularis*

Caneva's Blenny, *Lipophrys canevae*

Canton Mudskipper, *Periophthalmus cantonensis*

Cape Conger, *Conger wilsoni*

Cape Mono, *Monodactylus falciformes*

Cape Verde Chromis, *Chromis cauta?*

Cape Verde Snapper, *Apsilus fuscus*

Cardinal Scorpionfish, *Scorpaena cardinalis*

Cardinal Soldierfish, *Plectrypops retrospinis*

Cardinalfin Goby, *Kellogella cardinalis*

Cardinalfish 1, *Apogon* sp. 1

Cardinalfish 2, *Siphamia* sp.

Cardinalfish 3, *Cheilodipterus* sp.

Caribbean Garden Eel, *Heteroconger halis*

Caribbean Tonguefish, *Symphurus arawak*

Caribbean Trumpetfish, *Aulostomus maculatus*

Carpenter's Wrasse, *Paracheilinus carpenteri*

Casamance River Goby, *Gobius casamancus*

Castelnau's Golden Wrasse, *Dotalabrus aurantiacus*

Castelnau's Soldierfish, *Myripristis amaenus*

Cat Shark, *Chiloscyllium confusum*

Catalina Goby, *Lythrypnus dalli*

Cave Goby, *Fusigobius* sp. 1

Celebes Butterfly Bream, *Nemipterus celebicus*

Celebes Sweetlips, *Plectorhinchus celebicus*

Celebes Wrasse, *Cheilinus celebicus*

Ceram Blenny, *Salarias ceramensis*

Ceram Cardinalfish, *Apogon ceramensis*

Chain Moray, *Echidna catenata*

Chain Pipefish, *Syngnathus louisianae*

Chalk Bass, *Serranus tortugarum*

Chameleon Tilefish, *Hoplolatilus chlupatyi*

Chameleon Wrasse, *Halichoeres dispilus*

Charcoal Damsel, *Pomacentrus brachialis*

Checkerboard Wrasse, *Halichoeres hortulanus*

Checkered Blenny, *Starksia ocellata*

Checkered Snapper, *Lutjanus decussatus*

Cheeklined Wrasse, *Cheilinus digrammus*

Cherry Bass, *Sacura margaritacea*

Cherubfish, *Centropyge argi*

Chestnut Blenny, *Cirripectes castaneus*

Chevron Barracuda, *Sphyraena obtusata*

Chevron Butterflyfish, *Chaetodon trifascialis*

China Rockfish, *Sebastes nebulosus*

Chinaman Leatherjacket, *Nelusetta ayraudi*

Chinese Zebra Goby, *Ptereleotris zebra*

Chiseltooth Wrasse, *Pseudodax moluccanus*

Choat's Wrasse, *Macropharyngodon choati*

Chocolate Surgeonfish, *Acanthurus pyroferus*

Christmas Island Scorpionfish, *Scorpaenopsis* sp. 3

Christmas Wrasse, *Thalassoma trilobatum*

Chub Mackerel, *Scomber japonicus*

Chub, unidentified, *Kyphosus* sp.

Chubby Soldierfish, *Myripristis melanostictus*

Cigar Wrasse, *Cheilio inermis*

Cinnamon Flounder, *Pseudorhombus cinnamoneus*

Citron Butterflyfish, *Chaetodon citrinellus*

Citron Goby, *Gobiodon citrinus*

Clarion Angelfish, *Holacanthus clarionensis*

Clarion Damselfish, *Stegastes redemptus*

Clark's Anemonefish, *Amphiprion clarkii*

Clark's Triplefin, *Norfolkia clarkei*

Cleaner Mimic, *Plagiotremus rhinorhynchos*
Clearnose Guitarfish, *Rhinobatos armatus*
Clingfish, unidentified, Gobiesocidae sp. 2
Cloister Blenny, *Omobranchus elongatus*
Cloudy Dory, *Zenopsis nebulosa*
Clown Anemonefish, *Amphiprion percula*
Clown Coris, *Coris aygula*
Clown Grouper, *Pogonoperca punctata*
Clown Surgeonfish, *Acanthurus lineatus*
Clown Triggerfish, *Balistoides conspicillum*
Clown Wrasse, *Coris gaimard*
Clubhead Blenny, *Acanthemblemaria balanorum*
C-O Turbot, *Pleuronichthys coenosus*
Coachwhip Ray, *Himantura uarnak*
Cobalt Chromis, *Chromis flavicauda*
Cobia, *Rachycentron canadum*
Cockatoo Waspfish, *Ablabys taenianotus*
Cockburn Sound Clingfish, *Cochleoceps spatula*
Cocktail Fish, *Pteragogus flagellifer*
Coco Frillgoby, *Bathygobius cocosensis*
Cocoa Damselfish, *Stegastes variabilis*
Cod Scorpionfish, *Peristrominous dolosus*
Coelacanth, *Latimeria chalumnae*
Coin-bearing Frogfish, *Antennarius nummifer*
Collared Sea Bream, *Gymnocranius bitorquatus*
Collette's Blenny, *Ecsenius collettei*
Colon Goby, *Coryphopterus dicrus*
Colonial Salmon, *Eleutheronema tetradactylum*

Colorado Snapper, *Lutjanus colorado*
Comb Grouper, *Mycteroperca rubra*
Comber, *Serranus cabrilla*
Combfish, *Coris picta*
Comb-toothed Mudskipper, *Boleophthalmus pectinirostris*
Comet, *Calloplesiops altivelis*
Commerson's Cornetfish, *Fistularia commersonii*
Commerson's Frogfish, *Antennarius commersoni*
Common Buffalo Bream, *Kyphosus sydneyanus*
Common European Bass, *Dicentrarchus labrax*
Common Mediterranean Triplefin, *Tripterygion tripteronotus*
Common Sea Bass, *Lateolabrax japonicus*
Common Slipmouth, *Leiognathus equulus*
Common Sweeper, *Parapriacanthus dispar*
Common Triplefin, *Enneapterygius etheostomus*
Common Tripodfish, *Triacanthus biaculeatus*
Compressed Sharp-nosed Puffer, *Canthigaster compressus*
Compressed Sweeper, *Pempheris compressa*
Condolle's Lepadogaster, *Lepadogaster candollei*
Coney, *Cephalopholis fulva*
Connie's Wrasse, *Conniella apterygia*
Convict Blenny, *Pholidichthys leucotaenia*
Convict Goby, *Lythrypnus phorellus*
Convict Surgeonfish, *Acanthurus triostegus*
Cook's Black-banded Cardinalfish, *Apogon cookii*
Cook's Cardinalfish, *Apogon robustus*
Cook's Scorpionfish, *Scorpaena cooki*
Copper Rockfish, *Sebastes caurinus*

Copper Sweeper, *Pempheris schomburgki*

Copperband Butterflyfish, *Chelmon rostratus*

Coral Beauty, *Centropyge bispinosus*

Coral Demoiselle, *Neopomacentrus nemurus*

Coral Hogfish, *Bodianus axillaris*

Coral Rockcod, *Cephalopholis miniata*

Coral Scorpionfish, *Scorpaena albifimbria*

Coral Sea Gregory, *Stegastes gascoynei*

Coral Toadfish, *Sanopus splendidus*

Coral-dwelling Cleaning Goby, *Gobiosoma illecebrosum*

Cortez Angelfish, *Pomacanthus zonipectus*

Cortez Chromis, *Stegastes rectifraenum*

Cortez Chub, *Kyphosus elegans*

Cortez Garden Eel, *Taenioconger digueti*

Cortez Hovering Goby, *Ioglossus* sp.

Cortez Rainbow Wrasse, *Thalassoma lucasanum*

Cortez Soapfish, *Rypticus bicolor?*

Cottonwick, *Haemulon melanurum*

Cow-nose Ray, *Rhinoptera neglecta*

Cowtail Ray, *Dasyatis sephen*

Creole Wrasse, *Clepticus parrae*

Creole-fish, *Paranthias furcifer*

Crescent Snapper, *Lutjanus lunulatus*

Crescent Tailed Grouper, *Variola albomarginata*

Crescent Throat Blenny, *Glyptoparus delicatulus*

Crescent Triggerfish, *Rhinecanthus lunula*

Crescent-tail Bigeye, *Priacanthus hamrur*

Crested Flounder, *Limanda schrenki*

Crested Triplefin, *Norfolkia cristata*

Crested Weedfish, *Cristiceps australis*

Crevalle Jack, *Caranx hippos*

Crimson Cleaner Wrasse, *Suezichthys* sp.

Crimson Goby, *Trimma* sp. 3

Crimson Sea-Bream, *Evynnis cardinalis*

Crimson-eyed Coral Goby, Gobiidae sp. 1

Crosseyed Cardinalfish, *Fowleria aurita*

Crown Squirrelfish, *Sargocentron diadema*

Crowned Seahorse, *Hippocampus coronatus*

Crown-of-thorns Cardinalfish, *Siphamia fuscolineata*

Cuban Goby, *Microgobius signatus*

Cubbyu, *Equetus umbrosus*

Cubera Snapper, *Lutjanus cyanopterus*

Cunner, *Tautogolabrus adspersus*

Cupido Wrasse, *Thalassoma cupido*

Curry-comb Tripodfish, *Tripodichthys strigilifer*

Cusk-eel, unidentified, *Lepophidium* sp.

Cut-ribbon Rainbowfish, *Stethojulis interrupta*

Cuvier's Jawfish, *Opistognathus cuvieri*

Cuvier's Surgeonfish, *Acanthurus nigroris*

Cuvier's Tamarin, *Anampses cuvier*

Cyclops Electric Ray, *Diplobatos ommata*

Dalmatian Blenny, *Lipophrys dalmatinus*

D

Dampier Stonefish, *Dampierosa daruma*

Damselfish 1, *Stegastes* sp.

Damselfish 2, *Chromis* sp.

Damselfish 3, *Pomacentrus* sp.

Damselfish 4, Pomacentridae sp.

Dandy Blenny, *Petroscirtes breviceps*

Dane, *Porcostoma dentata*

Dapple Coris, *Coris variegata*

Dappled Cardinalfish, *Phaeoptyx pigmentaria*

Dark Dottyback, *Pseudochromis melas*

Dark-backed Snake Blenny, *Ophiclinus graclis*

Darkbar Hogfish, *Bodianus speciosus*

Dark-fin Chromis, *Chromis atripes*

Darkfin Sleeper, *Oxymetopon* sp.

Dark-spotted Moray, *Gymnothorax fimbriatus*

Darnley Cardinalfish, *Apogon darnleyensis*

Darwin Jawfish, *Opistognathus darwiniensis*

Darwin's Damselfish, *Stegastes imbricata*

Dash Goby, *Gobionellus saepepallens*

Dash-dot Blenny, *Salarias irroratus*

Dash-dot Goatfish, *Parupeneus barberinus*

Decorated Butterflyfish, *Chaetodon semeion*

Decorated Firefish, *Nemateleotris decora*

Decorated Goby, *Istigobius decoratus*

Decorated Guitarfish, *Rhinobatos hynnicephalus*

Decorated Sole, *Soleichthys heterorhinos*

Decorated Spinefoot, *Siganus puellus*

Decorated Warbonnet, *Chirolophis decoratus*

Deep-bodied Emperor, *Lethrinus haematopterus*

Deep-bodied Silver-Biddy, *Gerres poeti*

Deep-reef Chromis, *Chromis* sp. "D"

Deepreef Scorpionfish, *Scorpaenodes tredecimspinosus*

Deepsea Bonefish, *Pterothrissus gissu*

Deepsea Fairy Basslet, *Pseudanthias rubrizonatus*

Deepwater Sculpin, *Nautichthys oculofasciatus*

Delicate Triplefin, *Enneanectes sexmaculatus*

Delicate-fin Shrimp Goby, *Amblyeleotris* sp. aff *periophthalmus*

Derjugin's Ronquil, *Bathymaster derjugini*

Desjardin's Sailfin Tang, *Zebrasoma desjardinii*

Devil Frogfish, *Halophryne diemensis*

Devil Lionfish, *Pterois miles*

Devil Scorpionfish, *Scorpaenopsis diabolus*

Devilfish, *Inimicus* sp.

Diadem Basslet, *Pseudochromis diadema*

Diagonal Bar Prawn Goby, *Amblyeleotris diagonalis*

Diagonal Striped Goby, *Cryptocentroides insignis*

Diamond Blenny, *Malacoctenus boehlkei*

Diamond Turbot, *Hypsopsetta guttulata*

Diamond Wrasse, *Halichoeres nigrescens*

Diana's Hogfish, *Bodianus diana*

Diaphanous Cowfish, *Lactoria diaphanus*

Disappearing Wrasse, *Pseudocheilinus evanidus*

Divided Wrasse, *Macropharyngodon bipartitus*

Doctorfish, *Acanthurus chirurgus*

Doderlein's Cardinalfish, *Apogon doederleini*

Doderlein's Wrasse, *Symphodus doderleini*

Dog Snapper, *Lutjanus jocu*

Dogtooth Tuna, *Gymnosarda unicolor*

Dolphinfish, *Coryphaena hippurus*
Dot-dash Blenny, *Ecsenius melarchus*
Dot-dash Cusk-eel, *Lepophidium prorates*
Dot-dash Goby, *Exyrias puntang*
Dotted Cardinalfish, *Vincentia punctatus*
Dotted Gizzard Shad, *Konosirus punctatus*
Dotted Goatfish, *Pseudupeneus maculatus*
Dotted-reef Eel, *Muraena lentiginosa*
Dotty Triggerfish, *Balistoides viridescens*
Double Banded Forktail Blenny, *Meiacanthus anema*
Double Bar Chromis, *Chromis opercularis*
Double Blotch Spinecheek, *Scolopsis bimaculatus*
Doublebar Snapper, *Lutjanus biguttatus*
Double-barred Spinefoot, *Siganus virgatus*
Doubleline Clingfish, *Lepadichthys lineatus*
Doubleline Toby, *Canthigaster rivulata*
Doubleline Tongue Sole, *Paraplagusia bilineata*
Double-lined Mackerel, *Grammatorcynus bicarinatus*
Double-spot Blenny, *Ecsenius bimaculatus*
Doublespot Cardinalfish, *Archamia biguttata*
Doublespot Goby, *Eviota bimaculata*
Double-spot Wrasse, *Halichoeres bimaculatus*
Doublespotted Queenfish, *Scomberoides lysan*
Double-striped Goby, *Arenigobius bifrenatus*
Doublesword Sea Robin, *Dixiphichthys hoplites*

Doublewhip Threadfin Bream, *Nemipterus nematophorus*
Doubtful Flounder, *Hippoglossoides dubius*
Dowager Angelfish, *Pomacanthus striatus*
Downy Blenny, *Labrisomus kalisherae*
Dragon Moray Eel, *Muraena pardalis*
Dragon Wrasse, *Novaculichthys taeniourus*
Dragonet 1, *Synchiropus* sp.
Dragonet 2, *Neosynchiropus* sp.
Dragonfish, *Stomias* sp.
Duboulay's Angelfish, *Chaetodontoplus duboulayi*
Dusky Blenny, *Atrosalarias fuscus*
Dusky Brotulid, *Brotulina fusca*
Dusky Cardinalfish, *Apogon melas*
Dusky Cherub, *Centropyge multispinis*
Dusky Finned Goby, *Callogobius snelliusi*
Dusky Fusegoby, *Fusigobius neophytus*
Dusky Goby, *Chaenogobius urotaenia*
Dusky Gregory, *Stegastes nigricans*
Dusky Jawfish, *Opistognathus whitehursti*
Dusky Morwong, *Dactylophora nigricans*
Dusky Parrotfish, *Scarus niger*
Dusky Ray, *Trygonoptena testaceus*
Dusky Spinefoot, *Siganus fuscescens*
Dusky Tilefish, *Hoplolatilus cuniculus*
Dusky Wrasse, *Halichoeres marginatus*
Duskyfin Rockcod, *Cephalopholis nigripinnis*
Dusky-finned Bullseye, *Heteropriacanthus cruentatus*
Dusky-tail Anemonefish, *Amphiprion fuscocaudatus*

Dusky-winged Flying Fish,
Cheilopogon atrisignis
Dussumier's Halfbeak,
Hyporhamphus dussumieri
Dussumier's Surgeonfish,
Acanthurus dussumieri
Dutorti, *Pseudochromis dutoiti*
Dwarf Blenny, *Ecsenius minutus*
Dwarf Leatherjacket, *Rudarius minutus*
Dwarf Scorpionfish, *Scorpaena elachys*
Dwarf Surfperch, *Micrometrus minimus*
Dwarf Wrasse, *Doratonotus megalepis*
Dwarfed Frogfish, *Antennarius pauciradiatus*

E

Eagle Ray, *Myliobatis tobijei*
Eared Eel-blenny, *Congrogadus subducens*
Earnshaw's Hawkfish,
Amblycirrhitus earnshawi
Earspot Coral Goby, *Gobiodon atrangulatus*
East Indian Red Hind, *Epinephelus corallicola*
East Pacific Banded Butterflyfish,
Chaetodon humeralis
East Pacific Red Snapper, *Lutjanus peru*
East Pacific Triggerfish, *Balistes polylepis*
Easter Island Demoiselle,
Chrysiptera rapanui
Easter Island Pygmy Angelfish,
Centropyge hotumatua
Ecklonia Weedfish, *Heteroclinus eckloniae*
Eclipse Hogfish, *Bodianus mesothorax*
Eel Clingfish, *Alabes parvulus*
Eel Pipefish, *Bulbonaricus brauni*

Eibl-Eibesfeldt's Pygmy Angelfish,
Centropyge eibli
Eight-armed Snailfish, *Liparis ochotensis*
Eightbar Goby, *Cryptocentrus cryptocentrus*
Eightline Wrasse, *Paracheilinus octotaenia*
Eight-lined Cardinalfish,
Cheilodipterus macrodon
Eightstripe Wrasse, *Pseudocheilinus octotaenia*
Eight-striped Butterflyfish,
Chaetodon octofasciatus
Eight-toothed Parrotfish, *Scarus oktodon*
Electric Ray, *Torpedo marmoratus*
Elegant Blenny, *Omobranchus elegans*
Elegant Coris, *Coris venusta*
Elizabeth's Fairy Basslet,
Odontanthias elizabethae
Elongate Hardyhead, *Atherinosoma elongata*
Elongate Scute Herring, *Ilisha elongata*
Elongate Sea Moth, *Parapegasus* sp.
Elongate Tonguefish, *Symphurus elongatus*
Elongate Wrasse, *Pseudojuloides elongatus*
Elusive Blenny, *Emblemaria walkeri*
Ember Blenny, *Cirripectes stigmaticus*
Ember Parrotfish, *Scarus rubroviolaceus*
Emery's Gregory, *Stegastes emeryi*
Emperor Angelfish, *Pomacanthus imperator*
Emperor Snapper, *Lutjanus sebae*
Engelhard's Goldie, *Pseudanthias engelhardi*
Englishman, *Chrysoblephus anglicus*
Enjambre, *Cephalopholis panamensis*
Epaulette Shark, *Hemiscyllium ocellatum*

Eschmeyer's Scorpionfish,
Rhinopias eschmeyeri
European Eel, *Anguilla anguilla*
European Flounder, *Platichthys
flesus*
European Scorpionfish, *Scorpaena
porcus*
European Seahorse, *Hippocampus
hippocampus*
Evermann's Cardinalfish, *Apogon
evermanni*
Evileye Pufferfish, *Amblyrhynchotes
honckenii*
Exquisite Wrasse, *Cirrhilabrus
exquisitus*
Exuma Cleaning Goby, *Gobiosoma
atronasum*
Eyed Flounder, *Bothus ocellatus*
Eyespot Blenny, *Ecsenius oculatus*

F

Fairy Basslet, unidentified,
Pseudanthias sp.
Fairy Cardinalfish, *Apogon
leptocaulus*
Falco Hawkfish, *Cirrhitichthys falco*
Fall Parrotfish, *Scarus frenatus*
False Batfish, *Parachaetodon
ocellatus*
False Blackfin Sweeper, *Pempheris*
sp. cf *japonicus*
False Bluestreak Wrasse, *Labroides
pectoralis*
False Cleanerfish, *Aspidontus
taeniatus*
False Combfish, *Coris pictoides*
False Lined Butterflyfish, *Chaetodon
oxycephalus*
False Longfinned Trevally,
Carangoides sp. (*?armatus*)
False Pacific Hagfish, *Eptatretus* sp.
(*stouti?*)
False Senator Wrasse, *Pictilabrus* sp.
False Skunk-striped Anemonefish,
Amphiprion perideraion

False Variegated Cardinalfish,
Fowleria sp. cf *variegata*
False-eye Damsel, *Abudefduf
sparoides*
False-eye Toby, *Canthigaster solandri*
Fan-bellied Leatherjacket,
Monacanthus chinensis
Fancytail Goby, *Cryptocentrus* sp. 2
Fantail Filefish, *Pervagor spilosoma*
Fantail Sole, *Xystreurys liolepis*
Fat Barracuda, *Sphyraena pinguis*
Feather Star Clingfish, *Discotrema
crinophila*
Featherfin Scorpionfish,
Hipposcorpaena sp. 1
Feathertail Sleeper, *Oxymetopon
cyanoctenosum*
Ferocious Goby, *Ctenogobiops
feroculus*
Few-rayed Dragonet, *Diplogrammus
pauciradiatus*
Few-rayed Hardyhead,
Craterocephalus pauciradiatus
Fiddler Ray, *Trygonorhina fasciata*
Fifteen-spine Stickleback, *Spinachia
spinachia*
Fiji Blenny, *Ecsenius fijiensis*
Filament Blenny, *Cirripectes
filamentosus*
Filament Devilfish, *Inimicus
filamentosus*
Filamented Sand Eel, *Trichonotus
setigerus*
Filament-fin Wrasse, *Paracheilinus
filamentosus*
Filamentous Blenny, *Labrisomus
filamentosus*
Filefish, unidentified,
Paramonacanthus sp.
Finescale Wrasse, *Hologymnosus
annulatus*
Fine-spotted Flounder,
Pleuronichthys cornutus
Fingered Dragonet, *Dactylopus
dactylopus*

Fingerfish, *Monodactylus sebae*
Firefish, *Nemateleotris magnifica*
Fisher's Angelfish, *Centropyge fisheri*
Fishgod Blenny, *Malacoctenus ebisui*
Fishing Frog, *Antennarius hispidus*
Five-banded Parrotfish, *Scarus venosus*
Fivefinger Wrasse, *Xyrichthys pentadactylus*
Five-line Snapper, *Lutjanus quinquelineatus*
Fivelined Threadfin Bream, *Nemipterus tambuloides*
Fivesaddle Parrotfish, *Scarus dimidiatus*
Five-spotted Wrasse, *Symphodus roissali*
Fivestripe Wrasse, *Thalassoma quinquevittatum*
Five-striped Pigface, *Rhyncopelates oxyrhynchus*
Flag Cabrilla, *Epinephelus labriformis*
Flag Rockfish, *Sebastes rubrovinctus*
Flag Squirrelfish, *Sargocentron vexillarium*
Flag-fin Cardinalfish, *Apogon elliotti*
Flag-finned Goby, *Pterogobius zonoleucus*
Flame Angelfish, *Centropyge loriculus*
Flame Basslet, *Mirolabrichthys ignitus*
Flame Goby, *Trimma macrophthalma*
Flameback Angelfish, *Centropyge aurantonotus*
Flamefish, *Apogon maculatus*
Flashlight Fish, *Photoblepharon palpebratus*
Flat Needlefish, *Ablennes hians*
Flathead, unidentified, *Platycephalus* sp.
Floating Blenny, *Aspidontus dussumieri*
Floral Blenny, *Petroscirtes mitratus*
Floral Wrasse, *Cheilinus chlourus*

Florida Pompano, *Trachinotus carolinus*
Flowerspot Flounder, *Pseudorhombus* sp.
Flowing Finned Goby, *Exyrias belissimus*
Flying Fish, Exocoetidae sp.
Flying Gurnard, *Dactylopterus volitans*
Fontanes Goby, *Amblyeleotris fontanesii*
Footballer, *Chrysiptera annulata*
Footballer Trout, *Plectropomus maculatus*
Forktail Blenny, *Meiacanthus atrodorsalis*
Forktail Dottyback, *Pseudochromis dixurus*
Forktailed Sea Bass, *Mirolabrichthys pascalus*
Formosan Goby, *Parioglossus formosus*
Forsskal's Goatfish, *Parupeneus forsskalii*
Forster's Hawkfish, *Paracirrhites forsteri*
Forster's Parrotfish, *Scarus japanensis*
Four-banded Cardinalfish, *Apogon fasciatus*
Four-bar Porcupinefish, *Lophodiodon calori*
Four-barred Toby, *Canthigaster coronata*
Four-barred Weever, *Parapercis tetracantha*
Four-eyed Butterflyfish, *Chaetodon capistratus*
Four-lined Terapon, *Pelates quadrilineatus*
Four-lined Wrasse, *Pseudocheilinus tetrataenia*
Fourmanoir's Blenny, *Ecsenius fourmanoiri*

Four-spot Butterflyfish, *Chaetodon quadrimaculatus*
Four-spot Wrasse, *Halichoeres trispilus*
Foxface, *Lo vulpinus*
Foxfish, *Bodianus frenchii*
Freckleback Grouper, *Epinephelus epistictus*
Freckleback Sharp-nosed Puffer, *Canthigaster callisterna*
Freckled Cardinalfish, *Phaeoptyx conklini*
Freckled Grouper, *Epinephelus cyanopodus*
Freckled Hawkfish, *Cirrhitus pinnulatus*
Freckled Head Midshipman, *Porichthys margaritatus*
Freckled Soapfish, *Rypticus bistrispinus*
Freckled Waspfish, *Paracentropogon vespa*
Freckletail Lyretail Angelfish, *Genicanthus lamarck*
French Angelfish, *Pomacanthus paru*
French Butterflyfish, *Chaetodon marcellae*
French Grunt, *Haemulon flavolineatum*
Freshwater Demoiselle, *Neopomacentrus taeniurus*
Frigate Tuna, *Auxis thazard*
Frill Shark, *Chlamydoselachus anguineus*
Fringed Blenny, *Chirolophis japonicus*
Fringed Filefish, *Monacanthus ciliatus*
Fringe-finned Trevally, *Pantolabus radiatus*
Fringe-lipped Flathead, *Thysanophrys otaitensis*
Fringyfin Goby, *Yongeichthys criniger*
Fringyhead Scorpionfish, *Scorpaena neglecta*

Fusilier Damsel, *Lepidozygus tapeinosoma*

G

Gafftopsail Pompano, *Trachinotus rhodopus*
Gag, *Mycteroperca microlepis*
Galapagos Hogfish, *Bodianus eclancheri*
Galapagos Sawtail, *Prionurus laticlavius*
Gardner's Butterflyfish, *Chaetodon gardneri*
Garibaldi, *Hypsypops rubicunda*
Garnet Red Parrotfish, *Scarus sordidus*
Garrett's Slender Filefish, *Pseudomonacanthus garretti*
Garrupa, *Cephalopholis aurantia*
Gecko Goby, *Chriolepis zebra*
Gecko Shark, *Galeus sauteri*
Genie's Cleaning Goby, *Gobiosoma genie*
Geoffroy's Wrasse, *Macropharyngodon geoffroyi*
Geometric Moray, *Siderea grisea*
Germain's Blenny, *Omobranchus germaini*
Ghana Butterflyfish, *Chaetodon robustus*
Ghana Goby, Gobiidae sp. 2
Ghana Rockcod, *Cephalopholis taeniops*
Ghana Yellowtail Damselfish, *Microspathodon frontalis*
Ghost Cardinalfish, *Apogon savayensis*
Ghost Goby, *Ptereleotris monoptera*
Ghost Mojarra, *Gerres baconensis*
Ghost Pipefish, *Solenostomus paegnis*
Ghost Shark, *Chimaera phantasma*
Ghost Snake Eel, *Brachysomophis cirrhocheilus*
Giant Chilean Clingfish, *Sicyases sanguineus*

Giant Damselfish, *Microspathodon dorsalis*

Giant Hawkfish, *Cirrhitus rivulatus*

Giant Jawfish, *Opistognathus rhomaleus*

Giant Kelpfish, *Heterostichus rostratus*

Giant Moray, *Gymnothorax javanicus*

Giant Sea Bass, *Stereolepis gigas*

Giant Sleepy Shark, *Nebrius concolor*

Giant Trevally, *Caranx ignobilis*

Gibert's Irish Lord, *Hemilepidotus gilberti*

Gilbert's Cardinalfish, *Apogon gilberti*

Gilbert's Sanddab, *Citharichthys gilberti*

Gilded Pipefish, *Corythoichthys schultzi*

Gill-rakered Needlefish, *Platybelone argalus*

Gill's Blenny, *Malacoctenus gilli*

Girard's Wrasse, *Pseudojulis girardi*

Girdled Blenny, *Malacoctenus zonifer*

Girdled Cardinalfish, *Archamia zosterophora*

Girdled Goby, *Priolepis cincta*

Girdled Moray, *Echidna pozyzona*

Glass Blenny, *Coralliozetus diaphanus*

Glassfish, *Chanda ranga*

Glassy Bombay Duck, *Harpadon translucens*

Glover's Frogfish, *Rhycherus gloveri*

Glowlight Bullseye, *Pempheris adspersa*

Gluttonous Goby, *Chasmichthys gulosus*

Gnomefish, *Scombrops boops*

Gobbleguts, *Apogon ruppelli*

Goblin Devilfish, *Inimicus sinensis*

Goblinfish, *Glyptauchen panduratus*

Goby 1, *Coryphopterus* sp.

Goby 2, *Gnatholepis* sp. 2

Goby 3, Gobiidae sp. 3

Goby 4, *Fusigobius* sp. (*neophytus?*)

Goddess Rainbowfish, *Xyrichthys dea*

Gold-band Fusilier, *Pterocaesio lativittata*

Goldband Trevally, *Carangoides dinema*

Gold-bar Snapper, *Tropidinus zonatus*

Goldbar Wrasse, *Thalassoma hebraicum*

Goldcheek Devilfish, *Plesiops oxycephalus*

Golden Angelfish, *Centropyge heraldi*

Golden Butterflyfish, *Chaetodon auripes*

Golden Dottyback, *Pseudochromis luteus*

Golden Dottyback, *Pseudochromis aureus*

Golden Goby, *Gobius auratus*

Golden Gregory, *Stegastes aureus*

Golden Grubfish, *Parapercis aurantica*

Golden Hawkfish, *Paracirrhites xanthus*

Golden Mimic Blenny, *Plagiotremus laudandus flavus*

Golden Mini-Grouper, *Assessor flavissimus*

Golden Orange Rockcod, *Cephalopholis analis*

Golden Pollock, *Pollachius pollachius*

Golden Puffer, *Arothron meleagris*

Golden Rainbowfish, *Halichoeres chrysus*

Golden Redfish, *Sebastes marinus*

Golden Sea Bream, *Pagrus auratus*

Golden Snapper, *Lutjanus fulviflammus*

Golden Spotted Rock Bass, *Paralabrax auroguttatus*

Golden Threadfin Bream, *Nemipterus virgatus*

Golden Trevally, *Gnathanodon speciosus*

Golden Wrasse, *Pseudojulis melanotus*

Golden Yellow Hawkfish, *Cirrhitichthys aureus*

Golden-girdled Coralfish, *Coradion chrysozonus*

Golden-lined Spinefoot, *Siganus lineatus*

Golden-spotted Goatfish, *Parupeneus luteus*

Goldentail Moray, *Muraena miliaris*

Gold-eyed Snapper, *Lutjanus adetii*

Gold-lined Sea Bream, *Gnathodentex aurolineatus*

Gold-lined Threadfin Bream, *Nemipterus gracilis*

Goldmann's Goby, *Istigobius goldmanni*

Goldmann's Sweetlips, *Plectorhinchus goldmanni*

Goldribbon Soapfish, *Aulacocephalus temmincki*

Gold-rimmed Surgeonfish, *Acanthurus glaucopareius*

Goldsaddle Hogfish, *Bodianus perdito*

Gold-spangled Angelfish, *Apolemichthys xanthopunctatus*

Goldspot Goby, *Gobiosoma saucrum*

Gold-spotted Puffer, *Chelonodon laticeps*

Goldspotted Snake Eel, *Myrichthys oculatus*

Gold-spotted Spinefoot, *Siganus chrysospilos*

Gold-spotted Sweetlips, *Plectorhinchus multivittatus*

Goldstripe Fusilier, *Pterocaesio chrysozona*

Gold-striped Cardinalfish, *Apogon cyanosoma*

Gold-striped Sardine, *Sardinella* sp. cf *jussieu*

Gold-striped Snapper, *Lutjanus erythropterus*

Goldtail Angelfish, *Pomacanthus chrysurus*

Goodlad's Dragonet, *Callionymus goodladi*

Goose Scorpionfish, *Rhinopias frondosa*

Goosehead Scorpionfish, *Scorpaena bergi*

Gopher Rockfish, *Sebastes carnatus*

Gorgeous Goby, *Lythrypnus pulchellus*

Gosline's Mimic Blenny, *Plagiotremus goslinei*

Gossamer Blenny, *Omobranchus ferox*

Gowan's Clingfish, *Lepadogaster lepadogaster*

Graceful Lizardfish, *Saurida gracilis*

Graceful Pearlfish, *Encheliophis gracilis*

Grammistes Blenny, *Meiacanthus grammistes*

Grass Puffer, *Takifugu niphobles*

Grass Scorpionfish, *Scorpaena grandicornis*

Gray Angelfish, *Pomacanthus arcuatus*

Gray Chromis, *Chromis axillaris*

Gray Cod, *Gadus macrocephalus*

Gray Demoiselle, *Chrysiptera glauca*

Gray Lance, *Haliophis guttatus*

Gray Moray, *Gymnothorax nubilis*

Gray Reef Shark, *Carcharhinus amblyrhynchos*

Gray Snapper, *Lutjanus griseus*

Gray Triggerfish, *Balistes capriscus*

Graybar Grunt, *Haemulon sexfasciatum*

Gray's Crested Flounder, *Samaris cristatus*

Gray's Pipefish, *Halicampus grayi*

Grayshy, *Cephalopholis cruentatus*

Greasy Rock Cod, *Epinephelus tauvina*

Great Barracuda, *Sphyraena barracuda*

Greater Amberjack, *Seriola dumerili*

Greater Soapfish, *Rypticus saponaceus*

Green Clingfish, *Lepadichthys frenatus*

Green Hata, *Epinephelus awoara*

Green Jack, *Caranx caballus*

Green Jobfish, *Aprion virescens*

Green Moray, *Gymnothorax funebris*

Green Parrotfish, *Scarus dubius*

Green Razorfish, *Xyrichthys splendens*

Green Wrasse, *Halichoeres chloropterus*

Greenback Fairy Wrasse, *Cirrhilabrus scottorum*

Greenback Yellowtail, *Seriola lalandi dorsalis*

Greenband Goby, *Gobiosoma multifasciatum*

Greenband Wrasse, *Halichoeres bathyphilus*

Greenbeak Parrotfish, *Scarus prasiognathos*

Greenblotch Parrotfish, *Sparisoma atomarium*

Greengar, *Strongylura anastomella*

Greenling, *Pleurogrammus azonus*

Grenadier, *Hymenocephalus gracilis*

Grey Grunter, *Pomadasys furcatus*

Grigorjew's Halibut, *Eopsetta grigorjewi*

Grigorjew's Shanny, *Stichaeus grigorjewi*

Grooved Razorfish, *Centriscus scutatus*

Grouper, unidentified, *Epinephelus* sp.

Grunt Sculpin, *Rhamphocottus richardsoni*

Guamanian Damsel, *Pomachromis guamensis*

Guenther's Butterflyfish, *Chaetodon guentheri*

Guichenot's Hawkfish, *Cirrhitichthys guichenoti*

Guinea Fowl Wrasse, *Macropharyngodon meleagris*

Guineafowl Moray, *Gymnothorax meleagris*

Gulf Coney, *Epinephelus acanthistius*

Gulf Damsel, *Pristotis jerdoni*

Gulf Opaleye, *Girella simplicidens*

Gunther's Rainbowfish, *Pseudolabrus guntheri*

Gunther's Sculpin, *Cottus pollux*

Gurnard Scorpion Perch, *Neosebastes pandus*

Guyana Butterflyfish, *Chaetodon guyanensis*

H

Haacke's Flathead, *Platycephalus haackei*

Hairy Blenny, *Labrisomus nuchipinnis*

Hairy Stingfish, *Scorpaenopsis oxycephala*

Hakeling, unidentified, *Physiculus* sp.

Half Yellow Pygmy Angelfish, *Centropyge joculator*

Half-and-Half Chromis, *Chromis dimidiata*

Half-and-Half Thicklip, *Hemigymnus melapterus*

Half-banded Goby, *Amblygobius seminctus*

Half-banded Lyretail Angelfish, *Genicanthus semifasciatus*

Half-banded Seaperch, *Ellerkeldia hunti*

Half-banded Snake Eel, *Leiuranus semicinctus*

Half-banded Snapper, *Lutjanus semicinctus*
Half-blue Demoiselle, *Chrysiptera hemicyanea*
Half-dotted Hawkfish, *Paracirrhites hemistictus*
Halfmoon, *Medialuna californiensis*
Halfmoon Rock Cod, *Epinephelus rivulatus*
Halfmoon Triggerfish, *Sufflamen chrysopterus*
Halfspotted Grouper, *Cephalopholis hemistiktos*
Half-striped Lyretail Angelfish, *Genicanthus semicinctus*
Half-tail Blenny, *Xiphasia setifer*
Half-yellow Butterflyfish, *Chaetodon hemichrysus*
Hamilton's Anchovy, *Thryssa hamiltoni*
Hamilton's Manta, *Manta hamiltoni*
Hamilton's Puffer, *Sphoeroides hamiltoni*
Hamlet, *Hypoplectrus unicolor*
Hancock's Blenny, *Acanthemblemaria hancocki*
Harlequin Bass, *Serranus tigrinis*
Harlequin Filefish, *Oxymonacanthus longirostris*
Harlequin Fish, *Othos dentex*
Harlequin Pipefish, *Micrognathus ensenadae*
Harlequin Rockcod, *Cephalopholis polleni*
Harlequin Sweetlips, *Plectorhinchus chaetodonoides*
Harlequin Tuskfish, *Lienardella fasciata*
Hartzfeld's Cardinalfish, *Apogon hartzfeldii*
Hartzfeld's Wrasse, *Halichoeres hartzfeldii*
Hasselt's Goby, *Callogobius hasselti*
Hasselt's Round Herring, *Dussumieria hasselti*

Hass's Garden Eel, *Taenioconger hassi*
Hawaiian Aholehole, *Kuhlia sandvicensis*
Hawaiian Basslet, *Pikea aurora*
Hawaiian Bicolor Chromis, *Chromis hanui*
Hawaiian Blood Red Wrasse, *Verriculus sanguineus*
Hawaiian Blue Parrotfish, *Scarus perspicillatus*
Hawaiian Bristletooth, *Ctenochaetus hawaiiensis*
Hawaiian Chromis, *Chromis ovalis*
Hawaiian Cleaner Wrasse, *Labroides phthirophagus*
Hawaiian Fairy Basslet, *Pseudanthias ventralis hawaiiensis*
Hawaiian Gold-barred Butterflyfish, *Chaetodon excelsa*
Hawaiian Hawkfish, *Cyprinocirrhites* sp.
Hawaiian Lionfish, *Pterois sphex*
Hawaiian Pearlyscale Angelfish, *Apolemichthys arcuatus*
Hawaiian Redtail Filefish, *Pervagor aspricaudus*
Hawaiian Rock Damsel, *Plectroglyphidodon sindonis*
Hawaiian Soapfish, *Suttonia lineata*
Hawaiian Squirrelfish, *Sargocentron ensifer*
Hawaiian Stingray, *Dasyatis hawaiiensis*
Hawaiian Threespot Damsel, *Dascyllus albisella*
Hector's Goby, *Amblygobius hectori*
Helfrich's Firefish, *Nemateleotris helfrichi*
Heliotrope Eyeband Wrasse, *Coris batuensis*
Hellmuth's Threadfin Bream, *Pentapodus hellmuthi*
Hemphill's Blenny, *Stathmonotus hemphilli*

Herald's Pipefish, *Heraldia nocturna*
Herre's Samoan Prawn Goby,
  Ctenogobiops aurocingulus
Herring Cale, *Odax cyanomelas*
Herring, unidentified, *Jenkinsia* sp.
Herzenstein's Righteye Flounder,
  Cleisthenes pinetorum herzensteini
Hifin Goby, *Myersina macrostoma*
Higheye Blenny, *Dialommus fuscus*
High-hat, *Equetus acuminatus*
High-middle Pretty Fish, *Hypomesus pretioses japonicus*
Hilgendorf Saucord, *Helicolenus hilgendorfi*
Hinalea, *Coris flavovittata*
Hinalea Luahine, *Thalassoma ballieui*
Hirai's Flying Fish, *Cypselurus hiraii*
Hoefler's Butterflyfish, *Chaetodon hoefleri*
Hogchoker, *Trinectes maculatus*
Hogfish, *Lachnolaimus maximus*
Honey Damselfish, *Stegastes mellis*
Honeycomb Filefish, *Cantherhines pardalis*
Honeycomb Grouper, *Epinephelus merra*
Honeycomb Moray, *Gymnothorax favagineus*
Honeycomb Podge, *Pseudogramma polycantha*
Honeycomb Rockfish, *Sebastes umbrosus*
Honeycomb Toby, *Canthigaster janthinoptera*
Honeycomb Trunkfish, *Lactophrys polygonia*
Honey-head Damsel, *Dischistodus prosopotaenia*
Horn Shark, *Heterodontus francisci*
Horned Squirrelfish, *Sargocentron melanospilus*
Hornyhead Turbot, *Pleuronichthys verticalis*
Horse Crevalle, *Carangoides equula*

Horse Mackerel, *Trachurus japonicus*
Horse-eye Jack, *Caranx latus*
Horseshoe Leatherjacket,
  Meuschenia hippocrepis
Houndfish, *Tylosurus crocodilus*
Hovering Goby, *Ioglossus helenae*
Humpback Scorpionfish,
  Scorpaenopsis gibbosa
Humpback Snapper, *Lutjanus gibbus*
Humpback Unicornfish, *Naso brachycentron*
Humphead Bannerfish, *Heniochus varius*
Humphead Parrotfish, *Bolbometopon muricatus*
Humphead Wrasse, *Cheilinus undulatus*
Hump-headed Blenny, *Istiblennius gibbifrons*
Humpnose Unicornfish, *Naso tuberosus*
Hunchback Scorpionfish, *Scorpaena* sp. 3

I

Ijima's Dragonet, *Neosynchiropus ijimai*
Imelda's Seaperch, *Mirolabrichthys imeldae*
Impostor Boxfish, *Ostracion trachys*
Incised Triplefin, *Norfolkia incisa*
Indian Driftfish, *Ariomma indica*
Indian Frogfish, *Antennarius indicus*
Indian Goatfish, *Parupeneus indicus*
Indian Golden-barred Butterflyfish,
  Chaetodon jayakari
Indian Little Fish, *Apogon semilineatus*
Indian Mackerel, *Trachurus indicus*
Indian Ocean Bannerfish, *Heniochus pleurotaenia*
Indian Ocean Chevron
  Butterflyfish, *Chaetodon madagascariensis*

Indian Scad, *Decapterus russelli*
Indian Threadfin, *Alectis indicus*
Indian Triggerfish, *Melichthys indicus*
Indies Gizzard Shad, *Anodontostoma chacunda*
Indigo Hamlet, *Hypoplectrus indigo*
Indo-Pacific Leatherjacket, *Cantherhines fronticinctus*
Indo-Pacific Squirrelfish, *Sargocentron tieroides*
Inshore Lizardfish, *Synodus foetens*
Island Goby, *Lythrypnus nesiotes*
Iso Blenny, *Ecsenius isos*

J

Jackknife-fish, *Equetus lanceolatus*
Jansen's Wrasse, *Thalassoma jansenii*
Janss' Pipefish, *Doryrhamphus janssi*
Japanese Anchovy, *Engraulis japonicus*
Japanese Angel Shark, *Squatina japonica*
Japanese Aulopus, *Aulopus japonicus*
Japanese Basslet, *Liopropoma susumi*
Japanese Bigeye, *Pristigenys niphonia*
Japanese Black Rockfish, *Sebastes inermis*
Japanese Blue Sprat, *Spratelloides japonicus*
Japanese Bluefish, *Scombrops gilberti*
Japanese Blue-striped Pipefish, *Doryrhamphus japonicus*
Japanese Boarfish, *Pentaceros japonicus*
Japanese Bonyhead, *Ostichthys japonicus*
Japanese Butterfish, *Hyperoglyphe japonica*
Japanese Butterfly Ray, *Gymnura japonica*
Japanese Cardinalfish, *Apogon ishigakiensis*
Japanese Coral Trout, *Plectranthias japonicus*

Japanese Dragonet, *Calliurichthys japonicus*
Japanese Eel, *Anguilla japonica*
Japanese Fairy Basslet, *Pseudanthias taira*
Japanese Feather-finned Flying Fish, *Cheilopogon pinnatibarbatus japonicus*
Japanese Filefish, *Rudarius ercodes*
Japanese Flathead, *Bembras japonicus*
Japanese Fringehead Blenny, *Neoclinus bryope*
Japanese Golden-barred Butterflyfish, *Chaetodon modestus*
Japanese Halfbeak, *Hyporhamphus sajori*
Japanese Horned Shark, *Heterodontus japonicus*
Japanese Leopard Shark, *Triakis scyllia*
Japanese Numbfish, *Narke japonica*
Japanese Parrotfish, *Calotomus japonicus*
Japanese Pygmy Angelfish, *Centropyge interruptus*
Japanese Sand Bass, *Chelidoperca hirundinacea*
Japanese Scorpionfish, *Sebastes baramenuke*
Japanese Sculpin, *Dasycottus japonicus*
Japanese Sea Bream, *Evynnis japonica*
Japanese Seahorse, *Hippocampus japonicus*
Japanese Seaperch, *Ditrema temmincki*
Japanese Silver Bream, *Acanthopagrus latus*
Japanese Slimefish, *Gephyroberyx japonicus*
Japanese Smoothhound, *Mustelus manazo*

Japanese Sole, *Heteromycteris japonicus*

Japanese Spanish Mackerel, *Scomberomorus niphonius*

Japanese Squirrelfish, *Sargocentron ittodai*

Japanese Stargazer, *Uranoscopus japonicus*

Japanese Stingray, *Dasyatis akajei*

Japanese Threadfin Bream, *Nemipterus japonicus*

Japanese Threadfin Shad, *Nematalosa japonica*

Japanese Tilefish, *Branchiostegus japonicus*

Japanese Tonguefish, *Paraplagusia japonica*

Japanese Torpedo Ray, *Torpedo tokionis*

Japanese Tubesnout, *Aulichthys japonicus*

Japanese Velvetfish, *Aploactis aspera*

Japanese Wobbegong, *Orectolobus japonicus*

Japanese Wrasse, *Pseudolabrus japonicus*

Japanese Yellowtail, *Seriola quinqueradiata*

Java Parrotfish, *Scarus javanicus*

Jelly Cardinalfish, *Pseudamia gelatinosa*

Jenkin's Parrotfish, *Calotomus zonarchia*

Jenyns' Flounder, *Pseudorhombus jenynsii*

Jet Black Dottyback, *Pseudochromis paranox*

Jewel Damsel, *Paraglyphidodon lacrymatus*

Jewel Goby, *Stiphodon* sp.

Jewelfish, *Microspathodon chrysurus*

Jewelled Rockskipper, *Salarias fasciatus*

Jewfish, *Epinephelus itajara*

Jock Stuart Scorpionfish, *Helicolenus papillosus*

John Dory, *Zeus japonicus*

Johnston Damsel, *Plectroglyphidodon johnstonianus*

Jolthead Porgy, *Calamus bajonado*

Jordan's Fairy Wrasse, *Cirrhilabrus jordani*

Jordan's Squirrelfish, *Neoniphon scythrops*

Jordan's Tuskfish, *Choerodon jordani*

Joyner's Stingfish, *Sebastes joyneri*

Joyner's Tonguefish, *Cynoglossus joyneri*

Jumping Blenny, *Lepidoblennius marmoratus*

K

Kallochrome Wrasse, *Halichoeres kallochroma*

Kawakawa, *Euthynnus affinis*

Kawamebar's Bass, *Coreoperca kawamebari*

Kazika Sculpin, *Cottus kazika*

Kellogg's Scorpionfish, *Scorpaenodes kelloggi*

Kelp Greenling, *Hexagrammos decagrammus*

Kelp Grouper, *Epinephelus moara*

Kelp Rockfish, *Sebastes atrovirens*

Kelpfish, *Chironemus marmoratus*

Key Blenny, *Starksia starcki*

Key Brotutid, *Ogilbia cayorum*

Keyhole Angelfish, *Centropyge tibicin*

Keys Jawfish, *Opistognathus* sp. 1

Khaki Blenny, *Ecsenius aroni*

Kidako Moray Eel, *Gymnothorax kidako*

King Angelfish, *Holacanthus passer*

King Coris, *Coris auricularis*

King Demoiselle, *Chrysiptera rex*

King George's Salmon, *Arripes georgianus*

King Spanish Mackerel, *Scomberomorus guttatus*

Klausewitz's Blenny, *Ecsenius lineatus*

Klein's Butterflyfish, *Chaetodon kleini*

Klunzinger's Chromis, *Chromis klunzingeri*

Klunzinger's Wrasse, *Thalassoma klunzingeri*

Knight-fish, *Cleidopus gloriamaris*

Knobbed Porgy, *Calamus nodosus*

Knobby Seahorse, *Hippocampus breviceps*

Koningsberg's Herring, *Harengula koningsbergeri*

Koran Angelfish, *Pomacanthus semicirculatus*

Kosi Rockskipper, *Pereulixia kosiensis*

Kuiter's Tall Triplefin, *Apopterygion alta*

L

Laboute's Fairy Wrasse, *Cirrhilabrus labouti*

Ladder Glider, *Valenciennea sexguttata*

Ladd's Goby, *Bathygobius laddi*

Ladyfish, *Elops machnata*

Lagoon Damsel, *Hemiglyphidodon plagiometopon*

Lake Goby, *Rhinogobius brunneus*

Lance-like Smelt, *Spirinchus lanceolatus*

Lancer, *Lethrinus nematacanthus*

Lancer Dragonet, *Paradiplogrammus bairdi*

Lancetail Fairy Wrasse, *Cirrhilabrus blatteus*

Lane Snapper, *Lutjanus synagris*

Lantern Bass, *Serranus baldwini*

Lantern-eye, *Anomalops katoptron*

Lanternfish, unidentified, Myctophidae sp.

Large-bodied Silver-Biddy, *Gerres macrosoma*

Largehead Hairtail, *Trichiurus lepturus*

Large-headed Scorpionfish, *Pontinus macrocephalus*

Large-scaled Grinner, *Saurida undosquamis*

Large-scaled Mullet, *Liza oligolepis*

Large-scaled Whiting, *Sillago macrolepis*

Laterally Banded Grouper, *Epinephelus latifasciatus*

Lattice Cardinalfish, *Apogon margaritiphora*

Lattice Seahorse, *Hippocampus* sp.

Lattice Soldierfish, *Myripristis violacea*

Latticed Blenny, *Ecsenius opsifrontalis*

Latticed Monocle Bream, *Scolopsis cancellatus*

Lavender Sculpin, *Leiocottus hirundo*

Leafy Seadragon, *Phycodurus eques*

Leaping Bonito, *Cybiosarda elegans*

Leaping Lizardfish, *Synodus jaculum*

Lea's Cardinalfish, *Archamia leai*

Leather Bass, *Epinephelus dermatolepis*

Leatherjacket, *Oligoplites saurus*

Lemon Shark, *Negaprion brevirostris*

Lemon Sweetlips, *Plectorhinchus flavomaculatus*

Lemonpeel, *Centropyge flavissimus*

Lennard's Tamarin, *Anampses lennardi*

Leonard's Common Goby, *Pomatoschistus microps*

Leopard Blenny, *Ecsenius pardus*

Leopard Flounder, *Bothus leopardus*

Leopard Frogfish, *Antennarius pardalis*

Leopard Grouper, *Plectropomus leopardus*

Leopard Moray, *Gymnothorax undulatus*
Leopard Pufferfish, *Canthigaster leopardus*
Leopard Rockcod, *Cephalopholis leopardus*
Leopard Rockskipper, *Exallias brevis*
Leopard Searobin, *Prionotus scitulus*
Leopard Shark, *Triakis semifasciatus*
Leopard Snake-eel, *Callechelys marmoratus*
Lesser Devil Ray, *Mobula diabolis*
Lesser Electric Ray, *Narcine brasiliensis*
Lesser Spotted Dogfish, *Scyliorhinus canicula*
Libertad Thread Herring, *Opisthonema libertate*
Lieutenant Surgeonfish, *Acanthurus tennenti*
Limbaugh's Chromis, *Chromis limbaughi*
Lined Basslet, *Liopropoma lineata*
Lined Blenny, *Ecsenius aequalis*
Lined Butterflyfish, *Chaetodon lineolatus*
Lined Cardinalfish, *Apogon lineatus*
Lined Chromis, *Chromis lineata*
Lined Cichlops, *Labracinus lineatus*
Lined Fairy Wrasse, *Cirrhilabrus lineatus*
Lined Goby, *Parioglossus lineatus*
Lined Silver Grunt, *Pomadasys hasta*
Lined Sole, *Achirus lineatus*
Lined Tamarin, *Anampses lineatus*
Linespot Hogfish, *Bodianus oxycephalus*
Lingcod, *Ophiodon elongatus*
Lionfish, *Pterois volitans*
Little Fairy Basslet, *Sacura parva*
Little Pineapplefish, *Sorosichthys ananassa*
Little Rainbow Wrasse, *Dotalabrus* sp.

Little Ruby Red Cardinalfish, *Apogon coccineus*
Little Spinefoot, *Siganus spinus*
Little Weed Whiting, *Neodax balteatus*
Little-scaled Parma, *Parma microlepis*
Littlewing Stonefish, *Synanceia alula*
Livingstone's Eel Blenny, *Notograptus livingstonei*
Lizard Island Goby, *Ctenogobiops pomastictus*
Lizard Triplefin, Tripterygiidae sp. 1
Lizardfish, unidentified, *Synodus* sp.
Lobed Stingaree, *Urolophus lobatus*
Lofty Triplefin, *Enneanectes altivelis*
Long Oarfish, *Regalecus glesne*
Long-barbeled Goatfish, *Parupeneus macronema*
Longfin Blenny, *Labrisomus haitiensis*
Longfin Damselfish, *Stegastes diencaeus*
Longfined Trevally, *Carangoides armatus*
Long-finned Batfish, *Platax pinnatus*
Longfinned Boarfish, *Zanclistius elevatus*
Long-finned Dottyback, *Pseudochromis longipinnis*
Long-finned Eel, *Anguilla dieffenbachii*
Long-finned Goby, *Favonigobius lateralis*
Long-finned Grouper, *Epinephelus megachir*
Long-finned Silver-Biddy, *Pentaprion longimanus*
Long-finned Snake Eel, *Pisodonophis cancrivorus*

Long-horned Cowfish, *Lactoria cornuta*

Longjaw Squirrelfish, *Holocentrus adscensiones*

Longlure Frogfish, *Antennarius multiocellatus*

Longnose Batfish, *Ogcocephalus corniger*

Longnose Hawkfish, *Oxycirrhites typus*

Longnose Sailfin Tang, *Zebrasoma rostratum*

Longnose Spiny Dogfish, *Squalus mitsukurii*

Longnose Trevally, *Carangoides chrysophrys*

Long-nosed Butterflyfish, *Forcipiger flavissimus*

Long-nosed Wrasse, *Symphodus rostratus*

Long-rayed Silver-Biddy, *Gerres filamentosus*

Longsnout Emperor, *Lethrinus rostratus*

Longsnout Seahorse, *Hippocampus reidi*

Long-snouted Boarfish, *Pentaceropsis recurvirostris*

Long-snouted Bream, *Lethrinus miniatus*

Longspine Cardinalfish, *Apogon leptacanthus*

Longspine Combfish, *Zaniolepis latipinnis*

Longspine Sailfin Scorpionfish, *Hypodytes longispinis*

Longspine Sea Bream, *Argyrops spinifer*

Longspine Snipefish, *Macrorhamphosus scolopax*

Longspine Squirrelfish, *Holocentrus rufus*

Long-spined Perchlet, *Ambassis interruptus*

Longtail Seamoth, *Pegasus volitans*

Longtail Tuna, *Thunnus tonggol*

Long-tailed Dragonfish, *Parapegasus natans*

Lookdown, *Selene vomer*

Loosetooth Parrotfish, *Nicholsina denticulata*

Lopez's Unicornfish, *Naso lopezi*

Lord Howe Butterflyfish, *Chaetodon tricinctus*

Lord Howe Leatherjacket, *Cantherhines howensis*

Lord Howe Scorpionfish, *Scorpaenopsis* sp. 2

Lord Howe Tamarin, *Anampses elegans*

Low Spot Sharpnosed Pufferfish, *Canthigaster inframacula*

Lowe's Beardfish, *Polymixia lowei*

Lubbock's Fairy Wrasse, *Cirrhilabrus lubbocki*

Lumpsucker, *Cyclopterus lumpus*

Lunar Fusilier, *Caesio lunaris*

Luther's Goby, *Cryptocentrus lutheri*

Lyretail Coralfish, *Pseudathias squamipinnis*

Lyretail Fairy Grouper, *Odontanthias fuscipinnis*

Lyretail Grouper, *Variola louti*

Lyretail Hogfish, *Bodianus anthioides*

Lyretail Wrasse, *Thalassoma lunare*

M

Macleay's Crested Pipefish, *Histiogamphelus cristatus*

MacNeill's Mini-grouper, *Assessor macneilli*

Madder Seaperch, *Mirolabrichthys dispar*

Madeira Sharp-nosed Puffer, *Canthigaster capistratus*

Madras Yellow-lined Snapper, *Lutjanus madras*

Magenta Dottyback, *Pseudochromis porphyreus*

Magnificent Foxface, *Lo magnifica*
Magnus' Goby, *Amblyeleotris sungami*
Magpie Morwong, *Cheilodactylus gibbosus*
Mahogany Snapper, *Lutjanus mahogoni*
Mahsena?, *Lethrinus mahsena?*
Mahsena, *Lethrinus mahsena*
Mailed Butterflyfish, *Chaetodon reticulatus*
Malabar Rockcod, *Epinephelus malabaricus*
Malabar Snapper, *Lutjanus malabaricus*
Malabar Trevally, *Carangoides malabaricus*
Malamalama, *Coris ballieui*
Mandarin Fish, *Pterosynchiropus splendidus*
Maned Goby, *Oxyurichthys microlepis*
Mangrove Goby, *Acentrogobius gracilis*
Manta Ray, *Manta birostris*
Many Banded Angelfish, *Centropyge multifasciatus*
Many Banded Pipefish, *Doryrhamphus multiannulatus*
Many Banded Scat, *Selenotoca multifasciata*
Many-banded Goatfish, *Parupeneus multifasciata*
Many-lined Blenny, *Istiblennius lineatus*
Many-lined Cardinalfish, *Apogon chrysotaenia*
Manylined Coral Goby, *Gobiodon multilineatus*
Many-spotted Blenny, *Laiphognathus multimaculatus*
Many-striped Sweetlips, *Plectorhinchus polytaenia*
Manytooth Parrotfish, *Cryptotomus roseus*

Many-toothed Blenny, *Ecsenius dentex*
Maomao, *Abudefduf abdominalis*
Maori Wrasse, *Ophthalmolepis lineolatus*
Marbled Blenny, *Paraclinus marmoratus*
Marbled Cat Shark, *Atelomycterus macleayi*
Marbled Dragonet, *Synchiropus marmoratus*
Marbled Electric Ray, *Torpedo sinuspersici*
Marbled Grouper, *Epinephelus inermis*
Marbled Parrotfish, *Leptoscarus vaigiensis*
Marbled Puffer, *Sphoeroides marmoratus*
Marbled Rockfish, *Sebasticus marmoratus*
Marblefish, *Aplodactylus meandritus*
Margarita Blenny, *Malacoctenus margaritae*
Margate, *Haemulon album*
Marginate Damselfish, *Dascyllus marginatus*
Marianas Squirrelfish, *Neoniphon marianus*
Market Anchovy, *Anchovia commersoniana*
Marley's Butterflyfish, *Chaetodon marleyi*
Marlin, *Makaira* sp.
Marquesan Sharpnosed Pufferfish, *Canthigaster marquesensis*
Marquesas Butterflyfish, *Chaetodon declevis*
Marshall's Dottyback, *Pseudochromis marshallensis*
Martinique Fairy Basslet, *Holanthias martinicensis*
Mashed Goby, *Gobionellus stigmaticus*

Masked Bannerfish, *Heniochus monoceros*

Masked Butterflyfish, *Chaetodon larvatus*

Masked Cardinalfish, *Apogon thermalis*

Masked Goby, *Coryphopterus personatus*

Masked Lyretail Angelfish, *Genicanthus personatus*

Masked Moray, *Gymnothorax panamensis*

Masked Puffer, *Arothron diadematus*

Masked Rainbowfish, *Halichoeres melanochir*

Masuda's Hogfish, *Bodianus masudai*

Mata Tang, *Acanthurus mata*

Matsubara's Red Rockfish, *Sebastes matsubarae*

Matted Leatherjacket, *Acreichthys tomentosus*

McCulloch's Anemonefish, *Amphiprion mccullochi*

McCulloch's Dottyback, *Pseudochromis mccullochi*

McCulloch's Scalyfin, *Parma mccullochi*

McCulloch's Seaperch, *Ellerkeldia maccullochi*

Mediterranean Banded Goby, *Gobius vittatus*

Mediterranean Black Goby, *Gobius niger*

Mediterranean Brown Wrasse, *Labrus merula*

Mediterranean Damselfish, *Chromis chromis*

Mediterranean Gray Wrasse, *Symphodus cinereus*

Mediterranean Ocellated Wrasse, *Symphodus ocellatus*

Mediterranean Rainbow Wrasse, *Coris julis*

Mediterranean Seahorse, *Hippocampus ramulosus*

Mediterranean Wrasse, *Symphodus mediterraneus*

Melbourne Mojarra, *Parequula melbournenses*

Melon Butterflyfish, *Chaetodon trifasciatus*

Merlet Scorpionfish, *Rhinopias aphanes*

Mertens's Butterflyfish, *Chaetodon mertensii*

Messmate Pipefish, *Corythoichthys haematopterus*

Metallic Demoiselle, *Neopomacentrus metallicus*

Mexican Blenny, *Paraclinus mexicanus*

Mexican Goatfish, *Mulloides dentatus*

Mexican Hogfish, *Bodianus diplotaenia*

Mexican Horned Shark, *Heterodontus mexicanus*

Mexican Lookdown, *Selene brevoortii*

Mexican Parrotfish, *Scarus perrico*

Meyer's Butterflyfish, *Chaetodon meyeri*

Micronesian Wrasse, *Labropsis micronesica*

Midas Blenny, *Ecsenius midas*

Midget Chromis, *Chromis acares*

Midnight Angelfish, *Centropyge nox*

Midnight Parrotfish, *Scarus coelestinus*

Midwater Rockcod, *Epinephelus undulosus*

Milk Trevally, *Lactarius lactarius*

Milkfish, *Chanos chanos*

Miller's Damselfish, *Pomacentrus milleri*

Millet-seed Butterflyfish, *Chaetodon miliaris*

Mimic Blenny, *Hemiemblemaria simulus*

Mimic Cardinalfish, *Apogon phenax*
Mimic Surgeonfish, *Acanthurus chronixis*
Mini-goby, unidentified, *Eviota* sp. 4
Mirror Butterflyfish, *Chaetodon speculum*
Miry's Damselfish, *Neopomacentrus miryae*
Misty Grouper, *Epinephelus mystacinus*
Mitsukuri Croaker, *Nibea mitsukurii*
Moens Parrotfish, *Scarus moensi*
Molly Miller, *Scartella cristata*
Molucca Damsel, *Pomacentrus moluccensis*
Moluccan Goatfish, *Upeneus moluccensis*
Monarch Damsel, *Dischistodus pseudochrysopoecilus*
Monkeyface Prickleback, *Cebidichthys violaceous*
Monkeyfish, *Erosa erosa*
Monogrammed Monocle Bream, *Scolopsis monogramma*
Monrovian Surgeonfish, *Acanthurus monroviae*
Moon Butterflyfish, *Chaetodon selene*
Moonfish, *Lampris guttatus*
Moonlighter, *Tilodon sexfasciatum*
Moore's Dottyback, *Pseudochromis moorei*
Moorish Idol, *Zanclus canescens*
Moray eel, unidentified, *Gymnothorax* sp.
Mosaic Leatherjacket, *Eubalichthys mosaicus*
Mosaic Moray, *Gymnothorax ramosus*
Mosshead Warbonnet, *Chirolophis nugator*
Mottled Goby, *Trimma* sp. 2
Mottled Jawfish, *Opistognathus* sp. 2
Mottled Moray, *Gymnothorax prionodon*

Moustache Jawfish, *Opistognathus lonchurus*
Mowbray's Cave Bass, *Liopropoma mowbrayi*
Mozambique Blenny, *Meiacanthus mossambicus*
Mozambique Cardinal, *Archamia mozambiquensis*
Mozambique Emperor, *Lethrinus hypselopterus*
Mudskipper, unidentified, Periophthalmidae sp.
Mueller's Butterflyfish, *Chelmon muelleri*
Mullet Snapper, *Lutjanus aratus*
Mullet, unidentified, *Liza* sp.
Multicolor Snake Blenny, *Sticharium dorsale*
Multicolorfin Rainbowfish, *Halichoeres poecilopterus*
Multispined Damsel, *Paraglyphidodon polyacanthus*
Multi-spotted Searobin, *Pterygotrigla multiocellatus*
Mushroom Scorpionfish, *Scorpaena inermis*
Mustard Surgeonfish, *Acanthurus guttatus*
Mutton Hamlet, *Alphestes afer*
Mutton Snapper, *Lutjanus analis*
Myers's Balloonfish, *Diodon myersi*
Myers's Goby, *Myersina nigrivirgata*
Mystery Butterflyfish, *Chaetodon nigropunctatus*
Mystery Wrasse, *Pseudocheilinus* sp.

N

Naked-head Sea Bream, *Gymnocranius griseus*
Nalolo, *Ecsenius nalolo*
Nannygai, *Trachichthodes affinis*
Narrow Snouted Tripodfish, *Tripodichthys angustifrons*
Narrow-bar Damsel, *Plectroglyphidodon dickii*

Narrow-barred Butterflyfish, *Amphichaetodon melbae*

Narrow-barred Spanish Mackerel, *Scomberomorus commerson*

Narrow-lined Toadfish, *Arothron manilensis*

Nassau Grouper, *Epinephelus striatus*

Natal Sharpnosed Pufferfish, *Canthigaster natalensis*

Natal Stumpnose, *Rhabdosargus sarba*

Needlefish, *Tylosurus leiurus*

Needlescaled Queenfish, *Scomberoides tol*

Neon Goby, *Gobiosoma oceanops*

Net Batfish, *Halieutaea retifera*

New Guinea Cardinalfish, *Apogon novaeguinea (?cyanosoma)*

New Guinea Tamarin, *Anampses neoguinaicus*

New Zealand Red Mullet, *Upeneichthys porosus*

New Zealand Spotty, *Pseudolabrus celidotus*

Nieuhof's Tripodfish, *Triacanthus nieuhofi*

Night Sergeant, *Abudefduf taurus*

Nine-banded Cardinalfish, *Apogon novemfasciatus*

Nineline Goby, *Ginsburgellus novemlineatus*

Nine-spined Batfish, *Zabidius novemaculeatus*

Nocturnal Goby, *Amblygobius nocturnus*

Noddlefish, *Salangichthys microdon*

Norfolk Cardinalfish, *Apogon norfolcensis*

Northern Anchovy, *Engraulis mordax*

Northern Bigscale Flathead, *Onigocia macrolepis*

Northern Buffalo Bream, *Kyphosus gibsoni*

Northern Clingfish, *Gobiesox maeandricus*

Northern Devilfish, *Plesiops cephalotaenia*

Northern Dragonet, *Diplogrammus xenicus*

Northern Lancetfish, *Alepisaurus borealis*

Northern Seahorse, *Hippocampus erectus*

Northern Searobin, *Prionotus carolinus*

Northern Wobbegong, *Sutorectus wardi*

Northwest Triplefin, *Norfolkia* sp. 2

Notched Threadfin Bream, *Nemipterus tolu*

Nurse Shark, *Ginglymostoma cirratum*

Nystrom's Conger Eel, *Gnathophis nystromi*

O

Oblique-banded Cardinalfish, *Apogon semiornatus*

Oblong Filefish, *Paramonacanthus oblongus*

Oblong Rockfish, *Sebastes oblongus*

Ocean Pout, *Macrozoarces americanus*

Ocean Sunfish, *Mola mola*

Ocean Surgeon, *Acanthurus bahianus*

Ocean Triggerfish, *Canthidermis sufflamen*

Ocean Whitefish, *Caulolatilus princeps*

Oceanic Seahorse, *Hippocampus kuda*

Oceanic Whitetip Shark, *Carcharhinus longimanus*

Ocellate Damselfish, *Pomacentrus vaiuli*

Ocellate Eel Blenny, *Blennodesmus scapularis*

Ocellate Goby, *Vanderhorstia ornatissima*
Ocellated Blenny, *Ecsenius oculus*
Ocellated Cardinalfish, *Apogonichthys ocellatus*
Ocellated Frogfish, *Antennarius ocellatus*
Ocellated Gudgeon, *Valenciennea longipinnis*
Ocellated Longfin, *Plesiops corallicola*
Ocellated Soapfish, *Grammistops ocellatus*
Ocellated Turbot, *Pleuronichthys ocellatus*
Ocellated Waspfish, *Apistus carinatus*
Ochreband Goatfish, *Upeneus sundaicus*
Oddscale Cardinalfish, *Apogon anisolepis*
Offshore Crocodilefish, *Satyrichthys laticephalus*
Offshore Lizardfish, *Synodus poeyi*
Offshore Pony Fish, *Leiognathus rivulatus*
Ogilby's Hardyhead, *Atherinomorus ogilbyi*
Ogilby's Puller, *Mecaenichthys immaculatus*
Oilfish, *Ruvettus pretiosus*
Old Wife, *Enoplosus armatus*
Olive Dottyback, *Pseudochromis olivaceous*
Olive Flounder, *Paralichthys olivaceus*
Olive Rockfish, *Sebastes serranoides*
Olive-spotted Monocle Bream, *Scolopsis xenochrous*
Olive-striped Snapper, *Lutjanus vitta*
One-band Wrasse, *Cheilinus unifasciatus*
One-eyed Blenny, *Ecsenius monoculus*
One-lined Cardinalfish, *Apogon exostigma*

One-spot Demoiselle, *Chrysiptera unimaculata*
One-spot Snapper, *Lutjanus monostigma*
One-spot Waspfish, *Liocranium praepositum*
One-spot Worm Eel, *Gunnelichthys monostigma*
One-stripe Snake Eel, *Aprognathodon platyventris*
Opaleye, *Girella nigricans*
Ophichthid eel, Ophichthidae sp.
Orange Anemonefish, *Amphiprion sandaracinos*
Orange Band Wrasse, *Stethojulis balteata*
Orange Eelpout, *Dinematichthys dasyrhynchus*
Orange Filefish, *Aluterus schoepfi*
Orange Gurnard, *Dactyloptena peterseni*
Orange Hamlet, *Hypoplectrus gummigutta*
Orange Wrasse, *Pseudolabrus luculentus*
Orange-axil Wrasse, *Stethojulis bandanensis*
Orangeback Bass, *Serranus annularis*
Orangeback Damsel, *Stegastes dorsopunicans*
Orange-banded Goby, *Eviota* sp. 1
Orangebar Weever, *Parapercis* sp.
Orangecheek Parrotfish, *Scarus* sp. 1
Orange-dotted Tuskfish, *Choerodon anchorago*
Orange-ear Goby, *Gnatholepis inconsequens*
Orange-epaulette Surgeonfish, *Acanthurus olivaceus*
Orange-finned Anemonefish, *Amphiprion chrysopterus*
Orangehead Goby, *Gobiosoma* sp.
Orangemarked Goby, *Amblygobius decussatus*

Orange-spine Unicornfish, *Naso lituratus*
Orangespot Goby, *Coryphopterus urospilus*
Orange-spotted Blenny, *Istiblennius chrysospilos*
Orange-spotted Dorsal Goby, *Amblyeleotris* sp. 3
Orange-spotted Filefish, *Cantherhines pullus*
Orange-spotted Goby, *Nes Longus*
Orange-spotted Prawn Goby, *Amblyeleotris guttata*
Orange-stripe Cardinalfish, *Apogon* sp. 2
Orangestripe Prawn Goby, *Amblyeleotris randalli*
Orange-striped Emperor, *Lethrinus ramak*
Orangethroat Pikeblenny, *Chaenopsis alepidota*
Orange-tipped Rainbowfish, *Halichoeres melanurus*
Orbiculate Batfish, *Platax orbicularis*
Orbiculate Burrfish, *Cyclichthys orbicularis*
Orbiculate Cardinalfish, *Sphaeramia orbicularis*
Orchid Dottyback, *Pseudochromis fridmani*
Oriental Flying Gurnard, *Dactyloptena orientalis*
Oriental Sweetlips, *Plectorhinchus picus*
Oriental Thornback, *Platyrhina sinensis*
Oriental Wrasse, *Labropsis manabei*
Ornate Butterflyfish, *Chaetodon ornatissimus*
Ornate Emperor, *Lethrinus ornatus*
Ornate Goby, *Istigobius ornatus*
Ornate Scorpionfish, *Scorpaenodes varipinnis*
Ornate Threadfin Bream, *Nemipterus hexodon*

Ornate Wrasse, *Macropharyngodon ornatus*
Otakii's Greenling, *Hexagrammos otakis*
Oualan Forktail Blenny, *Meiacanthus oualanensis*
Outer-reef Damsel, *Pomacentrus emarginatus*
Oval Soldierfish, *Myripristis hexagonatus*
Owen's Threadfin Bream, *Nemipterus ovenii*
Owstonia, *Owstonia grammodon?*
Oxeye Herring, *Megalops cyprinoides*

P

Pacific Angel Shark, *Squatina californica*
Pacific Barracuda, *Sphyraena argentea*
Pacific Beardfish, *Polymixia berndti*
Pacific Black-eared Wrasse, *Halichoeres melasmopomas*
Pacific Burrfish, *Chilomycterus affinis*
Pacific Convict Goby, *Priolepis cinctus*
Pacific Creole-fish, *Paranthias colonus*
Pacific Dog Snapper, *Lutjanus novemfasciatus*
Pacific Flagfin Mojarra, *Eucinostomus gracilis*
Pacific Gregory, *Stegastes fasciolatus*
Pacific Hagfish, *Eptatretus stouti*
Pacific Halibut, *Hippoglossus stenolepis*
Pacific Hamlet, *Alphestes multiguttatus*
Pacific Herring, *Clupea harengus pallasi*
Pacific Ladyfish, *Elops hawaiiensis*
Pacific Lemon Shark, *Negaprion acutidens*

Pacific Lined Grunt, *Haemulon flaviguttatum*
Pacific Palometa, *Trachinotus kennedyi*
Pacific Pomfret, *Brama japonicus*
Pacific Porgy, *Calamus brachysomus*
Pacific Sailfin Tang, *Zebrasoma veliferum*
Pacific Sand Perch, *Diplectrum pacificum*
Pacific Saury, *Cololabias saira*
Pacific Seahorse, *Hippocampus ingens*
Pacific Sierra, *Scomberomorus sierra*
Pacific Soldierfish, *Myripristis* sp. 1
Pacific Spadefish, *Chaetodipterus zonatus*
Pacific Spotted Velvetfish, *Caracanthus maculatus*
Pacific Threadfin, *Polydactylus sexfilis*
Pacific Whitecheek Parrotfish, *Scarus bleekeri*
Pacific Yellowcheek Wrasse, *Halichoeres* sp.
Painted Blenny, *Ecsenius pictus*
Painted Comber, *Serranus scriba*
Painted Frogfish, *Antennarius pictus*
Painted Goldie, *Pseudanthias pictilis*
Painted Greenling, *Oxylebias pictus*
Painted Moki, *Cheilodactylus ephippium*
Painted Scorpionfish, *Scorpaena picta*
Painted Sculpin, *Pseudoblennius cottoides*
Painted Sweetlips, *Plectorhinchus pictus*
Painted Triggerfish, *Rhinecanthus aculeatus*
Painted Wrasse, *Symphodus tinca*
Painted-face Blenny, *Istiblennius meleagris*
Paintspotted Moray, *Siderea picta*
Pale Cardinalfish, *Apogon planifrons*
Pale Dottyback, *Pseudochromis pesi*

Pale Gudgeon, *Ptereleotris heteropterus*
Paleback Goby, *Gobulus myersi*
Paleband Spinecheek, *Scolopsis ghanam*
Palefin Basslet, *Liopropoma pallidum*
Palefinned Threadfin Bream, *Nemipterus marginatus*
Pale-headed Checkerboard Wrasse, *Halichoeres podostigma*
Palespot Goby, *?Eviota* sp. 2
Pale-tail Chromis, *Chromis xanthura*
Pale-tailed Damselfish, *Stegastes leucorus*
Pallas's Flounder, *Pleuronectes pallasii*
Pallid Goby, *Coryphopterus eidolon*
Palmate Horned Blenny, *Parablennius sanguinolentus*
Palometa, *Trachinotus goodei*
Panamanian Lizardfish, *Synodus lacertinus*
Panamic Clingfish, *Gobiesox adustus*
Panamic Fanged Blenny, *Ophioblennius steindachneri*
Panamic Frillfin Goby, *Bathygobius ramosus*
Panamic Green Moray, *Gymnothorax castaneus*
Panamic Porkfish, *Anisotremus taeniatus*
Panamic Sergeant Major, *Abudefduf troscheli*
Panamic Soldierfish, *Myripristis leiognathus*
Pancake Batfish, *Halieutichthys aculeatus*
Panther Flounder, *Bothus pantherinus*
Panther Goby, *Barbulifer pantherinus*
Panther Grouper, *Cromileptes altivelis*
Panther Puffer, *Takifugu pardalis*
Papuan Goby, *Oxyurichthys papuensis*
Parika, *Parika scaber*

Parrotfish, unidentified, *Scarus* sp. 2

Pastel Fairy Basslet, *Callanthias japonica*

Pastel Jawfish, *Lonchopisthus micrognathus*

Pastel Pygmy Angelfish, *Centropyge multicolor*

Pastel Tilefish, *Hoplolatilus fronticinctus*

Paving-tile Blenny, *Ecsenius tessera*

Paxton's Pipefish, *Corythoichthys paxtoni*

Peacock Blenny, *Salaria pavo*

Peacock Flounder, *Bothus lunatus*

Peacock Sole, *Pardachirus pavoninus*

Peacock Wrasse, *Xyrichthys pavo*

Peacockfin Goby, *Gnatholepis deltoides*

Pearl Blenny, *Entomacrodus nigricans*

Pearl Cardinalfish, *Apogon guamensis*

Pearl Sergeant, *Abudefduf margariteus*

Pearl Toby, *Canthigaster margaritata*

Pearlfish, *Carapus bermudensis*

Pearl-scaled Angelfish, *Centropyge vroliki*

Pearly Monocle Bream, *Scolopsis margaritifera*

Pearly Razorfish, *Xyrichthys novacula*

Pearly Rockskipper, *Entomacrodus striatus*

Pebbled Butterflyfish, *Chaetodon multicinctus*

Pelicier's Wrasse, *Halichoeres pelicieri*

Pen Shell Pearlfish, *Carapus homei*

Pencil Weed Whiting, *Siphonognathus beddomei*

Pennant Glider, *Valenciennea strigata*

Peppered Butterflyfish, *Chaetodon guttatissimus*

Peppered Chromis, *Chromis cinerascens*

Peppered Righteye Flounder, *Clidoderma asperrimum*

Peppermint Basslet, *Liopropoma rubre*

Peppermint Goby, *Coryphopterus lipernes*

Perch Sculpin, *Pseudoblennius percoides*

Perchlet, *Ambassis macracanthus*

Permit, *Trachinotus falcatus*

Peron's Butterfly Bream, *Nemipterus peronii*

Peter's Cave Bass, *Dinoperca petersi*

Pflueger's Goatfish, *Mulloides pflugeri*

Phallic Blenny, *Starksia spinipenis*

Phenax, *Plagiotremus phenax*

Philippine Bandfish, *Acanthocepola indica*

Philippine Blenny, *Ecsenius dilemma*

Philippine Chevron Butterflyfish, *Chaetodon xanthurus*

Philippine Damsel, *Pomacentrus philippinus*

Philippine Fairy Basslet, *Pseudanthias luzonensis*

Philippine Garden Eel, *Gorgasia preclara*

Philippine Pennant Coralfish, *Heniochus singularius*

Phoenix Damsel, *Plectroglyphidodon phoenixensis*

Piano Blenny, *Plagiotremus tapeinosoma*

Picarel, *Spicara maena flexuosa*

Picasso Triggerfish, *Rhinecanthus assasi*

Pickhandle Barracuda, *Sphyraena jello*

Picture Wrasse, *Halichoeres nebulosus*

Pile Surfperch, *Damalichthys vacca*

Pilot Fish, *Naucrates ductor*

Pineapplefish, *Monocentris japonicus*

Pinfish, *Lagodon rhomboides*

Pink Basslet, *Luzonichthys waitei*
Pink Cardinalfish, *Apogon pacifici*
Pink Maomao, *Caprodon longimanus*
Pink Speckled Wrasse, *Xenojulis margaritaceous*
Pink Weedfish, *Heteroclinus roseus*
Pink-and-Blue Spotted Goby, *Cryptocentrus leptocephalus*
Pinkbar Goby, *Amblyeleotris aurora*
Pink-eye Sea Whip Goby, *Bryaninops natans*
Pinktail Triggerfish, *Melichthys vidua*
Pipefish, unidentified, *Syngnathus* sp.
Pirate Blenny, *Emblemaria piratula*
Pixy Hawkfish, *Cirrhitichthys oxycephalus*
Plaice, *Pleuronectes platessa*
Plain Dottyback, *Pseudoplesiops inornatus*
Plain Grouper, *Trisotropis dermopterus*
Plain Parrotfish, *Scarus macrocheilus*
Plainfin Midshipman, *Porichthys notatus*
Planehead Filefish, *Monacanthus hispidus*
Pluma, *Calamus pennatula*
Plume Blenny, *Protemblemaria lucasana*
Pocket Butterflyfish, *Chaetodon wiebeli*
Poey's Butterflyfish, *Chaetodon aculeatus*
Pointed Sawfish, *Anoxypristis cuspidata*
Polka Dot Boxfish, *Ostracion cubicus*
Polka-dot Batfish, *Ogcocephalus radiatus*
Polka-dot Cardinalfish, *Sphaeramia nematoptera*
Pond Smelt, *Hypomesus olidus*
Pontif Blenny, *Parablennius incognitus*

Pony Fish, *Secutor ruconius*
Poor Knights Wrasse, *Pseudolabrus inscriptus*
Popeye Catalufa, *Pristigenys serrula*
Popeyed Sea Goblin, *Inimicus didactylus*
Porae, *Cheilodactylus douglasi*
Porcupine Ray, *Urogymnus africanus*
Porcupinefish, *Diodon hystrix*
Port Jackson Shark, *Heterodontus portusjacksoni*
Portenoy's Blenny, *Ecsenius portenoyi*
Portuguese Man-O-War Fish, *Nomeus gronovii*
Potato Bass, *Epinephelus tukula*
Potter's Angelfish, *Centropyge potteri*
Pouter, *Lycogramma zesta*
Powder-blue Tang, *Acanthurus leucosternon*
Prawn Goby 1, *Amblyeleotris* sp. 1
Prawn Goby 2, *Amblygobius* sp.
Pretty Cheek Goby, *Amblyeleotris* sp. cf *japonicus* 2
Pretty Prawn Goby, *Valenciennea puellaris*
Prettytail Cardinalfish, *Apogon menesemus*
Prickly Fanfish, *Pterycombus petersii*
Prickly Leatherjacket, *Chaetodermis penicilligerus*
Prickly-headed Goby, *Paragobiodon echinocephalus*
Princess Parrotfish, *Scarus taeniopterus*
Promethean Escolar, *Promethichthys prometheus*
Psychedelic Fish, *Synchiropus picturatus*
Pudding Wife, *Halichoeres radiatus*
Puffcheek Blenny, *Labrisomus bucciferus*
Puffer, unidentified, *Sphoeroides* sp.
Pufferfish, unidentified, *Amblyrhynchotes* sp.

Pugjaw Wormfish, *Cerdale floridana*
Pugnose Pipefish, *Bryx dunckeri*
Purple Blotch Basslet, *Pseudanthias pleurotaenia*
Purple Goatfish, *Parupeneus porphyreus*
Purple Queen, *Mirolabrichthys tuka*
Purple Reeffish, *Chromis scotti*
Purple Surgeonfish, *Acanthurus xanthopterus*
Purple Tilefish, *Hoplolatilus purpureus*
Purpleheaded Wrasse, *Halichoeres prosopeion*
Purplemouth Moray, *Gymnothorax vicinus*
Purple-spotted Bullseye, *Priacanthus tayenus*
Pygmy Goby, *Pandaka lidwilli*
Pygmy Leatherjacket, *Brachaluteres jacksonianus*
Pygmy Sea Bass, *Serraniculus pumilio*

Q

Queen Angelfish, *Holacanthus ciliaris*
Queen Coris, *Coris formosa*
Queen Parrotfish, *Scarus vetula*
Queen Snapper, *Etelis oculatus*
Queen Triggerfish, *Balistes vetula*
Queensland Spanish Mackerel, *Scomberomorus queenslandicus*
Quillback Rockfish, *Sebastes maliger*

R

Raccoon Butterflyfish, *Chaetodon lunula*
Radiant Wrasse, *Halichoeres iridis*
Raffle's Butterflyfish, *Chaetodon rafflesi*
Railway Glider, *Valenciennea helsdingenii*

Rainbow Fairy Basslet, *Pseudanthias ventralis ventralis*
Rainbow Fish, *Odax acroptilus*
Rainbow Monocle Bream, *Scolopsis temporalis*
Rainbow Parrotfish, *Scarus guacamaia*
Rainbow Runner, *Elagatis bipinnulata*
Rainbow Scorpionfish, *Scorpaenodes xyris*
Rainbow Surfperch, *Hypsurus caryi*
Rainbow Wrasse, *Thalassoma amblycephalum*
Rainford's Butterflyfish, *Chaetodon rainfordi*
Rainford's Perch, *Rainfordia opercularis*
Rake-gilled Mackerel, *Rastrelliger kanagurta*
Randall's Chromis, *Chromis randalli*
Randall's Fusilier, *Pterocaesio randalli*
Randall's Yellownose Goby, *Gobiosoma randalli*
Ransonet's Surfperch, *Neoditrema ransonneti*
Rapa Sharpnose Pufferfish, *Canthigaster rapaensis*
Recherche Clingfish, Gobiesocidae sp. 1
Rectangle Triggerfish, *Rhinecanthus rectangulus*
Red Anemonefish, *Amphiprion rubrocinctus*
Red and Green Dottyback, *Pseudochromis novaehollandiae*
Red and Green Snapper, *Lutjanus fulvus*
Red Blush Goatfish, *Parupeneus chrysopleuron*
Red Bonnetmouth, *Emmelichthyops ruber*
Red Brotula, *Brosmophycis marginata*

Red Bullseye, *Priacanthus macracanthus*
Red Butterfly Perch, *Caesioperca lepidoptera*
Red Checkerboard Wrasse, *Halichoeres ornatissimus*
Red Clingfish, *Arcos robiginosus*
Red Clown, *Amphiprion frenatus*
Red Coral Goby, *Eviota* sp. 3
Red Fairy Basslet, *Pseudanthias kashiwae*
Red Grouper, *Epinephelus morio*
Red Grubfish, *Parapercis schuinslandi*
Red Hake, *Urophycis chuss*
Red Hind, *Epinephelus guttatus*
Red Hogfish, *Decodon puellaris*
Red Indianfish, *Pataecus fronto*
Red Irish Lord, *Hemilepidotus hemilepidotus*
Red Lizardfish, *Synodus synodus*
Red Moki, *Cheilodactylus spectabilis*
Red Morwong, *Cheilodactylus fuscus*
Red Mumea, *Lutjanus bohar*
Red Opercled Snapper, *Lethrinus rubrioperculatus*
Red Scorpin, *Scorpaena sumptuosa*
Red Scorpionfish, *Scorpaena notata*
Red Sea Angelfish, *Apolemichthys xanthotis*
Red Sea Bannerfish, *Heniochus intermedius*
Red Sea Butterflyfish, *Chaetodon mesoleucos*
Red Sea Chevron Butterflyfish, *Chaetodon paucifasciatus*
Red Sea Clown Surgeon, *Acanthurus sohal*
Red Sea Fourline Wrasse, *Larabicus quadrilineatus*
Red Sea Melon Butterflyfish, *Chaetodon austriacus*
Red Sea Mimic Blenny, *Ecsenius gravieri*

Red Sea Multicolor Wrasse, *Minilabrus striatus*
Red Sea Raccoon Butterflyfish, *Chaetodon fasciatus*
Red Seaperch, *Ellerkeldia rubra*
Red Snapper, *Lutjanus campechanus*
Red Spotted Grouper, *Epinephelus akaara*
Red Spotted Rock Bass, *Paralabrax maculatofasciatus*
Red Tai, *Pagrus pagrus*
Red-and-Blue Goby, *Pterogobius virgo*
Redband Lizardfish, *Synodus engelmani*
Redband Parrotfish, *Sparisoma aurofrenatum*
Red-banded Wrasse, *Pseudolabrus biserialis*
Redbarred Cardinalfish, *Archamia fucata*
Redbelly Fusilier, *Caesio erythrogaster*
Redbreasted Wrasse, *Cheilinus fasciatus*
Red-brown Lizardfish, *Synodus ulae*
Redcheek Wrasse, *Thalassoma genivittatum*
Redchin Sea Robin, *Bellator gymnostethius*
Red-diamond Eel, *Quassiremus notochir*
Redeye Triplefin, *Enneanectes pectoralis*
Redeye Wrasse, *Pseudojuloides erythrops*
Redfilament Threadfin Bream, *Nemipterus mesoprion*
Redfin Fusilier, *Caesio xanthonotus*
Redfin Sailfin Scorpionfish, *Hypodytes rubripinnis*
Red-finned Fairy Wrasse, *Cirrhilabrus rubripinnis*
Redflush Rockcod, *Aethaloperca rogaa*

Redhead Goby, *Gobiosoma puncticulatus*

Redline Puffer, *Sphoeroides erythrotaenia*

Red-lined Butterflyfish, *Chaetodon semilarvatus*

Red-lined Parrotfish, *Hipposcarus harid*

Red-lined Wrasse, *Halichoeres biocellatus*

Redlip Cleaner Wrasse, *Labroides rubrolabiatus*

Redlip Rubberlip, *Plectorhinchus sordidus*

Red-lipped Morwong, *Cheilodactylus rubrolabiatus*

Red-marbled Lizardfish, *Synodus rubromarmoratus*

Redmouth Goby, *Gobius cruentatus*

Red-necked Goby, *Amblyeleotris fasciatus*

Redrump Blenny, *Xenomedia rhodopyga*

Red-saddled Sand Stargazer, *Dactyloscopus pectoralis*

Red-sided Fairy Wrasse, *Cirrhilabrus* sp. "D"

Redspine Threadfin Bream, *Nemipterus nemurus*

Red-spotted Emperor, *Lethrinus lentjan*

Redspotted Hawkfish, *Amblycirrhitus pinos*

Red-spotted Razorfish, *Xyrichthys verrens*

Redstripe Tilefish, *Hoplolatilus marcosi*

Red-striped Bigeye, *Pristigenys multifasciata*

Redtail Filefish, *Pervagor melanocephalus*

Redtail Parrotfish, *Sparisoma chrysopterum*

Redtail Surfperch, *Amphistichus rhodoterus*

Red-tailed Butterflyfish, *Chaetodon collare*

Red-tailed Dottyback, *Pseudochromis flammicauda*

Red-tailed Sea Bream, *Gymnocranius lethrinoides*

Red-tailed Tamarin, *Anampses chrysocephalus*

Red-tailed Yellow Bass, *Liopropoma* sp.

Red-tipped Fringed-blenny, *Cirripectes variolosus*

Red-tipped Rock Cod, *Epinephelus truncatus*

Redtooth Triggerfish, *Odonus niger*

Reef Bass, *Pseudogramma gregoryi*

Reef Butterflyfish, *Chaetodon sedentarius*

Reef Chromis, *Chromis agilis*

Reef Croaker, *Odontoscion dentex*

Reef Scorpionfish, *Scorpaenodes carribbaeus*

Reef Squirrelfish, *Sargocentron coruscum*

Reef Surfperch, *Micrometrus aurora*

Reefsand Blenny, *Ekemblemaria* sp. (*myersi?*)

Reeve's Shad, *Hilsa reevesii*

Regal Angelfish, *Pygoplites diacanthus*

Regal Demoiselle, *Neopomacentrus cyanomos*

Regal Mudskipper, *Periophthalmus regius*

Reid's Damsel, *Pomacentrus reidi*

Remora, unidentified, *Echeneis* sp.

Reticulated Bumphead Wrasse, *Semicossyphus reticulatus*

Reticulated Damselfish, *Dascyllus reticulatus*

Reticulated Emperor, *Lethrinus reticulatus*

Reticulated Moray, *Gymnothorax permistus*

Reticulated Pufferfish, *Arothron reticularis*
Rhino Leatherjacket, *Pseudalutarius nasicornis*
Ribbon Eel, *Rhinomuraena quaesita*
Ribbon-striped Soldierfish, *Myripristis vittatus*
Richardson's Dragonet, *Repomucenus richardsoni*
Richardson's Reef-damsel, *Pomachromis richardsoni*
Rifle Cardinalfish, *Apogon kiensis*
Ringed Puffer, *Omegophora armilla*
Ringed Seaperch, *Ellerkeldia annulata*
Ringed Wrasse, *Hologymnosus doliatus*
Rippled Coral Goby, *Gobiodon rivulatus*
Rippled Rockskipper, *Istiblennius edentulus*
Ritter's Catshark, *Centroscyllium ritteri*
Robust Boxfish, *Strophiurichthys robustus*
Robust Flounder, *Hippoglossoides robustus*
Rock Beauty, *Holacanthus tricolor*
Rock Cod, *Lotella rhacinus*
Rock Croaker, *Equetus viola*
Rock Flagtail, *Kuhlia rupestris*
Rock Gunnel, *Pholis gunnellus*
Rock Hind, *Epinephelus adscensionis*
Rock Sculpin, *Taurulus bubalis*
Rock Sole, *Lepidopsetta bilineata*
Rock Wrasse, *Halichoeres semicinctus*
Rockfish, *Sebastes* sp.
Rogue Fish, *Amblyapistus taenionotus*
Rolland's Demoiselle, *Chrysiptera rollandi*
Roosterfish, *Nematistius pectoralis*
Rose Devilfish, *Pseudoplesiops rosae*
Rosefinned Parrotfish, *Scarus rhodurupterus*
Rosy Blenny, *Malacoctenus macropus*

Rosy Dory, *Cyttopsis rosea?*
Rosy Grouper, *Mycteroperca rosacea*
Rosy Jobfish, *Pristipomoides filamentosus*
Rosy Rockfish, *Sebastes rosaceus*
Rosy Snapper, *Lutjanus lutjanus*
Rosy Triplefin, *Norfolkia* sp. 1
Rough Bullseye, *Pempheris klunzinger*
Rough Leatherjacket, *Scobinichthys granulatus*
Rough Scorpionfish, *Scorpaenodes scaber*
Rough Triggerfish, *Canthidermis maculatus*
Roughback, *Raja porosa*
Roughhead Blenny, *Acanthemblemaria aspera*
Roughhead Triplefin, *Enneanectes boehlkei*
Roughjaw Frogfish, *Antennarius avalonis*
Roughlip Cardinalfish, *Apogon robinsi*
Roughneck Grunt, *Pomadasys corvinaeformis*
Rough-peaked Soldierfish, *Myripristis trachyacron*
Rough-scaled Soldierfish, *Plectrypops lima*
Rough-scaled Squirrelfish, *Sargocentron lepros*
Roughy, *Trachichthys australis*
Round Herring, *Etrumeus teres*
Round Pompano, *Trachinotus blochii*
Round Scad, *Decapterus punctatus*
Round Stingray, *Urolophus halleri*
Round-faced Batfish, *Platax teira*
Roux's Blenny, *Parablennius rouxi*
Royal Basslet, *Lipogramma regia*
Royal Blue Pygmy Angelfish, *Centropyge flavicauda*
Royal Dottyback, *Pseudochromis paccagnellae*
Royal Gramma, *Gramma loreto*

Rudderfish, *Girella punctata*
Rufous Snapper, *Lutjanus rufolineatus*
Russell's One-spot Snapper, *Lutjanus russelli*
Russet Squirrelfish, *Sargocentron rubrum*
Rusty Angelfish, *Centropyge ferrugatus*
Rusty Goby, *Priolepis hipoliti*
Rusty Parrotfish, *Scarus ferrugineus*

S

Sablefish, *Anoplopoma fimbria*
Sabretooth Blenny, *Plagiotremus azaleus*
Saddle Anemonefish, *Amphiprion ephippium*
Saddle Stargazer, *Heteristius rubrocinctus*
Saddle Wrasse, *Thalassoma duperreyi*
Saddleback Butterflyfish, *Chaetodon ehippium*
Saddleback Clown, *Amphiprion polymnus*
Saddleback Hogfish, *Bodianus bilunulatus*
Saddled Blenny, *Malacoctenus triangulatus*
Saddled Butterflyfish, *Chaetodon ulietensis*
Saddled Rock-cod, *Epinephelus damelii*
Saddled-ocellated Sharpnosed Puffer, *Canthigaster ocellicincta*
Saffron Cod, *Eleginus gracilis*
Saffron Damsel, *Amblyglyphidodon aureus*
Sailfin Bass, *Rabaulichthys altipinnis*
Sailfin Blenny, *Emblemaria pandionis*
Sailfin Cardinalfish, *Pterapogon mirifica*
Sailfin Dottyback, *Pseudochromis veliferus*

Sailfin Leaffish, *Taenianotus triacanthus*
Sailfin Poacher, *Podothecus sachi*
Sailfin Sandfish, *Arctoscopus japonicus*
Sailfin Scorpionfish, *Hypodytes* sp.
Sailfin Sculpin, *Blepsias* sp.
Sailfish, *Istiophorus platypterus*
Sailor's Choice, *Haemulon parrai*
Saint Helena Butterflyfish, *Chaetodon sanctaehelenae*
Saint Helena Sharpnose Puffer, *Canthigaster sanctaehelenae*
Salmonet, *Upeneus bensasi*
Samoan Cardinalfish, *Foa fo*
Sand Dragonet, *Callionymus calcaratus*
Sand Perch, *Diplectrum formosum*
Sand Sole, *Psettichthys melanostictus*
Sand Steenbras, *Lithognathus mormyrus*
Sand Tiger, *Odontaspis taurus*
Sand Tilefish, *Malacanthus plumieri*
Sandager's Coris, *Coris sandageri*
Sandbar Shark, *Carcharhinus plumbeus*
Sanddiver Lizardfish, *Synodus intermedius*
Sandfish, *Crapatalus arenarius*
Sargassum Blenny, *Exerpes asper*
Sargassum Fish, *Histrio histrio*
Sargassum Triggerfish, *Xanthichthys ringens*
Sargo, *Anisotremus davidsonii*
Saucereye Porgy, *Calamus calamus*
Saupe, *Boops salpa*
Sawcheek Cardinalfish, *Apogon quadrisquamatus*
Saw-edged Perch, *Niphon spinosus*
Saw-jawed Monocle Bream, *Scolopsis ciliatus*
Sawtail Grouper, *Mycteroperca prionura*
Sawtooth Boxfish, *Kentrocapros aculeatus*

Saw-tooth Grouper, *Polyprion oxygeneiosus*

Sawtooth Pipefish, *Maroubra perserrata*

Scaled-cheek Brotula, *Brotula erythrea?*

Scalloped Hammerhead, *Sphyrna lewini*

Scaly Chromis, *Chromis lepidolepis*

Scaly Damsel, *Pomacentrus lepidogenys*

Scalyhead Sculpin, *Artedius harringtoni*

Scarbreast Tuskfish, *Choerodon azurio*

Scarlet Frogfish, *Antennarius coccineus*

Scarlet Soldierfish, *Myripristis pralinia*

Scarlet Wrasse, *Pseudolabrus milesi*

Scarletfin Blenny, *Coralliozetus micropes*

Scarlet-fin Squirrelfish, *Sargocentron spiniferum*

Scarletfin Wrasse, *Paracheilinus lineopunctatus*

Schlegel's Bream, *Acanthopagrus schlegeli*

Schlegel's Guitarfish, *Rhinobatos schlegeli*

Schlegel's Parrotfish, *Scarus schlegeli*

Schlegel's Red Bass, *Caprodon schlegeli*

Schoolmaster, *Lutjanus apodus*

Schroeder's Blenny, *Ecsenius schroederi*

Schultz's Triplefin, *Norfolkia brachylepis*

Schwanfeld's Six-bar Wrasse, *Thalassoma schwanefeldi*

Scintillating Damsel, *Pomacentrus pikei*

Scissor Damselfish, *Chromis atrilobata*

Scooter Dragonet, *Neosynchiropus ocellatus*

Scorpionfish 1, *Scorpaenopsis* sp. 1

Scorpionfish 2, *Scorpaenodes* sp. 2

Scorpionfish 3, *Sebastapistes* sp.

Scorpionfish 4, Scorpaenidae sp.

Scrawled Cowfish, *Lactophrys quadricornis*

Scribbled Boxfish, *Ostracion solorensis*

Scribbled Filefish, *Aluterus scriptus*

Scribbled Toadfish, *Arothron mappa*

Sculptured Goby, *Callogobius mucosus*

Scythe-marked Butterflyfish, *Chaetodon falcifer*

Sea Carp, *Dactylosargus* sp.

Sea Goldie, *Pseudanthias hutchi*

Sea Lamprey, *Petromyzon marinus*

Sea Moth, *Eurypegasus draconis*

Sea Perch, *Psammoperca waigiensis*

Sea Raven, *Hemitripteris americanus*

Sea Robin, unidentified, *?Peristedion* sp.

Sea Urchin Cardinalfish, *Siphamia mossambica?*

Sea Whip Goby, *Bryaninops amphis*

Seagrass Wrasse, *Novaculichthys macrolepidotus*

Seale's Cardinalfish, *Apogon sealei*

Seaweed Blenny, *Parablennius marmoreus*

Seba's Anemonefish, *Amphiprion sebae*

Seminole Goby, *Microgobius carri*

Senator Wrasse, *Pictilabrus laticlavius*

Senorita, *Oxyjulis californica*

Senorita De Cintas, *Pseudojulis notospilus*

Sepia Stingray, *Urolophus aurantiacus*

Sergeant Baker, *Aulopus purpurissatus*

Sergeant Major, *Abudefduf saxatilis*

Sevenband Wrasse, *Thalassoma septemfasciatum*

Seven-banded Grouper, *Epinephelus septemfasciatus*

Sevenbar Damsel, *Abudefduf septemfasciatus*

Seychelle Blenny, *Stanulus seychellensis*

Seychelles Emperor, *Lethrinus enigmaticus*

Seychelles Goby, *Ctenogobiops crocineus*

Seychelles Mudskipper, *Periophthalmus sobrinus?*

Seychelles Spinecheek, *Scolopsis frenatus*

Seychelles Squirrelfish, *Sargocentron seychellensis*

Shaggy Sea Raven, *Hemitripteris villosus*

Sharknose Goby, *Gobiosoma evelynae*

Sharphead Tripodfish, *Tripodichthys oxycephalus*

Sharpnose Puffer, *Canthigaster rostrata*

Sharpnose Sevengill Shark, *Heptranchias perlo*

Sharp-nosed Trench Sardine, *Amblygaster leiogaster*

Sharp-nosed Weever, *Parapercis hexophthalma*

Sharptail Eel, *Myrichthys aciminatus*

Sharp-tailed Pearlfish, *Onuxodon margaritiferae*

Sharptooth Cardinalfish, *Cheilodipterus quinquelineatus*

Shaw's Cowfish, *Aracana aurita*

Sheepshead, *Archosargus probatocephalus*

Sheepshead Bream, *Diplodus puntazzo*

Shepard's Angelfish, *Centropyge shepardi*

Shimmering Cardinalfish, *Siphamia versicolor*

Shiner Surfperch, *Cymatogaster aggregata*

Shore Rockfish, *Scorpaenodes littoralis*

Short Bigeye, *Pristigenys alta*

Shortbill Spearfish, *Tetrapturus angustirostris*

Short-bodied Ghost Pipefish, *Solenostomus paradoxus*

Short-browed Scorpionfish, *Scorpaenopsis brevifrons*

Shortfin Lionfish, *Dendrochirus brachypterus*

Shortfin Mako Shark, *Isurus oxyrinchus*

Short-finned Sculpin, *Myoxocephalus scorpius*

Shortheaded Lizardfish, *Trachinocephalus myops*

Shortjaw Lizardfish, *Saurida normani*

Shortnose Batfish, *Ogcocephalus nasutus*

Shortnose Blacktail Shark, *Carcharhinus wheeleri*

Shortspine Combfish, *Zaniolepis frenata*

Shortspine Porcupinefish, *Diodon liturosus*

Shortspine Thornyhead, *Sebastolobus alascanus*

Short-spined Scorpionfish, *Scorpaenodes parvipinnis*

Shortstripe Cleaning Goby, *Gobiosoma chancei*

Shorttail Pipefish, *Microphis brachyurus*

Short-tail Stingray, *Dasyatis brevicaudata*

Short-tooth Cardinalfish, *Apogon apogonides*

Shoulderbar Soldierfish, *Myripristis kuntee*

Shovelnose Guitarfish, *Rhinobatos productus*

Showy Snailfish, *Liparis pulchellus*
Shrimpfish, *Aeoliscus strigatus*
Shy Butterflyfish, *Hemitaurichthys polylepis*
Shy Filefish, *Cantherhines verrucundus*
Shy Hamlet, *Hypoplectrus guttavarius*
Shy Puffer, *Takifugu oblongus*
Sickle Butterflyfish, *Chaetodon falcula*
Sickle Pomfret, *Taractichthys steindachneri*
Sideburn Wrasse, *Pteragogus pelycus*
Signae Blenny, *Emblemaria hypacanthus*
Silhouette Goby, *Silhouettea insinuans*
Silk Snapper, *Lutjanus vivanus*
Silver and Black Butterflyfish, *Chaetodon argentatus*
Silver Conger Eel, *Muraenesox cinereus*
Silver Grunt, *Pomadasys argyreus*
Silver Grunter, *Mesopristes argenteus*
Silver Jenny, *Eucinostomus gula*
Silver Large-eye Bream, *Gymnocranius japonicus*
Silver Mono, *Monodactylus argenteus*
Silver Pomfret, *Pampas argenteus*
Silver Sillago, *Sillago sihama*
Silver Snapper, *Lutjanus argentimaculatus*
Silver Squirrelfish, *Neoniphon argenteus*
Silver Streak Goldie, *Pseudanthias taeniatus*
Silver Surfperch, *Hyperprosopon ellipticum*
Silver Sweeper, *Pempheris oualensis*
Silver-banded Sweetlips, *Plectorhinchus diagrammus*
Silverline Mudskipper, *Periophthalmus argentilineatus*
Silverspot, *Threpterius maculosus*

Silverspot Squirrelfish, *Sargocentron caudimaculatum*
Silverstripe Chromis, *Chromis altus*
Silvertip Shark, *Carcharhinus albimarginatus*
Similan Goby, *Pleurosicya* sp.
Singapore Silver-Biddy, *Gerres kapas*
Singleband Forktail Blenny, *Meiacanthus vittatus*
Sirm, *Amblygaster sirm*
Six-banded Angelfish, *Euxiphipops sexstriatus*
Six-banded Weever, *Parapercis sexfasciata*
Six-bar Wrasse, *Thalassoma hardwicke*
Six-barred Grouper, *Cephalopholis sexmaculata*
Six-barred Reef Grouper, *Epinephelus diacanthus*
Six-lined Trumpeter, *Pelates sexlineatus*
Sixstripe Soapfish, *Grammistes sexlineatus*
Sixstripe Wrasse, *Pseudocheilinus hexataenia*
Skilfish, *Erilepis zonifer*
Skipjack Tuna, *Katsuwonus pelamis*
Skipper, *Liza saliens*
Skunk Blenny, *Dictyosoma burgeri*
Skunk-striped Anemonefish, *Amphiprion akallopisos*
Sky Blue Parrotfish, *Scarus frontalis*
Slaty Goby, *Gobiosoma tenox*
Slaty Grenadier, *Coryphaenoides asper*
Slender Bombay Duck, *Harpadon microchir*
Slender Bullseye, *Parapriacanthus unwini*
Slender Cardinalfish, *Rhabdamia* sp.
Slender Filefish, *Monacanthus tuckeri*
Slender Glassy, *Ambassis natalensis*

Slender Grinner, *Saurida dermatogenys*

Slender Hardyhead, *Atherinomorus lacunosus*

Slender Mola, *Ranzania laevis*

Slender Reef-damsel, *Pomachromis exilis*

Slender Roughy, *Hoplostethus elongatus*

Slender Sandfish, *Creedia alleni*

Slender Shrimp Goby, *Amblyeleotris periophthalmus*

Slender Soldierfish, *Pristilepis oligolepis*

Slender Splitfin, *Luzonichthys microlepis*

Slender Sweeper, *Parapriacanthus ransonneti*

Slender Threadfin Bream, *Nemipterus metopias*

Slenderspine Pursemouth, *Gerres oyena*

Slender-spined Blenny, *Acanthemblemaria exilispinis*

Slide-mouth Bass, *Schultzea beta*

Slimy Mackerel, *Scomber australasicus*

Slimy Stingray, *Urolophus mucosus*

Slingjaw Wrasse, *Epibulus insidiator*

Slippery Dick, *Halichoeres bivittatus*

Sloan's Viperfish, *Chauliodus sloani*

Slopehead Parrotfish, *Scarus erythrodon*

Slope-head Sergeant, *Abudefduf declevifrons*

Slow Goby, *Aruma histrio*

Small Scorpionfish, *Scorpaena coniorta*

Small-headed Blind Goby, *Ctenotrypauchen microcephalus*

Smallmouth Grunt, *Haemulon chrysargyreum*

Small-mouthed Leatherjacket, *Oligoplites altus*

Smallmouthed Squirrelfish, *Sargocentron microstoma*

Smallscale Pursemouth, *Gerres acinaces*

Smallspot Pompano, *Trachinotus baillonii*

Smalltail Wrasse, *Pseudojuloides cerasinus*

Smalltooth Sawfish, *Pristis pectinata*

Small-toothed Threadfin Bream, *Pentapodus microdon*

Smiling Goby, *Mahidolia mystacina*

Smith's Blenny, *Meiacanthus smithii*

Smith's Coral Goby, *Eviota infulata*

Smith's Damsel, *Pomacentrus smithi*

Smith-vaniz's Fairy Basslet, *Mirolabrichthys smithvanizi*

Smoky Chromis, *Chromis fumea*

Smooth Hammerhead, *Sphyrna zygaena*

Smooth Pipefish, *Lissocampus caudalis*

Smooth Ronqiul, *Rathbunella hypoplecta*

Smooth Stargazer, *Kathetostoma averruncus*

Smooth Trunkfish, *Lactophrys triqueter*

Smooth Wolf Herring, *Chirocentrus nudus*

Snake eel, unidentified, *Muraenichthys* sp.

Snake Mackerel, *Gempylus serpens*

Snake Pipefish, *Entelurus aequoreus*

Snake-eyed Tartoor, *Opisthopterus tartoor*

Snakeskin Goby, *Chasmichthys dolichognathus*

Snook, *Centropomus undecimalis*

Snooty Wrasse, *Cheilinus oxycephalus*

Snow Bass, *Serranus chionaraia*

Snowflake Blenny, *Cirripectes alboapicalis*

Snowflake Moray, *Echidna nebulosa*

Snowpatch Goby, *Cryptocentrus niveatus*

Snowy Grouper, *Epinephelus niveatus*

Snowy Rockfish, *Sebastes nivosus*

Snubnose Pipefish, *Cosmocampus arctus*

Snyder's Weever, *Parapercis snyderi*

Soaring Hawkfish, *Cyprinocirrhites polyactus*

Social Fairy Wrasse, *Cirrhilabrus rubriventralis*

Soldierfish, unidentified, *Myripristis* sp. 2

Somali Butterflyfish, *Chaetodon leucopleura*

Sombre Sweetlips, *Plectorhinchus schotaf*

Sonara Blenny, *Malacoctenus gigas*

Sonora Brotulid, *Ogilbia* sp.

Sonora Clingfish, *Tomicodon humeralis*

Sonora Goby, *Gobiosoma chiquita*

Sonora Scorpionfish, *Scorpaena sonorae*

South Australian Numbfish, *Hypnos monopterygium*

South Seas Sergeant Major, *Abudefduf vaigiensis*

Southern Bluefin Tuna, *Thunnus maccoyii*

Southern Demoiselle, *Chrysiptera notialis*

Southern Eel, *Anguilla australis*

Southern Fusilier, *Paracaesio xanthura*

Southern Hulafish, *Trachinops noarlungae*

Southern Sea Garfish, *Hyporhamphus melanochir*

Southern Siphonfish, *Siphamia cephalotes*

Southern Stingray, *Dasyatis americana*

Spadefish, *Ephippus orbis*

Spaghetti Eel, *Moringa microchir*

Spangled Emperor, *Lethrinus nebulosus*

Spangled Gudgeon, *Ophiocara porocephala*

Spanish Bream, *Pagellus acarne*

Spanish Flag, *Gonioplectrus hispanus*

Spanish Grunt, *Haemulon macrostomum*

Spanish Hogfish, *Bodianus rufus*

Speckled Catshark, *Hemiscyllium trispeculare*

Speckled Damsel, *Pomacentrus bankanensis*

Speckled Fin Grouper, *Epinephelus summana*

Speckled Moray, *Gymnothorax obesus*

Speckled Sanddab, *Citharichthys stigmaeus*

Speckled Squirrelfish, *Sargocentron punctatissimum*

Specklefin Damsel, *Pomacentrus arenarius*

Speckle-finned Flathead, *Onigocia spinosa*

Spectacled Angelfish, *Chaetodontoplus conspicillatus*

Spectacled Blenny, *Gilloblennius tripennis*

Sphinx Blenny, *Aidablennius sphynx*

Sphinx Goby, *Amblygobius sphynx*

Spinecheek Goby, *Oplopomus oplopomus*

Spine-cheeked Anemonefish, *Premnas biaculeatus*

Spinster Wrasse, *Halichoeres nicholsi*

Spiny Blenny, *Acanthemblemaria spinosa*

Spiny Butterflyfish, *Hemitaurichthys multispinus*

Spiny Chromis, *Acanthochromis polyacanthus*

Spinycheek Soldierfish, *Corniger spinosus*

Spiny-eyed Cardinalfish, *Apogon fraenatus*

Spinyhead Cardinalfish, *Apogon kallopterus*

Spiny-nosed Pipefish, *Halicampus spinirostris*

Spinythroat Scorpionfish, *Pontinus nematophthalmus*

Spinytooth Parrotfish, *Calotomus spinidens*

Splendid Hawkfish, *Cirrhitus splendens*

Splendid Perch, *Callanthias allporti*

Split-banded Cardinalfish, *Apogon compressus*

Split-banded Goby, *Gymneleotris seminudus*

Sponge Cardinalfish, *Phaeoptyx xenus*

Sported Flagtail, *Kuhlia marginata*

Spot-band Butterflyfish, *Chaetodon punctatofasciatus*

Spot-bearing Scorpionfish, *Iracundus signifier*

Spotbelly Rockfish, *Sebastes pachycephalus*

Spotcheek Blenny, *Labrisomus nigricinctus*

Spot-cheeked Surgeonfish, *Acanthurus nigrofuscus*

Spotfin Bandfish, *Acanthocepola limbata*

Spotfin Butterflyfish, *Chaetodon ocellatus*

Spotfin Croaker, *Roncador stearnsi*

Spotfin Flounder, *Verasper variegatus*

Spotfin Goby, *Gobionellus stigmalophius*

Spotfin Hogfish, *Bodianus pulchellus*

Spotfin Lionfish, *Pterois antennata*

Spotfin Mojarra, *Eucinostomus argenteus*

Spotfin Squirrelfish?, *Neoniphon sammara?*

Spotfin Squirrelfish, *Neoniphon sammara*

Spotfin Tonguefish, *Symphurus fasciolaris*

Spothead Grubfish, *Parapercis cephalopunctata*

Spotlight Goby, *Gobiosoma louisae*

Spotnape Pony Fish, *Leiognathus nuchalis*

Spot-tail Butterflyfish, *Chaetodon ocellicaudus*

Spottail Cardinalfish, *Pseudamia amblyuroptera*

Spottail Coris, *Coris caudimacula*

Spottail Gudgeon, *Ptereleotris evides*

Spottail Rainbowfish, *Halichoeres trimaculatus*

Spotted Batfish, *Zalieutes elator*

Spotted Blenny, *Ecsenius stictus*

Spotted Boxfish, *Ostracion meleagris*

Spotted Bristletooth, *Ctenochaetus strigosus*

Spotted Burrfish, *Chilomycterus atinga*

Spotted Cabrilla, *Epinephelus analogus*

Spotted Cardinalfish, *Apogon maculifera*

Spotted Conchfish, *Astrapogon punticulatus*

Spotted Crocodile Flathead, *Inegocia guttata*

Spotted Cusk-eel, *Chilara taylori*

Spotted Dragonet, *Foetorepus* sp.

Spotted Drepane, *Drepane punctata*

Spotted Drum, *Equetus punctatus*

Spotted Eagle Ray, *Aetobatus narinari*

Spotted Fantail Filefish, *?Aluterus* sp.

Spotted Flathead, *Cociella crocodila*

Spotted Garden Eel, *Gorgasia maculata*

Spotted Goatfish, *Parupeneus pleurospilus*

Spotted Grouper, *Epinephelus maculatus*

Spotted Grunter, *Pomadasys commersoni*

Spotted Hardyhead, *Atherinomorus endrachtensis*

Spotted Hawkfish, *Cirrhitichthys aprinus*

Spotted Javelin Fish, *Pomadasys maculatum*

Spotted Jewfish, *Johnius diacanthus*

Spotted Knifejaw, *Oplegnathus punctatus*

Spotted Moonfish, *Mene maculata*

Spotted Moray, *Gymnothorax moringa*

Spotted Oceania Boxfish, *Ostracion meleagris meleagris*

Spotted Prawn Goby, *Ctenogobiops maculosus*

Spotted Pug, *Tandya latitabunda*

Spotted Ratfish, *Hydrolagus collei*

Spotted Reef Moray, *Muraena melanotis*

Spotted Rose Snapper, *Lutjanus guttatus*

Spotted Sailfin Tang, *Zebrasoma gemmatum*

Spotted Scat, *Scatophagus argus*

Spotted Scorpionfish, *Scorpaena* sp. 2

Spotted Sharpnose Puffer, *Canthigaster punctatissimus*

Spotted Shrimpfish, *Aeoliscus punctulatus*

Spotted Snake Eel, *Ophichthus ophis*

Spotted Snapper, *Lutjanus stellatus*

Spotted Soapfish, *Rypticus subbifrenatus*

Spotted Spanish Mackerel, *Scomberomorus maculatus*

Spotted Tilefish, *Hoplolatilus fourmanoiri*

Spotted Toby, *Canthigaster ambionensis*

Spotted Trunkfish, *Lactophrys bicaudalis*

Spotted Unicornfish, *Naso brevirostris*

Spotted Wobbegong, *Orectolobus maculatum*

Spotted-finned Rainbowfish, *Halichoeres miniatus*

Spotted-tail Cardinalfish, *Gymnapogon urospilotus*

Spottyhead Goby, *Amblygobius bynoensis*

Spottysail Dottyback, *Labracinus cyclophthalmus*

Spotwing Goby, *Lythrypnus spilus*

Springer's Demoiselle, *Chrysiptera springeri*

Springer's Dottyback, *Pseudochromis springeri*

Squaretail Grouper, *Epinephelus areolatus*

Squaretail Leopard Grouper, *Plectropomus areolatus*

Squaretail Mullet, *Liza vaigiensis*

Sri Lanka Dottyback, *Pseudochromis dilectus*

St. Helena Moray, *Muraena helena*

St. Helena Scorpionfish, *Scorpaenodes insularis*

Staghorn Damsel, *Amblyglyphidodon curacao*

Starck's Demoiselle, *Chrysiptera starcki*

Stargazer, *Ichthyoscopus lebeck*

Stark's Tilefish, *Hoplolatilus starcki*

Starry Batfish, *Halieutaea stellata*

Starry Blenny, *Salarias* sp. 1

Starry Conchfish, *Astrapogon stellatus*

Starry Fin Goby, *Asterropteryx semipunctatus*

Starry Flounder, *Platichthys stellatus*

Starry Goby, *Asterropteryx* sp.

Starry Moray, *Gymnothorax nudivomer*

Starry Pufferfish, *Arothron stellatus*
Starry Rockfish, *Sebastes constellatus*
Starry Triggerfish, *Abalistes stellatus*
Stars and Stripes Leatherjacket, *Meuschenia venusta*
Stars and Stripes Puffer, *Arothron hispidus*
Startailed Cardinalfish, *Apogon dispar*
Steene's Scorpionfish, *Scorpaenodes steenei*
Steindachner's Moray Eel, *Gymnothorax steindachneri*
Steinitz's Goby, *Amblyeleotris steinitzi*
Stinging Stargazer, *Uranoscopus asper*
Stone Flounder, *Kareius bicoloratus*
Stone Scorpionfish, *Scorpaena plumieri mystes*
Stonefish, *Synanceia verrucosa*
Stoplight Parrotfish, *Sparisoma viride*
Stout-body Chromis, *Chromis chrysura*
Straight-lined Thornfish, *Terapon theraps*
Straight-tail Razorfish, *Xyrichthys martinicensis*
Strasburg's Dascyllus, *Dascyllus strasburgi*
Streaked Spanish Mackerel, *Scomberomorus lineolatus*
Streaked Spinefoot, *Siganus javus*
Streaky Rockskipper, *Istiblennius dussumieri*
Streamlined Spinefoot, *Siganus argenteus*
Striated Frogfish, *Antennarius striatus*
Striped Bass, *Morone saxatilis*
Striped Blenny, *Meiacanthus lineatus*
Striped Bonito, *Sarda orientalis*
Striped Bristletooth, *Ctenochaetus striatus*

Striped Burrfish, *Chilomycterus schoepfi*
Striped Coral Goby, *Trimma striata*
Striped Coral Trout, *Plectropomus oligacanthus*
Striped Dottyback, *Pseudochromis sankeyi*
Striped Filefish, *Paramonacanthus japonicus*
Striped Goby, *Tridentiger trigonocephalus*
Striped Grouper, *Stereolepis ischinagi*
Striped Grunt, *Haemulon striatum*
Striped Hawkfish, *Cheilodactylus vittatus*
Striped Jack, *Pseudocaranx dentex*
Striped Jobfish, *Pristipomoides multidens*
Striped Kelpfish, *Gibbonsia metzi*
Striped Marlin, *Tetrapturus audax*
Striped Mullet, *Mugil cephalus*
Striped Parrotfish, *Scarus croicensis*
Striped Scat, *Scatophagus tetracanthus*
Striped Scorpionfish, *Minous versicolor*
Striped Snapper, *Lutjanus carponotatus*
Striped Trevally, *Carangoides vinctus*
Striped Tripodfish, *Paratriacanthodes retrospinis*
Striped Wing Flying Fish, *Hirundichthys* sp.
Stripe-fin Sculpin, *Enophrys diceraus*
Stripehead Goby, *Priolepis semidoliatus*
Stripetail Damsel, *Abudefduf sexfasciatus*
Stripey, *Microcanthus strigatus*
Sturgeon Poacher, *Agonus acipenserinus*
Suckley's Spiny Dogfish, *Squalus suckleyi*
Sulphur Damselfish, *Pomacentrus sulfureus*

Sulphur Goatfish, *Upeneus sulphureus*

Sumbawa Threadfin Bream, *Nemipterus sumbawensis*

Summer Flounder, *Paralichthys dentatus*

Sunburst Basslet, *Serranocirrhitus latus*

Sunburst Butterflyfish, *Chaetodon aureofasciatus*

Sunrise Dottyback, *Pseudochromis flavivertex*

Sunset Butterflyfish, *Chaetodon pelewensis*

Sunset-tail Goby, *Cryptocentrus aurora*

Sunshinefish, *Chromis insolatus*

Super Klipfish, *Clinus superciliosus*

Surge Silverside, *Iso rhothophilus*

Surge Wrasse, *Thalassoma purpureum*

Sutton's Flying Fish, *Cheilopogon suttoni*

Swales's Basslet, *Liopropoma swalesi*

Sweetlip Emperor, *Lethrinus chrysostomus*

Sweetlips Snapper, *Lutjanus lemniscatus*

Swell Shark, *Cephaloscyllium ventriosum*

Swordfish, *Xiphias gladius*

T

TABL Scad, *Decapterus tabl*

Tadpole Sculpin, *Psychrolutes paradoxus*

Tahitian Butterflyfish, *Chaetodon trichrous*

Tahitian Fairy Wrasse, *Cirrhilabrus punctatus*

Tahitian Reef-damsel, *Pomachromis fuscidorsalis*

Tahitian Squirrelfish, *Sargocentron tiere*

Tailbar Lionfish, *Pterois radiata*

Tail-jet Frogfish, *Antennarius analis*

Tail-light Damselfish, *Chromis notata*

Tailring Surgeonfish, *Acanthurus blochii*

Tail-spot Blenny, *Ecsenius stigmatura*

Tailspot Cardinalfish, *Apogon dovii*

Taiwan Guitarfish, *Rhinobatos formosensis*

Talang Queenfish, *Scomberoides commersonnianus*

Talbot's Blenny, *Stanulus talboti*

Talbot's Demoiselle, *Chrysiptera talboti*

Tangaroa Goby, *Ctenogobiops tangaroai*

Tapiroid Grunter, *Mesopristes cancellatus*

Tarakihi, *Nemadactylus macropterus*

Target Goby, *Cryptocentrus strigilliceps*

Targetfish, *Terapon jarbua*

Tasmanian Blenny, *Parablennius tasmanianus*

Tasmanian Clingfish, *Aspasmogaster tasmaniensis*

Tasmanian Golden Eel, *Muraenichthys tasmaniensis*

Tassled Wobbegong, *Eucrossorhinus dasypogon*

Tattler, *Serranus phoebe*

Tattooed Parrotfish, *Scarus brevifilis*

Tawneygirdled Butterflyfish, *Coradion altivelis*

Teardrop Butterflyfish, *Chaetodon unimaculatus*

Teary Goby, *Stenogobius lachrymosus*

Temminck's Wrasse, *Cirrhilabrus temmincki*

Tentacled Blenny, *Parablennius tentacularis*

Tentacled dragonet, *Anaora tentaculata*

Ternate Chromis, *Chromis ternatensis*

Ternate Damsel, *Amblyglyphidodon ternatensis*

Tessellated Goby, *Redigobius tessellatus*

Thick-lipped Bream, *Diplodus cervinus*

Thielle's Anemonefish, *Amphiprion thiellei*

Thompson's Butterflyfish, *Hemitaurichthys thompsoni*

Thompson's Fairy Basslet, *Mirolabrichthys thompsoni*

Thompson's Goby, *Gnatholepis thompsoni*

Thornback, *Platyrhinoidis triseriata*

Thornback Cowfish, *Lactoria fornasini*

Thornbacked Boxfish, *Tetrosomus gibbosus*

Thread Sailfin Filefish, *Stephanolepis cirrhifer*

Threadfin Bream, *Pentapodus vitta*

Threadfin Butterflyfish, *Chaetodon auriga*

Threadfin Goldie, *Nemanthias carberryi*

Threadfin Shad, *Dorosoma petenense*

Thread-fin Silver-Biddy, *Gerres macracanthus*

Threadfin, unidentified, *Polydactylus* sp.

Threadnose Fairy Basslet, *Pseudanthias tenuis*

Thread-tailed Filefish, *Monacanthus filicauda*

Three Spot Damselfish, *Stegastes planifrons*

Three-band Demoiselle, *Chrysiptera tricincta*

Threeband Pennant Coralfish, *Heniochus chrysostomus*

Three-banded Anemonefish, *Amphiprion tricinctus*

Three-banded Sweetlip, *Plectorhinchus cinctus*

Threebeard Rockling, *Gaidropsarus ensis*

Three-eye Flounder, *Ancylopsetta dilecta*

Threeline Basslet, *Lipogramma trilineata*

Threeline Bream, *Pentapodus* sp. 4

Three-line Damsel, *Pomacentrus trilineatus*

Three-line Monocle Bream, *Scolopsis trilineatus*

Three-lined Blenny, *Ecsenius trilineatus*

Three-lined Fusilier, *Pterocaesio trilineata*

Three-lined Grunt, *Parapristipoma trilineatum*

Three-lined Wrasse, *Stethojulis trilineata*

Three-ribbon Rainbowfish, *Stethojulis strigiventer*

Threespine Toadfish, *Batrachomoeus trispinosus*

Three-spot Angelfish, *Apolemichthys trimaculatus*

Three-spot Cardinalfish, *Apogon trimaculatum*

Three-spot Chromis, *Chromis verater*

Three-spot Damsel, *Pomacentrus tripunctatus*

Threespot Damselfish, *Dascyllus trimaculatus*

Three-spot Frogfish, *Lophiocharon trisignatus*

Three-spot Spinefoot, *Siganus trispilos*

Three-spotted Weever, *Parapercis trispilota*

Threestripe Rockfish, *Sebastes trivittatus*

Threestripe Wrasse, *Pseudojuloides trifasciatus*

Three-striped Goatfish, *Parupeneus atrocingulatus*

Threetooth Puffer, *Triodon macropterus*
Throatspotted Blenny, *Malacoctenus tetranemus*
Tiger Blenny, *Ecsenius tigris*
Tiger Cardinalfish, *Cheilodipterus lineatus*
Tiger Goby, *Gobiosoma macrodon*
Tiger Grouper, *Mycteroperca tigris*
Tiger Puffer, *Takifugu rubripes*
Tiger Reef Eel, *Uropterygius tigrinus*
Tiger Rockfish, *Sebastes nigrocinctus*
Tiger Sea Whip Goby, *Bryaninops tigris*
Tiger Shark, *Galeocerdo cuvieri*
Tiger Snake Eel, *Myrichthys maculosus*
Tigersnout Seahorse, *Hippocampus angustus*
Tilefish, *Lopholatilus chameleonticeps*
Timor Snapper, *Lutjanus timorensis*
Timor Wrasse, *Halichoeres timorensis*
Tinker's Butterflyfish, *Chaetodon tinkeri*
Tinsel Squirrelfish, *Sargocentron suborbitalis*
Toadstool Groper, *Trachypoma macracantha*
Tobacco Fish, *Serranus tabacarius*
Tokien Grenadier, *Coelorhinchus tokiensis*
Tomato Rockcod, *Cephalopholis sonnerati*
Tomini Bristletooth, *Ctenochaetus tominiensis*
Tomiyama's Shrimp Goby, *Tomiyamichthys oni*
Tommy Fish, *Limnichthys fasciatus*
Tompot Blenny, *Parablennius gattorugine*
Tomtate, *Haemulon aureolineatum*
Toothbrush Leatherjacket, *Pencipelta vittiger*
Toothed Pony Fish, *Gazza minuta*

Topknot Flounder, *Zeugopterus punctatus*
Topsmelt, *Atherinops affinis*
Torpedo, *Gymnocaesio gymnopterus*
Torpedo Scad, *Megalaspis cordyla*
Townsend's Mimic Blenny, *Plagiotremus townsendi*
Tracey's Demoiselle, *Chrysiptera traceyi*
Translucent Cardinalfish, *Apogon hyalosoma*
Translucent Goby, *Chriolepis fisheri*
Treefish, *Sebastes serriceps*
Trevally, unidentified, *Carangoides* sp.
Triangle Butterflyfish, *Chaetodon triangulum*
Triplefin 1, *Enneanectes* sp.
Triplefin 2, Tripterygiidae sp. 2
Triplefin 3, *Helcogramma* sp.
Triplefin 4, *Tripterygion* sp.
Triplefin 5, *Vauclusella* sp. 2
Tripletail, *Lobotes surinamensis*
Tripletail Wrasse, *Cheilinus trilobatus*
Tropical Sand Goby, *Favonigobius reichei*
Trout Reef Cod, *Epinephelus fario*
Truncate Coralfish, *Chelmonops truncatus*
Truncate Fairy Basslet, *Pseudanthias truncatus*
Trunk-eyed Moray, *Gymnothorax margaritophorus*
Trunkfish, *Lactophrys trigonus*
Tubelip Wrasse, *Labrichthys unilineatus*
Tubemouth, *Siphonognathus argyrophanes*
Tuberculated Frogfish, *Antennarius tuberosus*
Tubesnout, *Aulorhynchus flavidus*
Twinspot Bass, *Serranus flaviventris*
Twin-spot Chromis, *Chromis elerae*
Twinspot Coris, *Coris dorsomacula*

Twinspot Lionfish, *Dendrochirus biocellatus*

Two-banded Anemonefish, *Amphiprion bicinctus*

Two-banded Bream, *Diplodus vulgaris*

Two-banded Dottyback, *Pseudochromis bitaeniata*

Two-banded Grouper, *Diploprion bifasciatus*

Two-banded Stargazer, *Uranoscopus bicinctus*

Twobar Seabream, *Acanthopagrus bifasciatus*

Two-barred Goatfish, *Parupeneus bifasciatus*

Two-belted Cardinalfish, *Apogon taeniatus*

Two-colored Angelfish, *Centropyge bicolor*

Two-colored Blenny, *Ecsenius bicolor*

Two-lined Monocle Bream, *Scolopsis bilineatus*

Two-saddle Goatfish, *Parupeneus cinnabarinus*

Twospot Cardinalfish, *Apogon pseudomaculatus*

Two-spot Coralfish, *Coradion melanopus*

Two-spot Demoiselle, *Chrysiptera biocellata*

Two-spot Goby, *Signigobius biocellatus*

Two-spot Hawkfish, *Amblycirrhites bimacula*

Twospot Hogfish, *Bodianus bimaculatus*

Two-spot Wrasse, *Cheilinus bimaculatus*

Two-striped Bream, *Pentapodus bifasciatus*

Two-striped Cardinalfish, *Apogon quadrifasciatus*

Two-striped Coral Goby, *Eviota bifasciata*

Two-tone Fingerfin, *Chirodactylus brachydactylus*

Tyler's Sharpnosed Pufferfish, *Canthigaster tyleri*

U

Unarmed Monocle Bream, *Scolopsis inermis*

Undulate Triggerfish, *Balistapus undulatus*

Unicorn Leatherjacket, *Aluterus monoceros*

Unicorn Tang, *Naso unicornis*

Uniform Reef Eel, *Uropterygius concolor*

Uspi Spinefoot, *Lo uspi*

V

Vagabond Butterflyfish, *Chaetodon vagabundus*

Vagabond Filefish, *Cantherhines dumerili*

Vaigeu Snapper, *Lutjanus vaigiensis*

Valencienne's Goby, *Valenciennea muralis*

Valenciennes's Morwong, *Nemadactylus valenciennesi*

Valentini Mimic, *Paraluterus prionurus*

Vanderbilt's Chromis, *Chromis vanderbilti*

Vanikolo Goatfish, *Mulloides vanicolensis*

Vanikoro Sweeper, *Pempheris vanicolensis*

Variable Parrotfish, *Scarus oviceps*

Variegated Cardinalfish, *Fowleria variegata*

Variegated Emperor, *Lethrinus variegatus*

Variegated Ghost Pipefish, *Solenostomus cyanopterus*

Variegated Lizardfish, *Synodus variegatus*

Velvet Leatherjacket, *Thamnaconus australis*
Velvetfish, *Caracanthus madagascariensis*
Vermiculated Angelfish, *Chaetodontoplus mesoleucus*
Vermiculated Spinefoot, *Siganus vermiculatus*
Vermillion Rockfish, *Sebastes miniatus*
Vermillion Snapper, *Rhomboplites aurorubens*
Victoria Cardinalfish, *Apogon victoriae*
Victorian Scalyfin, *Parma victoriae*
Village Belle, *Chrysiptera taupou*
Violet Demoiselle, *Neopomacentrus violascens*
Violet Sea Whip Goby, *Bryaninops* sp.
Violet Squirrelfish, *Sargocentron violaceum*
Viviparous Eelpout, *Zoarchias veneficus*
Vlaming's Unicornfish, *Naso vlamingi*

W

Wahoo, *Acanthocybium solandri*
Wakanoura Moray, *Gymnothorax eurostus*
Wakiya's Perch, *Malakichthys wakiyai*
Walleye Basslet, *Serranus luciopercanus*
Wall-eye Pollock, *Theragra chalcogramma*
Ward's Damsel, *Pomacentrus wardi*
Warthead Blenny, *Protemblemaria bicirrhis*
Warty Frogfish, *Antennarius maculatus*
Warty Poacher, *Occella verrucosa*
Waspfish, *Centropogon australis*

Watanabe's Lyretail Angelfish, *Genicanthus watanabei*
Wavy Grubfish, *Parapercis haackei*
Wavy-lined Blenny, *Entomacrodus decussatus*
Wavy-lined Grouper, *Epinephelus undulostriatus*
Web Burrfish, *Chilomycterus antillarum*
Weber's Chromis, *Chromis weberi*
Weber's Coralline Velvetfish, *Paraploactis obbesi?*
Weber's Tripodfish, *Trixiphichthys weberi?*
Wedgetail Filefish, *Paramonacanthus barnardi*
Weed Blenny, *Springeratus xanthosoma*
Weedfish, *Heteroclinus* sp. cf *roseus*
Weedy Seadragon, *Phyllopteryx taeniolatus*
Weeping Scorpionfish, *Rhinopias argoliba*
Weever, unidentified, *Parapercis* sp. (*cylindrica?*)
West African Sea Bass, *Serranus heterurus*
West African Squirrelfish, *Holocentrus hastatus*
West African Triggerfish, *Balistes forcipitus*
West Australian Jewfish, *Glaucosoma hebraicum*
West Indian Cleaning Goby, *Gobiosoma prochilus*
Western Blue Devil, *Paraplesiops meleagris*
Western Buffalo Bream, *Kyphosus cornelii*
Western Cardinalfish, *Vincentia* sp.
Western Footballer, *Neatypus obliquus*
Western Gregory, *Stegastes obreptus*
Western Rock Blackfish, *Girella tephraeops*

Western Scalyfin, *Parma occidentalis*
Western Sea Perch, *Ellerkeldia wilsoni*
Westralian Puller, *Chromis westaustralis*
Whale Shark, *Rhincodon typus*
Wheeler's Prawn Goby, *Amblyeleotris wheeleri*
Whiptail Threadfin Bream, *Pentapodus nemurus*
Whiskered Brotula, *Brotula multibarbata*
Whiskered Goby, *Callogobius* sp.
Whiskered Pipefish, *Halicampus macrorhynchus*
Whiskered Prowfish, *Neopataecus waterhausi*
White Bream, *Diplodus sargus*
White Croaker, *Genyonemus lineatus*
White Damsel, *Dischistodus perspicillatus*
White Goby, *Nesogobius* sp.
White Grunt, *Haemulon plumieri*
White Lined Triplefin, *Helcogramma striata*
White Remora, *Remorina albescens*
White Shoulder Parma, *Parma alboscapularis*
White Speckled Rock Cod, *Epinephelus hexagonatus*
White Surfperch, *Phanerodon furcatus*
White Tilefish, *Branchiostegus albus*
White-backed Goby, *Cryptocentrus albidorsus*
Whiteband Surgeonfish, *Acanthurus leucopareius*
White-banded Damselfish, *Chrysiptera leucopoma*
White-bar Gregory, *Stegastes albifasciatus*
Whitebar Weever, *Parapercis xanthozona*
White-barred Boxfish, *Anaplocapros lenticularis*

White-bellied Stripey, *Atypichthys latus*
White-belly Damsel, *Amblyglyphidodon leucogaster*
White-belly Puffer, *Omegophora* sp.
Whitebelly Sailfin Scorpionfish, *Hypodytes leucogaster*
White-blotched Rockcod, *Epinephelus multinotatus*
Whitecap Anemonefish, *Amphiprion leucokranos*
White-capped Prawn Goby, *Lotilia graciiiosa*
Whitecheek Monocle Bream, *Scolopsis vosmeri*
Whitecheek Prawn Goby, *Cryptocentrus* sp. 3
Whitecrescent Parrotfish, *Hipposcarus caudovittatus*
White-dot Blenny, *Ecsenius mandibularis*
White-dotted Grouper, *Epinephelus socialis*
White-eared Garden Eel, *Taenioconger* sp.
White-edge Soldierfish, *Myripristis axillaris*
White-eye Goby, *Bollmannia boqueronensis*
White-faced Surgeonfish, *Acanthurus japonicus*
Whitefin Sharksucker, *Echeneis naucratoides*
White-line Feather Star Clingfish, *Discotrema* sp.
White-lined Rockcod, *Anyperodon leucogrammicus*
White-marked Razorfish, *Xyrichthys aneitensis*
Whitenose Pigfish, *Perryena leucometopon*
Whiteribbon Spinecheek, *Scolopsis leucotaenia*
Whitespot Coral Goby, *Trimma* sp. 1

White-spot Damsel, *Dischistodus chrysopoecilus*

Whitespot Tuskfish, *Choerodon cyanodus*

Whitespotted Filefish, *Cantherhines macroceros*

White-spotted Rainbowfish, *Pseudolabrus gymnogenis*

White-spotted Reef Cod, *Epinephelus caeruleopunctatus*

White-spotted Shovelnose Ray, *Rhynchobatus djeddensis*

White-spotted Spinefoot, *Siganus canaliculatus*

White-spotted Squirrelfish, *Sargocentron lacteoguttatus*

Whitespotted Toby, *Canthigaster jactator*

Whitestar Cardinalfish, *Apogon lachneri*

White-tail Chromis, *Chromis leucura*

Whitetail Damsel, *Pomacentrus chrysurus*

Whitetail Goblinfish, *Minous quincarinatus*

Whitetail Triggerfish, *Sufflamen albicaudatus*

White-tailed Humbug, *Dascyllus aruanus*

White-tailed Remora, *Echeneis naucrates*

White-tip Butterflyfish, *Chaetodon litus*

Whitetip Reef Shark, *Triaenodon obesus*

White-tip Shark, *Triaenodon apicalis*

White-tongued Crevalle, *Uraspis helvola*

Whitley's Boxfish, *Ostracion whitleyi*

Whitley's Sergeant, *Abudefduf whitleyi*

Whitley's Weedfish, *Heteroclinus whitleyi*

Whole-scaled Dory, *Zenion hololepis*

Wide-band Anemonefish, *Amphiprion latezonatus*

Wideband Prawn Goby, *Amblyeleotris latifasciata*

Widebanded Cleaning Goby, *Gobiosoma* sp. aff *limbaughi*

Wide-bodied Pipefish, *Stigmatopora nigra*

Willemawillum, *Chelmon marginalis*

Windowpane, *Scopthalmus aquosus*

Wine Red Blenny, *Starksia* sp.

Winter Flounder, *Pseudopleuronectes americanus*

Wire-netting Leatherjacket, *Cantherhines sandwichensis*

Wirrah, *Acanthistius serratus*

Wolf Cardinalfish, *Cheilodipterus artus*

Wolf-eel, *Anarrhichthys ocellatus*

Wolffish, *Anarhichas lupus*

Wonder Chromis, *Chromis mirationis*

Woodward's Moray, *Gymnothorax woodwardi*

Woodward's Pomfret, *Schuettea woodwardi*

Woolman's Halibut, *Paralichthys woolmani*

Wooly Sculpin, *Clinocottus analis*

Worm Blenny, *Stathmonotus sinuscalifornici*

Wounded Wrasse, *Halichoeres chierchiae*

Wrasse Basslet, *Liopropoma eukrines*

Wrasse, unidentified, *Cirrhilabrus* sp.

Wrought-iron Butterflyfish, *Chaetodon daedalma*

Y

Yaeyama Blenny, *Ecsenius yaeyamaensis*

Yamashiro's Rainbowfish, *Pseudocoris yamashiroi*

Yellow Chromis, *Chromis analis*

Yellow Chub, *Kyphosus incisor*
Yellow Clingfish, *Aspasmichthys ciconiae*
Yellow Conger Eel, *Anago anago*
Yellow Coral Goby, *Gobiodon okinawae*
Yellow Crested Weedfish, *Cristiceps aurantiacus*
Yellow Goatfish, *Mulloides martinicus*
Yellow Goby, *Vomerogobius flavus*
Yellow Head Jawfish, *Opistognathus aurifrons*
Yellow Jack, *Carangoides bartholomei*
Yellow Jawfish, *Opistognathus gilberti*
Yellow Lateral Line Fusilier, *Pterocaesio digramma*
Yellow Prawn Goby, *Cryptocentrus cinctus*
Yellow Prickly-headed Coral Goby, *Paragobiodon xanthosomus*
Yellow Ribbon Wrasse, *Leptojulis chrysotaenia*
Yellow Sailfin Tang, *Zebrasoma flavescens*
Yellow Stingray, *Urolophus jamaicensis*
Yellow Tailband Wrasse, *Anampses melanurus melanurus*
Yellow Wrasse, *Thalassoma lutescens*
Yellowback Sea Bream, *Dentex tumifrons*
Yellowback Wrasse, *Labropsis xanthonota*
Yellowband Wrasse, *Pseudojuloides* sp.
Yellow-banded Perch, *Acanthistius cinctus*
Yellow-banded Sweetlips, *Plectorhinchus lineatus*
Yellow-barred Red Rockfish, *Sebasticus albofasciatus*
Yellow-bellied Anemonefish, *Amphiprion chrysogaster*
Yellow-bellied Blue Damsel, *Pomacentrus caeruleus*

Yellowblotch Moray, *Gymnothorax leucostigma*
Yellow-blotch Spinefoot, *Siganus guttatus*
Yellowbreasted Tamarin, *Anampses twistii*
Yellowcheek Wrasse, *Halichoeres cyanocephalus*
Yellow-crowned Butterflyfish, *Chaetodon flavocoronatus*
Yellow-edge Chromis, *Chromis multilineatus*
Yellow-edged Moray, *Gymnothorax flavimarginatus*
Yelloweye Bristletooth, *Ctenochaetus marginatus*
Yelloweye Rockfish, *Sebastes ruberrimus*
Yellow-eyed Blenny, *Ecsenius prooculus*
Yellow-eyed Mullet, *Aldrichetta forsteri*
Yellow-face Butterflyfish, *Chaetodon flavirostris*
Yellowface Pikeblenny, *Chaenopsis limbaughi*
Yellowface Red Tai, *Pagrus major*
Yellow-face Soapfish, *Diploprion drachi*
Yellowface Triggerfish, *Pseudobalistes flavimarginatus*
Yellow-faced Angelfish, *Euxiphipops xanthometopon*
Yellowfin Croaker, *Umbrina roncador*
Yellow-fin Demoiselle, *Chrysiptera flavipinnis*
Yellow-fin Dottyback, *Pseudochromis wilsoni*
Yellowfin Emperor, *Lethrinus crocineus*
Yellowfin Goby, *Acanthogobius flavimanus*
Yellowfin Greenling, *Hexagrammos* sp.

Yellowfin Grouper, *Mycteroperca venenosa*
Yellowfin Mojarra, *Gerres cinereus*
Yellowfin Parrotfish, *Scarus flavipectoralis*
Yellowfin Sleeper, *Eleotris* sp. cf *acanthopoma*
Yellowfin Tuna, *Thunnus albacares*
Yellowfinned Grunt, *Pomadasys* sp.
Yellow-finned Pygmy Angelfish, *Centropyge flavopectoralis*
Yellow-flank Damselfish, *Amblyglyphidodon flavilatus*
Yellow-freckled Reeffish, *Chromis flavomaculata*
Yellowhead Butterflyfish, *Chaetodon xanthocephalus*
Yellow-head Dottyback, *Pseudochromis tapeinosoma*
Yellowhead Wrasse, *Halichoeres garnoti*
Yellow-headed Angelfish, *Chaetodontoplus chrysocephalus*
Yellow-headed Blenny, *Cirripectes chelomatus*
Yellowline Bananafish, *Pristipomoides macrophthalmus*
Yellowline Goby, *Gobiosoma horsti*
Yellowlined Basslet, *Gramma linki*
Yellow-lined Fusilier, *Caesio varilineata*
Yellow-lined Moray, *Gymnothorax thyrsoideus*
Yellow-lined Sweetlips, *Plectorhinchus albovittatus*
Yellow-lined Tilefish, *Branchiostegus* sp.
Yellow-lined Wrasse, *Halichoeres hoeveni*
Yellow-lip Snapper, *Lethrinus xanthochilus*
Yellowmouth Grouper, *Mycteroperca interstitialis*
Yellownose Goby, *Stonogobiops xanthorhinica*

Yellow-red Squirrelfish, *Sargocentron xantherythrus*
Yellowsaddle Goatfish, *Parupeneus cyclostomus*
Yellow-sided Parrotfish, *Scarus psittacus*
Yellowspotted Burrfish, *Chilomycterus spilostylus*
Yellow-spotted Crevalle, *Carangoides ferdau*
Yellow-spotted Emperor, *Lethrinus kallopterus*
Yellowspotted Kingfish, *Carangoides fulvoguttatus*
Yellow-spotted Scorpionfish, *Sebastapistes cyanostigma*
Yellow-spotted Tripodfish, *Tripodichthys* sp.
Yellowspotted Wrasse, *Macropharyngodon negrosensis*
Yellowstreak Whiptail, *Pentapodus porosus*
Yellow-streaked Monocle Bream, *Scolopsis dubiosus*
Yellowstripe Cleaning Goby, *Gobiosoma xanthiprora*
Yellowstripe Clingfish, *Diademichthys lineatus*
Yellowstripe Goatfish, *Mulloides flavolineatus*
Yellowstripe Hulafish, *Trachinops* sp.
Yellowstripe Ruby Snapper, *Etelis marshi*
Yellowstripe Scad, *Selaroides leptolepis*
Yellowstripe Threadfin Bream, *Pentapodus* sp. 1
Yellow-striped Blackfish, *Girella mezina*
Yellow-striped Butterfish, *Labracoglossa argentiventris*
Yellowstriped Fusilier, *Caesio pisang*
Yellow-striped Leatherjacket, *Meuschenia flavolineata*

Yellow-striped Squirrelfish, *Neoniphon aurolineatus*

Yellow-striped Whiptail, *Pentapodus macrurus*

Yellow-striped Worm Eel, *Gunnelichthys curiosus*

Yellowtail, *Seriola lalandi lalandi*

Yellowtail Barred Grunt, *Hapalogenys mucronatus*

Yellowtail Black Angelfish, *Apolemichthys xanthurus*

Yellowtail Damsel, *Pomacentrus trichourus*

Yellowtail Demoiselle, *Neopomacentrus azysron*

Yellow-tail Dottyback, *Pseudochromis xanthochir*

Yellowtail Goldie, *Mirolabrichthys evansi*

Yellowtail Hamlet, *Hypoplectrus chlorurus*

Yellowtail Parrotfish, *Sparisoma rubripinne*

Yellowtail Reeffish, *Chromis enchrysura*

Yellowtail Rockfish, *Sebastes flavidus*

Yellowtail Sailfin Tang, *Zebrasoma xanthurus*

Yellowtail Sawtail, *Prionurus punctatus*

Yellowtail Scad, *Atule mate*

Yellowtail Sergeant, *Abudefduf notatus*

Yellowtail Snapper, *Ocyurus chrysurus*

Yellow-tail Tamarin, *Anampses meleagrides*

Yellow-tailed Barracuda, *Sphyraena flavicauda*

Yellow-tailed Blenny, *Ecsenius namiyei*

Yellow-tailed Cleaner, *Diproctacanthus xanthurus*

Yellow-tailed Dascyllus, *Dascyllus flavicauda*

Yellow-tailed Fusilier, *Caesio cuning*

Yellow-tailed Snake Eel, *Sphagebranchus flavicauda*

Yellow-tailed Surgeonfish, *Acanthurus thompsoni*

Yellowtip Soldierfish, *Myripristis xanthacrus*

Yellow-winged Flying Fish, *Cypselurus poecilopterus*

Z

Zander's Cottid, *Argyrocottus zanderi*

Zanzibar Butterflyfish, *Chaetodon zanzibariensis*

Zebra Blenny, *Istiblennius zebra*

Zebra Goby, *Lythrypnus zebra*

Zebra Horn Shark, *Heterodontus zebra*

Zebra Lionfish, *Dendrochirus zebra*

Zebra Lyretail Angelfish, *Genicanthus caudovittatus*

Zebra Moray, *Gymnomuraena zebra*

Zebra Shark, *Stegostoma varium*

Zebra Sole, *Zebrias zebra*

Zebrafish, *Girella zebra*

Zebraperch, *Hermosilla azurea*

Zesty Blenny, *Petroscirtes xestus*

Zeylon Wrasse, *Halichoeres zeylonicus*

Zigzag Wrasse, *Halichoeres scapularis*

Zvonimiri's Blenny, *Parablennius zvonimiri*